ALLE ZEIT WACH
1842

Wolf Engels (Ed.)

Social Insects

An Evolutionary Approach to Castes and Reproduction

Contributors

A. Buschinger · R. M. Crewe · W. Engels · K. Hartfelder
C. G. J. van Honk · V. L. Imperatriz-Fonseca · C. D. Michener
C. Noirot · P.-F. Röseler · F. Ruttner · A. Strambi · H. H. W. Velthuis

Foreword by B. Rensch

Springer-Verlag Berlin Heidelberg NewYork
London Paris Tokyo HongKong

Prof. Dr. WOLF ENGELS
Universität Tübingen
Zoologisches Institut
Lehrstuhl Entwicklungsphysiologie
Auf der Morgenstelle 28
7400 Tübingen 1, FRG

With 41 Figures, One in Color

Typesetting and Layout: Bertram Engels, Tübingen

ISBN-13:978-3-642-74492-1 e-ISBN-13:978-3-642-74490-7
DOI: 10.1007/978-3-642-74490-7

Library of Congress Cataloging-in-Publication Data. Social insects: an evolutionary approach to castes and reproduction / W. Engels (editor); contributors, A. Buschinger ... [et al.]. p. cm. Includes bibliographical references. ISBN-13:978-3-642-74492-1 (U.S.) 1. Insects – Behavior. 2. Insects – Reproduction. 3. Social hierarchy in animals. I. Engels, Wolf. II. Buschinger, A. (Alfred) QL496.S62 1990 595.7'05248 – dc20 89-26246

Softcover reprint of the hardcover 1st edition 1990

2131/3130 (3020)-543210 – Printed on acid-free paper

Contents

Foreword

Evolution of Castes in Insects

BERNHARD RENSCH[1]

The evolution of organisms in the succession of generations has been primarilly achieved through the appearance of new hereditary variants with minor deviations along with the development of stable structures and functions. Harmful mutants were eradicated either immediately or in the subsequent generations through the process of natural selection while unharmful or advantageous ones usually survived. As this process continued, many new races and species developed which possessed structures and functions of a rational nature and which became increasingly independent from their environment. These evolutionary tendencies were also promoted through either geographical, ecological, physiological or genetical isolation of populations.

In some phylogenetic lines, there occurred an increase in the number of cells due to additional cellular divisions in the morphogenesis of structures. This opened up the possibility for the development of the division of labour within cellular groups leading to more complicated and consequently more successful functions of tissues and organs. A further advancement in the adaptation of a species to its environment was made possible through specialization for certain functions of whole individuals or groups of individuals. This occurred mainly in Hymenopteran species which produce anually more than one and overlapping generations and which live closely together in populous families. Because of the increase in the advantageous division of labour and the specialization in reproduction, special castes developed, the structures of which often deviate considerably from those of "normal" individuals. This evolution required a sequence of group selection processes, which eventually lead to the development of very morphologically and physiologically distinct castes.

In this way, termites, ants, bees, wasps and even aphids have evolved social structures. Specific patterns of social behavior

[1] *Zoologisches Institut der Universität Münster, BRD*

developed accordingly, for example in the procurement of certain foods, nest construction, the feeding of offspring, cultivation of funghi, the defense of the community etc. At the same time intraspecific communication systems became necessary. Already at an early stage in this process, there emerged differences between the few reproductive and the many productive individuals in the social groups: the castes of the fertile sexuals (queens) and the infertile workers.

Over 10,000 different species of social insects exist today at different evolutionary levels of social organization, making the development of castes and the analysis of their special functions, in particular concerning reproduction, a still insufficiently studied, but physiologically and sociobiologically most interesting area of research.

Preface

Perception of Caste Problems

WOLF ENGELS[1]

Social life is a matter of course for all human beings. Recent research has shown that sociality is fundamental not only for primates, but for many warm-blooded animals as well. Insect societies, on the other hand, have been studied since ancient times. Beekeeping was already developed in antiquity and in several countries (Fraser 1951), for instance in Lower Egypt which from the first dynasty onwards was officially called Bee-land. Honey hunters and beekeepers had to carefully observe several aspects of the social organization of bee colonies, and the ancient naturalists usually quoted from their knowledge. The earliest known scientific treatise on beekeeping and a honey bee colony's life history is found in Vol. V of Aristotle's famous History of Animals, probably written 344-342 B.C. He reported that the queen (which he called "ruler") is the mother of the colony, and that queenless hives are only capable of producing drones and not workers. This probably represents the first study dedicated to problems of the castes and reproduction of social insects.

Only about 25 years ago was a theory presented, which attempted to explain the evolution of social life at least in the Hymenoptera (Hamilton 1964). Other hypotheses have been added to the discussion since then which has stimulated the investigation into many aspects of socioevolution, the most of which have been recently reviewed by Dietz (1986) with special attention to bees.

One of the unexpected recent conclusions reached concerning bees was that even in a "society" of two females, which in basically solitary species may occasionally share a nest, reproduction is monopolized by one of them, the dominant female (Sakagami & Maeta 1985). So castes apparently already differentiate even during initial steps of the establishment of social life. The availability of modern analytical techniques has stimulated physiological and biochemical research on problems of caste development and caste-specific functions. As it has turned out, despite much basic conformity, there exists a surprising diversity in the ways in which the differentiation of caste-specific characters and the participation of the castes in reproduction is realized and regulated in various groups of social insects.

---

[1] *Zoologisches Institut der Universität Tübingen, BRD*

The present book is an attempt to bring together the contributions of several authors, all of whom work in the field of social insect evolution, castes and reproduction, but who of course specialize in different groups of social insects. Consequently, this multi-author volume is not a mere sequence of chapters, but a series of individual contributions which try to focus and, at the same time, to review our present knowledge on castes and reproduction.

We have attempted in this book to use a consistent terminology and to define new or diversely used terms. For instance, caste determination is only used in accordance with the meaning of "determination" in genetics or in cellular and developmental biology: a programming of gene activation in order to control the following steps of development which in our context represent caste-specific differentiation. Caste development may be induced by trophogenic stimuli which are transformed into an endocrine reaction. The resulting juvenile hormone titer will control worker- or queen-like morphogenesis. Another example is the sense of the term pheromone. The original definition by Karlson and Lüscher (1959) was oriented on both that which was known about the female silkmoth, which by means of volatile chemicals calls males over long distances, and about the domineering effects of the queen in termite and honey bee colonies. In these cases, which concern releaser or primer effects of pheromones, it is clear that the message was meant and not a single chemical compound which made up one of the components of the transmitted signal. I would like to thank all the authors for observing this and I hope that we can in this way avoid some of the misunderstandings which often result from inexact usage of terminology in social insect literature.

The first step towards this publication resulted from a symposium which dealt with similar topics and was organized for an International Congress of Entomology. More authors joined the project, and I would like to thank all of them for their cooperation. I am also very grateful for the assistance and suggestions of Dr. Dieter Czeschlik of the Springer Verlag during the production of the book.

References

Dietz A (1986) Evolution. In: Rinderer TE (ed) Bee genetics and breeding. Academic Press, London, pp 3-21

Fraser HM (1951) Beekeeping in Antiquity. 2nd Ed Univ London Press, London, 142 pp.

Hamilton WD (1964) The genetical theory of social behaviour I, II. J Theoret Biol 7: 1-16, 17-52

Karlson P, Lüscher M (1959) Pheromone. Ein Nomenklaturvorschlag für eine Wirkstoff-klasse. Naturwiss 46:63-64

Sakagami SF, Maeta Y (1985) Multifemale nests and rudimentary castes in the normally solitary bee *Ceratina japonica* (Hymenoptera: Xylocopinae). J Kansas Entomol Soc 57: 639-656

Chapter 1

Sexual Castes and Reproductive Strategies in Termites

CHARLES NOIROT[1]

CONTENTS

1 Introduction

In ISOPTERA, the polymorphism differs from that observed in the social HYMENOPTERA principally in two ways: the bisexual composition of the society and the hemimetabolous development. The former relates to (and perhaps is a consequence of) the mechanism of sex determination, which differs from the haplo-diploid system of the HYMENOPTERA. The latter allows the immature stages to play an active role in the social organization and to be subject to different social influences. This, when combined with the versatility of the post-

[1] *Laboratoire de Zoologie, Université de Dijon, France*

embryonic development, promotes a higher complexity in the caste formation. This is especially noticeable in the extraordinary diversity of the replacement or secondary reproductives, which may be either true imagines, as in HYMENOPTERA, or have differentiated from nymphal instars or from the workers. In this review, three main topics will be examined: the differentiation of the reproductives (both primary and secondary), their maturation and behaviour, and finally the evolution of the reproductive strategies.

2 Differentiation of the Reproductive Castes

2.1 Sex and Genetics

Termites differ from HYMENOPTERA in having a more classical sex determination, with male heterogamety. Recent cytogenetical observations demonstrated in several species extensive sex-linked chromosomal translocations in the heterozygous condition in males, a result of successive interchanges between sex chromosomes and autosomes. This was observed in several KALOTERMITIDAE (Luykx and Syren 1979; Syren and Luykx 1977, 1981), in *Reticulitermes* (family RHINOTERMITIDAE) (Fontana 1980) and in 34 species of TERMITIDAE (Vincke and Tilquin 1978, 1983), and this may well be a general character of the ISOPTERA, although it was not found in *Mastotermes* (Bedo 1987). A substantial increase of sex-linked characters (previously autosomal) may thus be expected, and this assumption is supported by some recent findings (Luykx 1981, 1985). Increase of sex-linkage, introducing a dissymmetry between males and females of the same society, was believed to be important in emergence of eusociality in termites (Lacy 1980; 1984; but see Leinhaas 1983 for an opposite interpretation). However, this chromosomal peculiarity appears as a derived condition (Crozier and Luykx 1985; Bedo 1987). In *Incisitermes schwartzi* where interchange is maximal, the sex ratio is perfectly equal (Luykx 1986; the slight excess of males in alate producing colonies is related to earlier development and dispersal of the females), and the sexes are distributed randomly within the colony, both within castes and between castes (Luykx et al. 1986). An exact symmetry between males and females was thus the primitive condition, retained only in some primitive species. In the course of termite evolution, many assymmetries appeared, either in the sex ratio, or by the way of sexual dimorphism, or both, and these differences may be variously expressed in the separate castes of the same species (Noirot 1989). Only a few examples are given here, because the assymetries are much more important and varied in the neuter castes than in the reproductives.

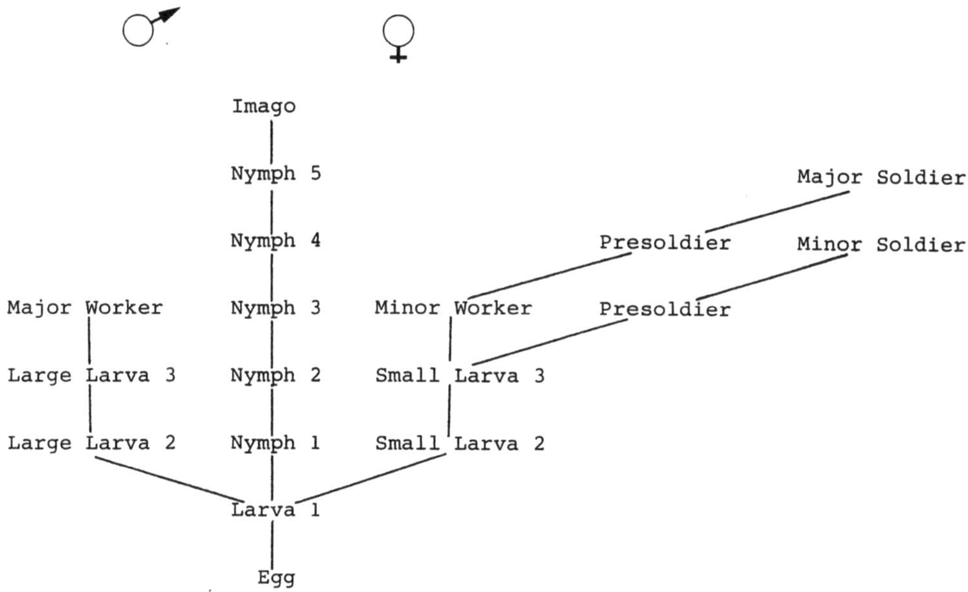

Fig. 1 - Caste development in *Macrotermes bellicosus*. All the first instar larvae are morphologically similar. They give rise, after the first molt, either to nymphs (with wing anlage) or to second instar larvae, which exhibit a sexual dimorphism in size. These larvae develop (through two additional molts) into workers (male major workers and female minor workers). The two kinds of soldiers differentiate from small third instar larvae and minor (unpigmented) workers respectively, and are all females.

Even in a primitive species such as *Kalotermes flavicollis*, the sexual inhibition is not the same in male and female reproductives (see 2.3). In higher termites (family TERMITIDAE), a sexual dimorphism in workers frequently occurs, and the soldiers are all of one sex. In *Macrotermes* for example, the large workers are males, the small females, and the two types of soldiers develope from the female neuter line (Fig. 1). In some extreme cases, as in *Trinervitermes*, each sex becomes specialized into one neuter caste, the soldiers being males, the workers females (Noirot 1955, 1969). However, it is possible, in experimental conditions, to obtain soldiers of the "wrong " sex, either in homosexual groups (Noirot 1955), or by application of JH analogs (Okot-Kotber 1980; Luamba 1980). These observation, with the frequent reversibility of the caste differentiation, strongly suggest epigenetic mechanisms in the caste formation: all the individuals in a given species possess the complete set of genes coding for the characters of the different castes, but their differential expression may be affected by some sex-linked genes giving rise to the sexual bias so frequently

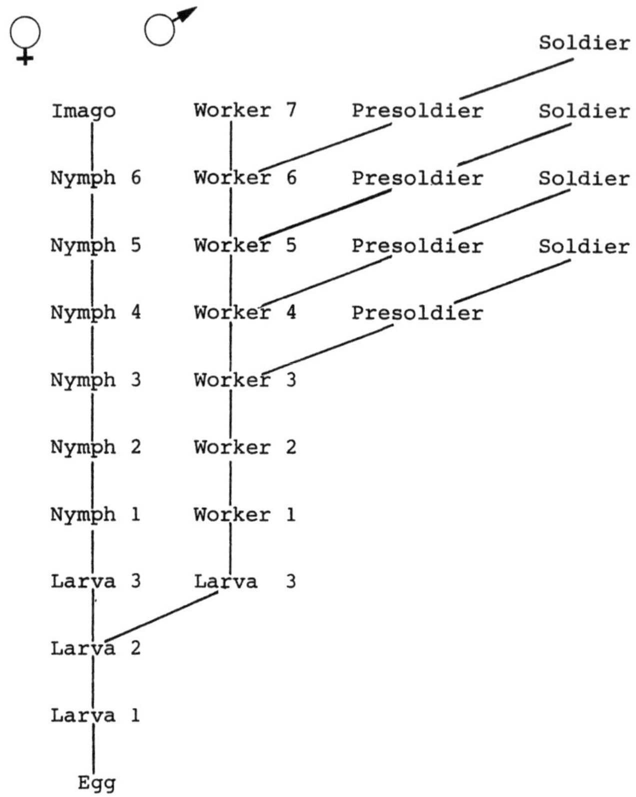

Fig. 2 - Caste development in *Anacanthotermes ahngerianus*. The sexual line, giving rise to the imagines through the 3 larval and 6 nymphal instars, includes both male and female individuals, whereas the neuters, morphologically recognizable after the second molt, are all males. The soldiers develop from several worker instars. From the data of Mednikova (1977).

observed. Meanwhile, in some genera, this bias is so striking that all the neuter castes are of one sex only. The first, *Anacanthotermes*, is a subdesert genus, belonging to the family HODOTERMITIDAE, and two species were studied with the same conclusions: *A. ochraceus* from Sahara (Clement 1953) and *A. ahngerianus* from Turkestan (Zhuzikov et al. 1972; Mednikova 1977). The differentiation between the larvae (neuter line) and nymphs[2] (sexual line) is visible after the second larval molt, but only the

(2) It is the custom, in termites studies, to call "larvae" all immature instars without wingpads and "nymphs" those with wingpads and thus engaged in the differentiation towards the sexual alates.

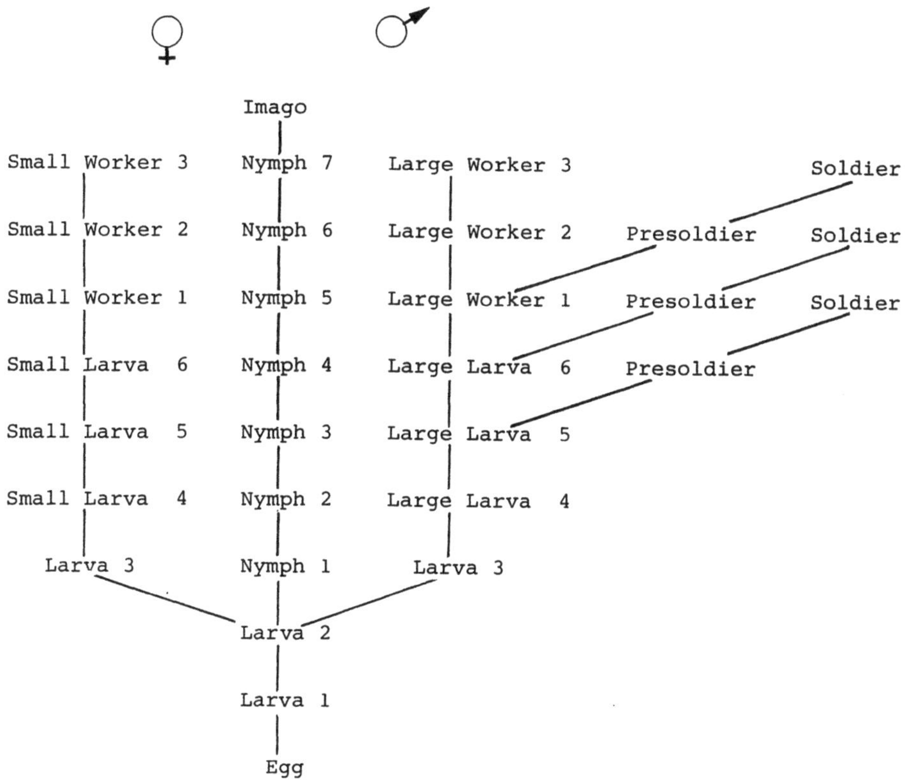

Fig. 3 - Caste development in *Hodotermes mossambicus*. This figure is to be compared with the Fig. 2, for the two genera are closely related. The main difference is the existence of two worker lines in *Hodotermes*, which is the expression of a sexual dimorphism. The large (male) workers are fully comparable with the workers of *Anacanthotermes*, and likewise are the origins of the soldiers. From the data of Luamba (1980).

male individuals can become workers and soldiers; consequently all the neuters are of the male sex, the sexuals beeing either male or female (Fig. 2). The polymorphism of another HODOTERMITIDAE, *Hodotermes*, very close to *Anacanthotermes* is, at first sight, completely different (Fig. 3) from the observations of Luamba (1980): here the neuter line, which differentiates after the second or third molt, comprizes both males and females, with a sexual dimorphism giving rise to two types of workers, the larger (males) and the smaller (females), each with several successive instars. However, the soldiers develop from the large workers and thus are all males. Comparing the development of castes in these two genera (Fig. 2 and

3), it is possible to hypothesize that *Anacanthotermes* originated from the *Hodotermes* type by suppression of the small female worker line. However, the mechanisms remain completely unknown. The second genus is *Schedorhinotermes* (family RHINOTERMITIDAE), two African species of which were studied by Renoux (1976). The situation is reverse of the preceeding, as all the neuters are of the female sex; Renoux noted furthermore a higher proprotion of females in the sexual caste (alates) in most (but not all) colonies examined, which further increased the disproportion in the numbers of the two sexes. Another Rhinotermitid of the same subfamily (RHINOTERMITINAE) was recently investigated by Roisin (1988): in *Parrhinotermes browni* too, all the neuters are females. From my observations, made on a limited number of specimens (Noirot 1989), *Cornitermes cumulans* (TERMITIDAE, NASUTITERMITINAE) may perhaps be added: all the neuters examined were found to be males. In all these termites the sex-ratio is completely abnormal. During most of the year, only neuters are produced, all of the same sex, and even at the season of sexual differentiation, the production of neuters continues uninterrupted. It seems necessary to postulate a special mechanism of sex and caste determination. Preliminary observations made by Renoux (1976) in *Schedorhinotermes* indicated a different karyotype in somatic (2n = 42) and germinal (2n = 48) cells, which suggests a chromosomal rearrangement during embryogenesis, perhaps in relation to sex and caste determination.

As pointed out by Crozier (1980), additional researches in the field of termite genetics are urgently needed.

2.2 Developmental Pathways

Assuming that each egg contains the same genetic information, the caste differentiation can be seen as a developmental process. By reference to the non-social insects, normal development ends in the formation of sexual alates; the other morphological castes are deviations occurring at some "decision points" and are either reversible or not. Due to the hemimetabolous development, it is easier in termites to follow the developmental steps and to identify the decision points, or more precisely, the first visible expression of the differentiation. In termites, the first anlage of wing pads seems the apparent indication of the sexual differentiation and appears at one or more definite developmental instar, and is expressed at the molt in some individuals which are then orientated towards sexual differentiation, whereas other individuals, without wing-pads, differentiate in the neuter castes. The time of this overt differentiation between neuters and sexuals is now known for a number of species; however, this is the morphological evidence of a commitment, a physiological induction, which may have occurred earlier. The first distinction between the commitment and its first morphological expression (overt differentiation) seems, in my opinion, an important one, not always taken into account.

As regards the time of the overt differentiation, a general trend of evolution seems apparent: the more evolved the species, the earlier the differentiation (for a different interpretation, see Watson and Sewell 1981). In several well studied primitive species (*Kalotermes flavicollis, Zootermopsis angusticollis, Prorhinotermes simplex*) nearly all immature individuals (with the exception of the youngest and the pre-soldiers) may perform several types of molt and develop into either a sexual (development of wing-pads, eyes ...), a soldier (development of the mandibles), a pseudergate or pseudo-worker[3] (regressive or stationary molt) or a replacement reproductive (review in Miller 1969; Lüscher 1974; Noirot 1985a; Noirot and Pasteels 1987). The wing pads appear very late (not before the fourth molt and often later), and the differentiation is completely reversible until the last nymphal instar, inclusively. No evidence exists about a commitment before the overt differentiation; it would be, at the best, very weak as indicated by the developmental versatility. *Reticulitermes lucifugus* seems more evolved; the nymphs are differentiated from the larvae after the second molt, and a worker caste (with several successive instars) is morphologically distinct. However, the variability of the developmental pathway remains very great, as shown by the extensive research of Buchli (1958). In the higher termites (family TERMITIDAE), the differences between the larvae (neuter line) and the nymphs (sexual line) are always visible after the first molt (Noirot 1955), and seem to be irreversible in normal conditions. The commitment occurs, at the latest, during the first larval instar. An earlier determination (before hatching) remains a possibility, supported by some evidence, especially the observations of Mensa Bonsu (1978) in my laboratory. In *Amitermes evuncifer*, the gonads of the young nymphs are much larger than in the larvae, and this difference was already observed in the first instar, externally undifferentiated larvae, even in the newly hatched individuals. Unfortunately, only a small number of specimens were available, and the observations point to, but do not demonstrate a determination prior to the eclosion of the egg. However, in *Macrotermes*, the differences in the gonadal development were only observed in the late first instar larvae, not in the newly-hatched ones. In another MACROTERMITINAE, *Odontotermes stercorivorus*, the difference in gonad development was observed by

(3) No true worker caste exists in these societies. Its role is performed by pseudo-workers, which are large larvae arrested (reversibly) in their development or nymphs having undergone a regressive molt (Grassé and Noirot 1947).

Truckenbrodt (1986) to occur early in the first larval instar and "seems to be determined in the embryo or in phase 1 of the 1st larval stadium". Thus the exact timing of commitment remains unclear and may vary somewhat depending on the species.

The caste differentiation is integrated within a fairly definite seasonal cycle, especially in the higher termites. The neuters are produced during the whole year [4], but the young nymphs only appear during a limited period (may be several weeks), frequently at the end of the rainy season in tropical countries. In a few species, two batches of nymphs may be produced in a year, in the same society, as in *Cephalotermes rectangularis* (Noirot 1969), but this seems quite exceptional. This cyclic production of young nymphs (with the consequence of a cyclic production of alates and the seasonal swarming) seems well synchronized with the climatic events, but no information is available on the way in which the climatic changes may control the polymorphism. The seasonal cycle is as much evident in the species with well air-conditioned nests as in those nesting in more exposed conditions, a fact which precludes a direct climatic influence on the eggs or on the brood. Most probably, the external factors (temperature, rainfall, food etc.) modify the behaviour and physiology of the workers, which in turn may influence either the queen, the eggs or the youngest larvae. In addition, a circannual clock may not be dismissed, although no direct evidence of such a clock exists. However, from the limited observations done, the seasonal cycle, in tropical or subtropical climate, seems less defined or absent in the KALOTERMITIDAE, a primitive family, than in the higher termites, although the former, nesting in dead wood, are not protected against external influences.

A further complication in the caste development is introduced by the differentiation of supplementary, or secondary reproductives (review in Noirot 1956; 1985b), which in most species occurs after the death of the primary sexuals, but in a few species can even occur in the presence of the royal pair (i.e. *Reticulitermes*, Buchli 1958). The secondaries are of different types and different origin, depending on the species. Sometimes, there are simply imaginal alates (named "adultoids") maturing in their nest without swarming. A very special case is observed in *Nasutitermes princeps* with the differentiation of "microimagoes" which cannot fly but can mature into the nest as replacement reproductives, concurrently with normal adultoids (Roisin and Pasteels 1985; 1986). More commonly, they differentiate, through one or several special molts, from some developmental instars. When the secondaries originate from nymphs (nymphoid reproductives), they retain a nymphal morphology (with

---

[4] Except when the reproduction is interrupted in winter in temperate climates or at altitude.

variously modified wing pads), and are to be considered as *neotenics*. When they originate from workers, their morphology remains essentially worker-like (ergatoid reproductives), although in several cases a development of the meso- and metathoracal nota, and of the eyes and ocelli is evident. These ergatoids may also be considered as neotenics, for their morphology is far simpler (larval-like) than that of the imagines, and workers are not imagines (Noirot and Pasteels 1987). In the few species where the differentiation of supplementaries was well analysed, the transformation of a nymph requires one molt, but two molts at least are necessary for the transformation of a worker (three molts were recently evidenced by Roisin and Pasteels (1987), in *Nasutitermes novarumhebridarum*). Once differentiated, the supplementaries cannot molt again. It is to be noted that a supplementary reproductive is never identical to a normal nymphal or worker instar. Its morphology is not always a firm indication of its origin, for example, during the transformation molt of a nymph, the wing pads may diminish or even disappear, they may appear during the transformation of a worker. As a consequence, a great diversity of supplementary reproductives exists in termites, and several types may be observed in the same species (Noirot 1956; 1985c). In most of the lower termites (KALOTERMITIDAE: *Kalotermes*, *Neotermes*;TERMOPSIDAE: *Zootermopsis*; RHINOTERMITIDAE: *Reticulitermes*, *Prorhinotermes*) every developmental instar, with the exception on the youngest and the presoldiers, can become a neotenic supplementary (review in Miller 1969; Lüscher 1974; Noirot 1985a). However, in *Mastotermes* (MASTOTERMITIDAE), the neotenics develop (perhaps exclusively) from the workers (Watson and Abbey 1985) and in HODOTERMITIDAE only from the nymphs (Mednikova 1977; Luamba 1980). In the higher termites (TERMITIDAE) the situation differs considerably according to the species; detailed observations were made in very few cases, and for numerous genera no supplementaries were ever found. Adultoid supplementaries were observed in many species, and seem the only type present in the fungus-growing termites (MACROTERMITINAE). In some cases, several types may be observed in one species (but in different colonies), for example, adultoids and ergatoids in *Termes hospes*, nymphoids and ergatoids in *Microcerotermes amboinensis* (Noirot 1956). In relation to the more rigidly fixed developmental pathway in the higher termites, the deviation toward the differentiation into supplementaries seems restricted to a limited number of instars, or even suppressed.

2.3 Mechanisms

The process of caste differentiation appears regulated at two principal levels. In a given individual, the development is the result of a differential expression of the gene pool, each caste being determined by the activity of a special set of genes. This expression (and the repression of genes determining the other castes) seems modulated by the internal milieu, i.e. the hormones and neurohormones, as are the successive steps in the development of all insects (Noirot and Bordereau 1988). In the society, however, the activity of the neuroendocrine system is under the influence both of external factors (temperature, humidity...) and of the complex interactions between members of the colony.

The endocrinology of the caste formation will be only briefly examined, for it was the subject of several reviews (Lüscher 1974, 1976; Lanzrein et al. 1985; Lebrun 1985; Noirot and Bordereau 1988). Even if the importance of hormones, and especially JH, is well established, we are far from a complete understanding of their role in polymorphism, especially for the reproductive castes. Two points will be emphasized. First, the possible importance, at least in higher termites, of the hormones (JH and ecdysteroids) synthesized in the queen and stored in the eggs (Lüscher 1976; Delbecque et al. 1978; Lanzrein et al. 1985), which may influence the development, and possibly relate to the seasonal cycle. Second, the ability of an insect to react to hormonal activity (for example the application of exogenous hormones) seems to vary during the development, especially in the higher termites. In *Macrotermes*, an application of a JH analog to the larvae enhances the production of soldiers as in all the species so far studied (review in Howard and Haverty 1979). However, only the larvae of the third instar may be commited to soldier development. It is interesting to observe that, in normal conditions (Fig. 1) it is *this* instar which gives rise to the small soldiers, but only the small (female) larvae. With application of JH analog, some large (male) third instar larvae are also transformed (Okot-Kotber 1980). Although this example is not related to the differentiation of reproductives, it underlines the importance of some critical steps of the development where one or another factor may act preferentially, if not exclusively.

As regards the extra-individual factors, we have almost no information about the action of the temperature, humidity etc.. In most temperate species, the reproduction is interrupted during the winter, but we have no idea of a possible influence of the cold on the subsequent formation of caste (as frequently evidenced in ants: Buschinger, this volume), and in termites where no such arrest is observed, the mode of regulation of the seasonal cycle remains unknown. The social influences seem especially important, but our information remains fragmentary. The problem of the nutrition of the immature instars is a complicated one. The feeding of the brood underwent profound modifications during the evolution of the Isoptera, in relation both of the symbiotic associations (proctodeal feeding restricted to the lower termites, with intestinal flagellates) and a progressively more complete dependence of the brood (Noirot 1952). Thus, liquid secretions (believed to be produced by the salivary glands) seem of paramount importance especially in the higher termites, but the composition of the "saliva" remains unexplored. Indeed, we have no evidence of differential feeding of larvae of the same instar, but the question has never been studied carefully in any species. Although the importance of nutrition in the differentiation of reproductives was sometimes put forward (i.e. Buchli 1958, for *Reticulitermes*), the arguments are scanty and the problem deserves further research.

Other social interactions are probably more important, especially chemical information, circulated by food-exchange and grooming, activities especially developed in termite colonies (Bordereau 1985). It is of interest to remember that the first evidence of social pheromones was obtained in termites (Light 1944; Lüscher 1955) and the elaboration of the concept of pheromone, and of the word itself, was a direct consequence of these works (Karlson and Butenandt 1959; Karlson and Lüscher 1959). The pheromones implicated in caste determination ("primer" pheromones, Wilson and Bossert 1963) act through the neuroendocrine system. It is nor entirely clear whether they act primarily on specialized chemoreceptor sensilla or directly on the target organs (neurosecretory cell, endocrine gland, or even final effector), although, in my opinion, the former seems more probable. Whatever it may be, the study of the primer pheromones is especially difficult, because they are effective only after prolonged application, their action observable after long delay, and only on a part of the population. This is especially frustrating in termites, where development is slow in comparison to the HYMENOPTERA, and the pheromones seem quickly destroyed in the society. Thus, in most cases, the knowledge of these pheromones is founded on indirect evidence.

The best documented example is that of *Kalotermes flavicollis*, following mainly the research of Lüscher (reviews in Lüscher 1974, 1975, 1977). The royal pair normally has an inhibitory effect on all the members of the society, which prevents the maturation of nymphs and larvae and slows down their imaginal differentiation. Two inhibitory pheromones are involved, one produced by the queen, the other by the king; they are transmitted by proctodeal feeding, and appear to be complementary, as a complete inhibition is only obtained when the two are acting together[5]. Most probably, these pheromones modify the hormonal equilibrium, mainly by a stimulation of the

[5] Some minor discrepancies remain on the respective role of the female and male pheromone (see Grassé and Noirot (1960a) on the one hand, and Lüscher (1964) on the other), but they are not important in the present discussion.

corpora allata. This hypothesis explains both the inhibition of neotenic formation (which is initiated when the *corpora allata* are inactive) and the retardation of imaginal differentiation. The same pheromones may also explain the stimulation of soldier production by the royal pair (Springhetti 1970), for the soldier differentiation is triggered by JH. However, the story is certainly much more complicated and it is difficult to imagine the post-embryonic development conditioned only by *inhibitory* actions. A combination of both inhibitory *and* stimulatory actions (mainly but perhaps not exclusively pheromonal) seems more likely. Thus, when nymphs of *Kalotermes* were reared in isolation, the proportion of neotenics produced was much less than within groups from the same colonies (Grassé and Noirot 1960b). Likewise, Lüscher (1964) noticed the stimulation of female neotenics by extracts from male reproductive heads. This double conditioning is now well evidenced in soldier regulation. For example, in *Nasutitermes lujae*, the soldiers produce in the secretion of their frontal gland an inhibitory pheromone slowing down the differentiation of new soldiers (Lefeuve and Bordereau 1984). At the same time, the royal pair stimulates the soldier differentiation, probably also by another pheromone (Bordereau and Han 1986).

In other species of termites the inhibitory pheromones produced by the royal pair were evidenced in *Zootermopsis* (Light 1944) and *Neotermes* (Nagin 1972). The mechanism proposed for *Kalotermes* could be valid for these termites too, although Stuart (1979) cannot observe in *Zootermopsis* the proctodeal transmission of inhibition. A strange situation seems to exist in *Mastotermes* where the formation of neotenic reproductives is *stimulated* by the neotenics still existing and very rarely occurs in their absence (Watson et al. 1975; Watson and Abbey 1985). In *Reticulitermes lucifugus*, the inhibition seems very weak or even absent (Buchli 1958). In the higher termites, the experimental evidence is very limited. To my knowledge, the only example of replacement reproductives obtained in laboratory cultures was with the adultoids of *Astalotermes* (formerly *Anoplotermes*): in this termite, the alates isolated with 100 to 200 workers quickly casted their wings and matured; eggs appeared 8 to 36 days later. When nymphs (from the last or last-but-one instar) are isolated with workers, they continued their normal development and only matured after the imaginal molt (Noirot 1956). These experiments (as well as the appearance in other species of supplementaries in field colonies deprived of their royal pair) strongly support the idea of an inhibition, exerted by the reproductives. To sum up the differentiation of supplementary reproductives seems in most (but not all) species triggered by the suppression of an inhibition normally exerted by the functional reproductives (in addition, some stimulatory influences remain a possibility). In many lower and some higher termites, a predisposition for a premature sexual maturation, or neoteny, is evident, and may be expressed in one or several developmental instars, either in the nymphal or the neuter line or both. A reduction of the neotenic differentiation during the evolution of the phylum and its dissappearance in many higher termites may be hypothesized.

As regards the differentiation of **primary** reproductives, i.e the alates, much less information is available. A possible hypothesis again postulates an inhibition of the functional reproductives, favouring the development of the neuter castes. Such an inhibition is supported by some observations on *Kalotermes* and *Zootermopsis* (Lüscher 1977) but cannot explain the high frequency of regressive molts in individuals reared in isolation (Grassé and Noirot 1960 b). The experiments of Bordereau (1975b) give an interesting insight on higher termites: he suppressed the royal pair in several nests of *Macrotermes bellicosus* soon after the swarming season (march) and observed, 7 to 9 weeks later, the appearance of young nymphs in 8 out of 11 nests. In this species, the young nymphs normally appear only in september and the differentiation of nymphs four or five months in advance may be related to the suppression of an inhibition from the royal pair. However, the phenomenon does not occur very regularly. In the 8 nests the nymphs were numerous in only 4. Additionally, the same kind of experiments on *M. subhyalinus* gave negative results (Sieber and Darlington 1982). In my opinion, the commitment toward sexual vs. neuter development may be realized in two or more steps; for example, in *Macrotermes*, the first predetermination could take place in the eggs during the oogenesis (possibly by variations in the hormone titers). However this determination could be changed during the embryonic development and/or the first larval instar by social influences, especially inhibitory pheromones either directly or by modifying the behaviour of workers. At our present level of knowledge, this scheme remains largely a working hypothesis.

3 Maturation and Behaviour of the Sexual Alates

A normal adult society produces, once a year (rarely twice) at a definite season, a batch of alates which remain at rest in the nest until the swarming. The period between the imaginal molt and the swarming allows the alates to wait for the favourable weather conditions which trigger the flight behaviour (Nutting 1969). The stay in the nest usually does not exceed several weeks, but can be longer. For example, the alates of *Macrotermes subhyalinus* can remain in the nest as long as four month in the dry grassland of Kenya (Darlington 1977). *Anacanthotermes ochraceus* from the Sahara desert seems to be an extreme case, for the alates emerge in October-

November but their flight occurs from January to May of the next year, depending of the rainfall (Clement 1956). During their stay in the nest, the alates remain clustered in the peripheral chambers. They are motionless, their food intake seems very limited and their sexual behaviour, as well as the maturation of the gonads, is completely inhibited. This period ends by the exit of the alates, which results from a complex and coordinated sequence of events involving the whole society. For example, holes giving a temporary direct communication with the exterior are opened by the workers, with, in some species, the building of special structures. Very often, a number of workers and soldiers exit and move around the swarming holes. These changes in behaviour (both for the alates and the neuter castes) are controlled by weather conditions, and for each species the flight occurs after a definite pattern of temperature, rainfall, wind ... (review in Nutting 1969).

During the time between the imaginal molt and the swarming, the alates undergo a physiological change after which they are able to swarm. This "maturation" (not directly related with the development of the genital organs) may be tentatively subdivided in two periods. During the first, alates remain in the central parts of the nest, intermingled with the other castes, and afterwards they are clustered in peripheral chambers, with very little contact with the workers. The second period is probably the longest, and the more variable, allowing a wait for favourable weather conditions. The observations made on *Hodotermes mossambicus* in South Africa give some insight into this second period (Hewitt and Nel 1969; Hewitt et al. 1972; Watson et al. 1972). When groups of 30 or more alates were maintained in cultures (even after the exit from the nest), they retained their gregarious (non-aggressive) behaviour and their ovaries did not mature, in contrast to those reared in pairs, either homo- or heterosexual. The inhibition of maturation in grouped alates seems essentially determined by the number of body contacts perceived by the antennae, which were fewer in the pairs than in the groups. No pheromonal influences could be demonstrated. However, the conditions may be more complex because individuals held in solitary confinement were in a state intermediate between those maintained in pairs and those in groups. The inhibition in the groups originated in the alates themselves, not in the royal pair. The situation seems different during the first period, although no definite information is available. As stated in 2.2, the alates may, in some species, become replacement reproductives in their own society. This premature maturation is probably the result of the removal of an inhibition normally exerted by the royal pair. The preceding observations on *Hodotermes* preclude such an inhibiton during the second period, whereas there is some indirect evidence for its action during the first. Indeed, adultoid reproductives were found in several cases with incomplete pigmentation suggesting that they matured just at the moment of the imaginal molt (Noirot 1956). Thus we may suppose, as a working hypothesis, that the inhibition by the royal pair only affects the alates during a short period following the imaginal molt. Afterwards, when the fully pigmented alates are gathered together in special chambers, with few contacts with their nestmates, the inhibition could be maintained by the crowding itself. At the same time, the alates become responsive to the weather pattern which induces flight behaviour. Clearly, more information is needed about the changes in the physiology and behaviour of the alates inside the nest, between the imaginal molt and the time they are ready for flight.

With regard to the endocrinological events associated with the sexual maturation, our knowledge remains fragmentary. The development of the *corpora allata* is evident in all types of reproductives (Noirot 1969) and seems to be in proportion to the fecundity of the queen. For example, in large queens of *Macrotermes*, which lay about 40.000 eggs per day, the volume of the *corpora allata* increased nearly one hundred fold from the time of swarming (Lüscher 1976). This explains the enormous quantity of juvenile hormone present in the queen and especially in her haemolymph (Mayer et al. 1976; Lanzrein et al. 1985). Most probably, as in many other insects, JH acts as a gonadotropic hormone, but the storage of JH in the eggs may have other significance, as discussed previously. In the king (especially in *Macrotermes*), the enlargement of the *corpora allata* is also apparent, although limited, and the JH titer in the haemolymph very high (Lüscher 1976; Lanzrein et al. 1985). On the other hand, the prothoracic glands degenerate completely in both primary and secondary reproductives immediately after the last molt (Noirot 1969). Thus, the ecdysone demonstrated in the queen has another origin, most probably the follicle cells of the ovary. Unlike the JH levels, the ecdysone titer is low in the queen's haemolymph (it is mainly concentrated in the ovaries), and it is doubtful whether this hormone has any influence on ovarian maturation. As it is stored in the eggs, its main purpose is probably for embryonic development. The behaviour of flying alates and the pairing were well reviewed by Nutting (1969)[(6)]. Two points only will be discussed here. The first is the question of inbreeding versus outbreeding during the pairing, which may have important consequences on the genetical structure of the populations (Crozier 1980) and even on the evolution of eusociality in termites (Bartz 1979). Unfortunately, our information is still only anecdotal. The synchronized flights of many colonies in the same area without doubt favours the pairing between alates of

---

(6) See also Nutting (1979), for the possible adaptive significance of the timing of flights during the day-night cycle.

different colonies, as does the assemblee of alates before pairing at the top of the trees (hill-topping behaviour) observed in a few species (Noirot and Bodot 1964). On the other hand, the flying distances seem extremely variable according to the species; when this distance is very short inbreeding is much more likely (Grassé and Noirot 1951). Moreover, inbreeding is the rule in the foundation of new colonies by "budding", for the supplementary reproductives are brothers and sisters. Most probably, the degree of inbreeding is variable, and I am inclined to asssume, in most species, either a moderate or a large amount of outbreeding between the colonies of a given area. However, as the spatial distribution of species is often discontinous, exchanges between more distant areas (for example two savannas separated by a strip of forest) seem much more limited. This may leed to the formation of relatively isolated populations, local races and finally isolated species. The problem deserves much more investigation using population genetics techniques (see for example Clement 1981; Reilly 1987). Mention must be done here of the hypothesis put forward by Bartz (1979) according to which a regular alternation of outbreeding (by flying alates) and inbreeding (by replacement reproductives) was a key factor in the evolution of social life in termites. From the review of Myles and Nutting (1988) this hypothesis seems poorly supported.

The second point is related to the nutrition of the founding pair and the incipient colony. In ants, the founding queen rarely collects food outside the nest and in a large majority of species depends on its own reserves for her nutrition and that of the first brood. These reserves are not only in the fat body but also in the flight muscles which are quickly resorbed after the nuptial flight (Wilson 1971).

In termites, the collection of food by the founding pair is more frequent, and seems the general rule in wood-inhabiting species, where the nest is bored in the food itself (majority of the lower termites). When the incipient colony develops in a subterranean chamber (most of the higher termites), no food is available until the differentiation of the first workers, except perhaps for the humus feeders. Indeed, in *Cubitermes fungifaber* (such an humivorous species), the royal pair ingests only a limited amount of earth before the emergence of the first brood of workers, and thus undergoes a spectacular weight loss (about 40% in 8 months). Interestingly, the curves of weight variation of the king and the queen are strictly parallel during this period. Thus, a very precise optimisation is realized for the utilization of the nutritive reserves of both the members of the founding pair; the queen, beeing in charge of the oogenesis, seems less involved in the feeding of the brood, which seems the main task of the male (Han and Noirot 1983). Likewise, in *Odontotermes assmuthi*, a fungus grower where the founding pair cannot take external food, the degeneration of the flight muscles (a significant source of nutriment although not so important than in ants) occurs at the same rate in the two sexes (Basalingappa 1982). In this manner, the termite society takes advantage of the necessity of a pair (instead of a single female) for the foundation of a new colony.

4 Evolution of the Reproductive Strategies

In solitary species, the size and variation of the population are determined by three main elements: fecundity, longevity and mortality. The same is true for the eusocial insects, but these factors must be considered both at individual level (particularly for the reproductives) and for the society as a whole.

4.1 Fecundity

4.1.1. The fecundity of the primary queen varies tremendously among termite species, in relation to the structure and physiology of the ovaries. In a primitive genus such as *Kalotermes*, each ovary is composed of only 7 ovarioles and the queen can scarcely produce more than a few dozen eggs per day. There is a clear tendency for an increase in the ovariole number in more evolved species, culminating in the higher termites (family TERMITIDAE) where each ovary is composed of between several hundred and several thousand ovarioles (review in Weesner 1969). This tremendous increase is correlated with a special type of ovarian physiology without counterpart in other insects: the **successive** maturation of the ovarioles along a postero-anterior gradient. At the time of colony foundation, very few ovarioles become functional (for example, 6-11 out of more than 400 in *Tenuirostritermes tenuirostris*, according to Weesner 1955). In proportion to the increase in the colony population, more and more ovarioles become functional, but the process is very slow taking several years. For example, Fig. 4 is a transverse section of the ovary of a young queen of *Macrotermes bellicosus* of about 3.5 cm in lenght (the maximum in this species is nearly 12 cm), it was estimated to be at least two years old. At this level (anterior quarter of the ovary) not one of the numerous ovarioles around the oviduct shows any sign of maturation, and in the most advanced oocytes the follicular envelope is not yet complete (Fig. 5). Furthermore, during this maturation, the lenght of each ovariole increases progressively (until 9.4 mm) as does the number of maturing oocytes (Truckenbrodt 1973). The gradual development of the ovaries allows a progressive increase of the fecundity closely related to the growth of the society: very slow in incipient colonies, the physogastry develops more quickly after introduction of additional workers (Sieber and Leuthold 1982), and much more so during the development of adultoid reproductives, which occurs with a large worker population (Sieber and Darlington 1982; Roisin and Pasteels 1986). In most species, the enlargement of the ovaries exceeds by several orders of magnitude the initial abdominal volume. For example, an old queen of *Macrotermes* has an abdomen about five hundred times larger than that of an alate; the king is almost unmodified. The phenomenon in the queen, called physogastry, is of variable intensity, and involves a complex and coordinated growth of all the abdominal organs and tissues, which is very rare in adult insects. This has been investigated recently by Bordereau in my laboratory. It is especially noteworthy that the epidermis is reactivated, cell multiplication and cuticle secretion occurring troughout the life of the queen (Bordereau 1979; 1982) as does the tracheal epithelium (Bordereau 1975a). The enormous growth of the cuticle (both in the integument and the tracheal system) appears to be very complex, involving not only the secretion of new layers of the cuticle but also a remodelling of the previously secreted layers, including a true growth of the epicuticle. The internal organs are also conspicuously modified (Truckenbrodt 1973), especially the mid-gut (Noirot-Timothée and Noirot 1965) and the fat body (Han and Bordereau 1982). The various aspects of physogastry are expressions of an adaptation which has no equivalent in the insect world. However, the physogastry is highly variable. It does not exist in KALOTERMITIDAE, nor (probably) in TERMOPSIDAE, but is sometimes evident in other lower termites (HODOTERMITIDAE, RHINOTERMITIDAE). It is common in higher termites (TERMITIDAE) but with great variations even within the same subfamily (for example the abdominal volume is increased about 20 times in *Cubitermes* and 1.000 times in *Cephalotermes*, both genera of the subfamily TERMITINAE). The detailed studies of Bordereau were done in two species only, and showed some significant differences between them. Physogastry seems a general trend in ISOPTERA, but now it is not clear whether it appeared once, or several times during their evolution.

Our knowledge of the fecundity of the secondary or replacement reproductives is more limited. In *Kalotermes* the neotenic females possess the same type of ovaries (with 7 ovarioles) as the imaginal

---

Fig. 4 - Ovary of a young queen of *Macrotermes bellicosus*, cross-sectioned in the anterior quarter. No one of the numerous ovarioles ending in the oviduct are yet mature, whereas in the posterior part several hundred ovarioles are still fully mature (x 70).

Fig. 5 - Higher magnification of the proximal part of four ovarioles, at the same level as in Fig. 4. The most advanced oocytes have not yet completed their follicular envelope (x 510).

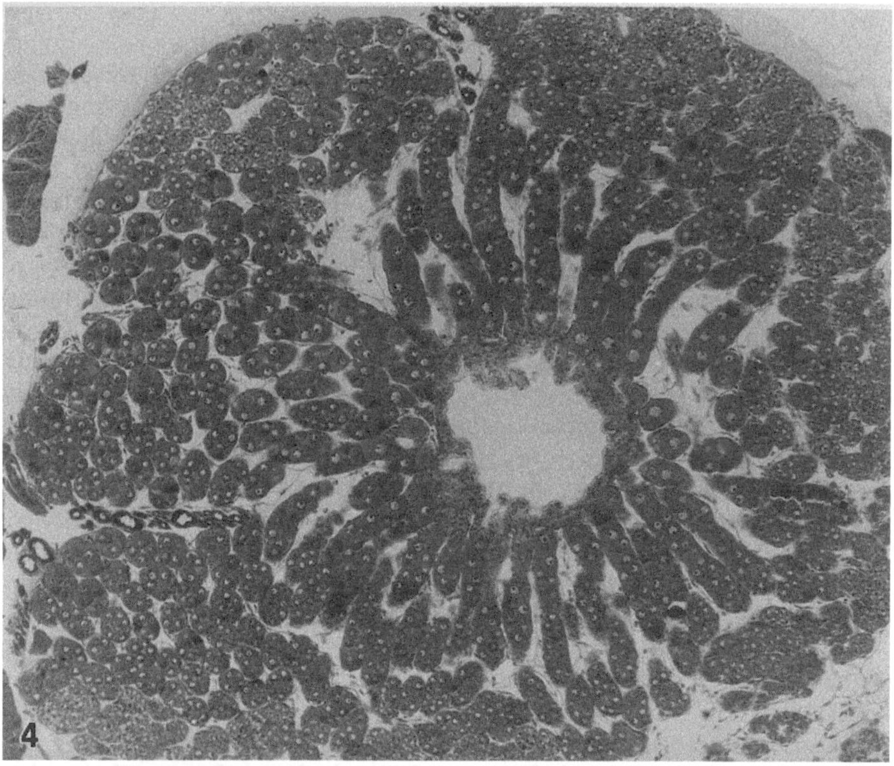

4

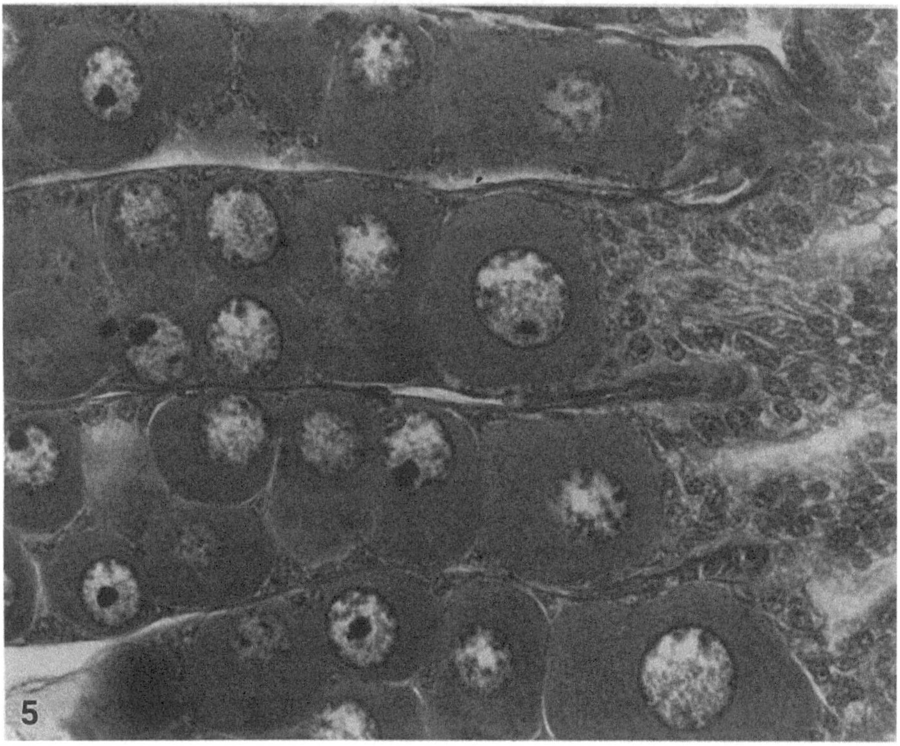

5

queens, and their fecundity seems similar. In more evolved species, the replacement reproductives may undergo a limited physogastry, which is very important only in adultoid secondaries (which are then sometimes difficult to identify as replacement reproductives). The ergatoid secondaries, especially in higher termites, do not enlarge very much their ovaries, develop a small number of functional ovarioles and their fecundity seems very low. Thus, the fecundity of replacement reproductives is, with the exception of some adultoids, generally far below that of primaries. Very often, however, this is compensated for by the presence of several, even numerous, secondaries, and in some species, as in *Reticulitermes lucifugus*, the differentiation of secondary reproductives seems a normal process during the development of the society, occurring probably in proportion to the increase in the population and to the extension of the nest. With adultoid replacement reproductives, the fecundity of the colony may be quickly restored, and even increased if more than one queen differentiates (Thorne 1984; Darlington 1985; Roisin and Pasteels 1985).

4.1.2. The eggs produced by the reproductives develop either in neuters or in sexual alates and these two categories are of different significance for the ergonomics of the society. Their production, in all eusocial insects, follows a definite temporal pattern. Using the terminology of Oster and Wilson (1978), it is useful to distinguish, in the life cycle of a colony, a **founding stage**, and (for perennial societies) an alternation of **ergonomic stages** and **reproductive stages**, where the energy investment is turned either to increasing ergonomic efficiency by production of a large number of workers and soldiers, or to the production of potential reproductives, whose number gives a measure of the fecundity at the colony level. The success of the reproductive stages is a direct function of the preceding ergonomic stage, especially in higher termites, where all developmental instars of alates are completely dependent on the workers for their alimentation. The number of alates produced each year (in comparison with the population of the nest) was counted in only a few species (review in Nutting 1969). All these counts gave high values (up to 25% or more of the population), and the investment appears even more important if the biomass (or energy content) of alates is taken into consideration, for in many species the size of alates (and even more their energy reserves) are greater than in non-reproductive castes (Nielsen and Josens 1978). Thus, in *Macrotermes subhyalinus* in the north of Senegal, the biomass of the alates is nearly equal to that of all the rest of the society (Lepage 1974). In *Nasutitermes corniger* (Panama), about 35% of the biomass is invested in alates (Thorne 1983).

So, the fecundity of a termite colony is very high, but the fertility (the proportion of alates founding a new colony) seems very low. According to species, the relative size of the alates varies greatly, and it seems that some species have chosen to invest in many small alates and other in fewer larger ones (see 4.4).

4.2 Longevity

For the workers and soldiers, the potential longevity seems high (compared to other social insects) up to several years, although precise informations are very scarce (review in Wilson 1971), but this factor is probably of limited importance in the energy budget, because the corpses are usually eaten and thus this energy recycled. More important is the longevity of the reproductives, which seems generally very great for the founding pair although our information is still mainly anecdotal. The longevity of the secondary reproductives is probably less, and seems especially short in the ergatoids of higher termites (Noirot 1956). In *Termes hospes*, my observations point to a continous production of short-living ergatoids, an expensive process energy-wise, but effective at any time of the seasonal cycle.

The societies of termites are often very long-lasting, may be several tens of years, but here again the very precise estimates are very few (review in Wilson 1971; Lee and Wood 1971). However, some observations indicate the possibility of large variations in the life span of the nests of different species in the same habitat. As for the analysis of population dynamics, this is probably the most neglected field of studies due to the need for long-term observations and experiments.

4.3 Mortality

In the absence of precise evaluation of mortality in natural conditions, we can only speculate about various adaptations which may minimize the energy loss caused by predation and adverse conditions. The most obvious is the appearance of the **soldier caste** very early in the evolution of the ISOPTERA. This caste is without equivalent in other social insects and is much more specialized, even in the most primitive termites, than the so-called soldiers present in some ants. This specialization is evident not only in the development and morphology but also in behaviour. Not only do the soldiers exhibit an extraordinary diversity of defense behaviour (Deligne et al. 1981) but they are so strictly adapted to this function that they are completely dependent on the workers for their feeding and take no part in the other tasks in the society (Grassé 1939). Thus, the only return for the investment is in defense[(7)]. The proportion of soldiers in the total population varies between narrow limits within a given species (although a change with aging of the colony may be observed) but differs enormously between different species. Some comprise in to 25% soldiers[8] but in several genera (the African species reviewed by Sands 1972) they are completely absent. A careful compilation of the available data (outside the scope of the present review) could possibly establish some correlation between the proportion of soldiers and the ecological niche. As an example, soldiers are always numerous in harvester termites (foraging in open field) and most, if not all, of the soldierless species are subterranean humus feeders. Furthermore, it would be necessary to consider the efficiency of different kinds of soldiers, and the various kinds of defensive behaviour exhibited by workers (especially in soldierless species) to have an estimation of the return-to-cost balance in soldier investment.

Another protective device is the architecture and situation of the nest, especially varied in termites (review in Noirot 1970). In all the termites, the nest is a closed space, without direct communication with the outer world and thus offers some protection against both predators and climatic hazards. However, depending on the species, the complexity of the nest and the energetic investment in its construction vary tremendously, and a correlation seems evident between this investment and the regulation of the microclimate of the nest on the one hand and the defensive efficacy on the other. It would be interesting to study whether some correlation exists between the nest structure and the proportion of soldiers, two main types of investment directed toward the protection of the society, and to apply to termite societies the distinction between *tychophobe* and *tychophile* species introduced by Oster and Wilson (1978) to underline the differences in defensive investments in ants.

4.4 Reproductive Strategies and Population Dynamics

The preceding considerations underline the interspecific variability of several important parameters, mainly the fecundity of the queen, the role and importance of replacement or secondary reproductives,

---

[7] However, a role of the soldiers in recruitment toward a new found source was evidenced in *Nasutitermes costalis* (Traniello 1982). See also Stuart (1981) for *N. corniger*.

[8] The proportion of soldiers is difficult to estimate with accuracy, due to their mobility and uneven distribution in the nest. A numbering of the whole society is necessary, and some values obtained by sampling the foraging population (i.e. Haverty 1977) seem overestimates.

and the investment in defense (nest construction and soldier formation). On the other hand, the investment in alate production seems less variable and is usually very high, the differences lying most probably in the chances of success of the flying alates. I have selected some examples where the main lines of population dynamics may be reasonably inferred from the available informations.

The colonies of *Kalotermes flavicollis* comprize a population from several hundred to a few thousand insects (the maximum I observed: 3.250), in relation to the low fecundity of the queen. Like the other KALOTERMITIDAE, it is a strictly wood-inhabiting species, nesting in the dead parts of trees, or in stumps, without any subterranean galleries which would allow extension of the nest into a neighbouring piece of wood. The investment in defense seems low: the nest is made simply of a network of galleries in the wood, without any actual construction, and the soldier caste represents only 2.9 + 0.92% of the active population(9). The society is thus not well protected and is subject to many dangers. The accidental elimination of the functional reproductives is not uncommon, as evidenced by the many replacementreproductives collected(10). The differentiation of secondaries (always neotenics) automatically occurs after the elimination of one or both the primaries and most of the nymphs and pseudergates which are receptive (the recently molted individuals) undergo the transformation in the following two or three weeks. Thus a great number of neotenics differentiate(11), most of which are eliminated by mutual aggression (Ruppli 1969) finally leaving only one pair to become functional. This process is very efficient and rapid, however it is very costly for the society due to the high mortality occasioned by the elimination of the surplus neotenics. On the other hand, it insures a potential indefinite life-span, and, in my opinion, the longevity of a colony of *Kalotermes* is mainly limited by the volume of the piece of dead wood it inhabits and from which it cannot escape. The production of alates may be very high, up to 30% of the population (excluding the young larvae), and the flying alates seem able to rapidly colonize every available niche. Thus, in the r-K scale defined by McArthur and Wilson (1967), *Kalotermes flavicollis* may be considered a r-strategist.

---

(9) Unpublished observations, resulting from counts of 28 colonies, made at Banyuls-sur-mer in different seasons of the year. The young larvae (until the 4th instar inclusively) were not taken into account, as well as the alates if present.

(10) In the same colonies, 35 reproductives were found, 20 primaries (8 males, 12 females) and 15 secondaries (8 males, 7 females).

(11) In laboratory cultures, Grassé and Noirot (1960b) obtained an average of 7.75% of neotenics, but the results are highly variable and probably follow a seasonal fluctuation.

Nasutitermes corniger was investigated by Thorne (1982, 1983, 1984, and personal communication) in Panama. It is an arboreal species, which constructs large carton-nests with a neuter population as high as 800.000 individuals. The physogastry of the queen is important (maximum wet weight recorded: 0.56 g, whereas the alate weight is approximately 0.012 g), and she has a high fecundity as evidenced by the number of the eggs collected in the nest (up to 155.000). The investment in soldier production is high (as in most *Nasutitermes*), and the soldiers represent 25% of the neuter population. Although well constructed and possessing a very hard royal chamber, the nest does not offer complete protection against ants, the small young nests seem especially vulnerable. About 1/3 of the colonies are headed by several (up to 33) physogastric imaginal queens and are thus polygynous (multiple kings were found less frequently). Neotenic reproductives were nerver observed. In only one colony, numerous ergatoids were found (Thorne and Noirot 1982), but it is doubtful whether they can reproduce effectively. This polygyny may be realized by two separate processes: either by pleometrosis (association of several swarming alates) or by "budding" (isolation of a part of the society, building a new nest at some distance of the parent nest, with maturation *in situ* of newly emerged alates). This strategy allows a faster growth of the new colonies, which spend less time in "founding stage" as defined by Oster and Wilson (1978), especially critical in this species. Although *Nasutitermes corniger* invests a large part of its energy in defense and protection, the constraint of a variable environment required further adaptations for the survival of the incipient colonies. However, these conclusions cannot be generalized for other species of the same genus: various reproductive strategies were evidenced in three sympatric species of New Guinea by Roisin (1987).

Cephalotermes rectangularis is frequently encountered in the rain forest of West Africa. It is a small species but has highly populated colonies (probably several million individuals), which relates to the extreme physogastry of the queen (see 3). The big carton-nest, constructed on the soil, generally leaning against a tree, is very elaborate (Grassé 1949; Noirot 1970) and seems one of the most effective in the protection against ant attacks: the wall (5 to 10 cm thick) is made with a very hard alveolar fecal material (which, in addition, gives a very good air conditioning), the royal pair is well protected in a compact nodule in the center of the nest. Moreover, soldiers are very scarce (less than 0.1% of the population) and only minimally larger than the workers. The investment in defense is almost completely turned on the nest construction. These colonies are very long-lasting, and although no estimate of their life-span is available it seems extremely long: in three years of observation in the Banco forest near Abidjan (where this species is frequent), I

Table 1 - Dry weight (in mg) for the different castes in two termites of the West African rain-forest. For *Cubitermes*, an humus-feeder, the weight is expressed after subtraction of the ash content, which in workers is near 60% (unpublished results, obtained from alcohol-preserved specimen).

	male alate	female alate	worker	soldier
Cephalotermes	0.75	0.81	0.14	0.20
Cubitermes	4.16	4.78	0.72	2.06

discovered only one young nest, and most of the observed nests exhibited only very limited growth during this period. The longevity is a consequence of the efficiency of the nest, both in its protection against predators and in the regulation of a favourable microclimate, shown by the fact that the imaginal pair was invariably discovered in the 15 nests explored by myself and later in the 62 opened by Bordereau (personal communication) in the same locality. The alates of *Cephalotermes* are among the smallest of all the ISOPTERA (Tab. 1). They are fragile and contain few nutritive reserves. Although produced in enormous quantities, twice a year, their chance of founding a new colony is extremely small. This species seems an extreme case, among the termites, of a K-strategist, it needs the stable environment of the tropical rain forest.

Cubitermes fungifaber adopted a very different strategy, although also inhabiting the forest of the west Africa. It is an humus feeding species, which erects a mushrom-shaped nest with populations of moderate size (rarely more than 30.000 workers). The physogastry of the queen is limited (twenty-fold increase at most, Bordereau 1975a), as is its fecundity[12].

Although I explored more than a hundred nests in Ivory Coast, I found no replacement reproductives. The nest, constructed with earth and fecal mortar, is relatively brittle, without a conspicious wall. It is made up of large, thin walled alveoles and lacks a special royal chamber. These poorly air-conditioned nests (Noirot 1970) are easily destroyed by accident or attacked by ants (Noirot et al.

(12) No records are available for this species, but the queen of *Cubitermes severus*, with populations of about the same size, lays between 200 and 600 eggs per day, according to Bodot (1969).

1986). The soldiers are not numerous. In 54 nests analyzed, I found an average of 1.60 + 0.17% soldiers (calculated in relation to workers). The investment in defense and protection is thus very small. The colonies are short-lived (probably no more than 5 years) and their turnover seems very fast in comparison with other termites. Accordingly, all stages of nest development, from the very small to the biggest construction, are easily observed in the field. The alates are produced at a small rate if only the number is considered (8,9% in proportion of the workers in one flourishing colony) but with a very high investment if biomass is taken into account (Tab. 1). These alates which are well provided with nutritive reserves seem much more successful than those of *Cephalotermes* in the same habitat. This species may certainly be classified as an r-strategist. Although frequently found in climactic rain forests, it can occupy more diverse and less permanent niches.

5 Conclusions

Whereas eusocial life has appeared several times during the evolution of the HYMENOPTERA, the ISOPTERA are the sole example in all the other insect orders. During the evolution of termites the diversification of the caste system was much more complicated than in other social group, especially for the reproductives. Therefore, a clear understanding of this evolution, by comparison with the social HYMENOPTERA, seems of paramount interest. However, despite the progress accomplished during the last decades, many large gaps remain. It seems necessary to increase study of the bisexual organization of the termite societies (sex ratio for the different castes, energetic investment for each sex in each caste, division of labour). Our knowledge of the genetics of termites is extremely poor, and the field of population genetics seems a very promising one. The time has arrived for the terminologists to critically apply to their material the modern theories on the origin and evolution of sociality, and to expand their research on the ergonomics of termite societies.

6 Acknowledgements

I thank Barbara Thorne and Christian Bordereau for the communication of unpublished informations, and Paula Green for the revision of the english manuscript.

Supported by the C.N.R.S. (U.A. 674).

7 References

Bartz SH (1979) Evolution of eusociality in termites. Proc Nat Acad Sci USA 76:5764-5768

Basalingappa S (1982) Degeneration of thoracic muscles in sexual forms of *Odontotermes assmuthi* Holmgren (Termitidae: Isoptera) following swarming and colony establishment. Entomon 7:63-64

Bedo DG (1987) Undifferentiated sex chromosomes in *Mastotermes darwiniensis* Froggatt (Isoptera: Mastotermitidae) and the evolution of eusociality in termites. Genome 29:76-79

Bodot P (1969) Composition des colonies des termites: ses fluctuations au cours du temps. Insect Soc 16:39-54

Bordereau C (1975a) Croissance des trachées au cours de l'évolution de la physogastrie chez la reine des termites supérieurs (Isoptera, Termitidae). Int J Insect Morphol & Embryol 4:431-465

Bordereau C (1975b) Déterminisme des castes chez les termites supérieurs: mise en évidence d'un contrôle royal dans la formation de la caste sexuée chez *Macrotermes bellicosus* Smeathman (Isoptera, Termitidae). Insect Soc 22:363-374

Bordereau C (1979) La cuticule de la reine de termite. Ultrastructure, composition chimique, étude de sa croissance en l'absence de mue. Thesis, Univ Dijon:252 p

Bordereau C (1982) Ultrastructure and formation of the physogastric termite queen cuticle. Tissue & Cell 14:371-396

Bordereau C (1985) The role of pheromones in termite caste differentiation. In: Watson JAL, Okot-Kotber BM, Noirot Ch (eds) Caste differentiation in social insects. Pergamon Press, Oxford, pp 221-226

Bordereau C, Han SH (1986) Stimulatory influence of the queen and king on soldier differentiation in the higher termites *Nasutitermes lujae* and *Cubitermes fungifaber*. Insect Soc 33:296-305

Buchli H (1958) L'origine des castes et les potentialités ontogéniques des termites européens du genre *Reticulitermes* Holmgren. Ann Sci Nat Zool 20:261-429

Clement G (1953) Sur la différenciation d'ovocytes dans les testicules des "neutres" d'*Anacanthotermes ochraceus*. CR Acad Sci 236:1095-1096

Clement G (1956) Observations sur l'essaimage d'*Anacanthotermes ochraceus* Burm. Bull Soc Entomol France 61:98-103

Clement JL (1981) Polymorphisme enzymatique des populations européennes des espèces du genre *Reticulitermes*. In: Howse PE, Clement JL (eds) Biosystematics of social insects. Academic Press, New York, pp 49-61

Crozier RH (1980) Genetical structure of social insect populations. In: Markl H (ed) Evolution and social behavior: Hypotheses and empirical tests. Verlag Chemie, Weinheim, pp 129-146

Crozier RH, Luykx P (1985) The evolution of termite eusociality is unlikely to have been based on a male-haploid analogy. Amer Nat 126:867-869

Darlington JPEC (1985) Multiple primary reproductives in the termite *Macrotermes michelseni* (Sjostedt). In: Watson JAL, Okot-Kotber BM, Noirot Ch (eds) Caste differentiation in social insects. Pergamon Press, Oxford, pp 54-58

Delbecque JP, Lanzrein B, Bordereau C, Imboden H, Hirn M, O'Connor JD, Noirot C, Lüscher M (1978) Ecdysone and ecdysterone in physogastric termite queens and eggs of *Macrotermes bellicosus* and *Macrotermes subhyalinus*. Gen Comp Endocrinol 36:40-47

Deligne J, Quennedey A, Blum MS (1981) The enemies and defense mechanisms in termites. In: Hermann HR (ed) Social insects. Academic Press, New York, Vol. II, pp 1-76

Fontana F (1980) Interchange complexes in Italian populations of *Reticulitermes lucifugus* Rossi (Isoptera, Rhinotermitidae). Chromosoma 81:169-175

Grassé PP (1939) Comportement et particularités physiologiques des soldats des termites. Bull Soc Zool France 64:251-262

Grassé PP (1949) Ordres des Isoptères, ou termites. In: Grassé PP (ed) Traité de zoologie, anatomie, systématique, biologie. Masson, Paris, Vol 9, pp 408-544

Grassé PP, Noirot C (1947) Le polymorphisme social du termite à cou jaune (*Calotermes flavicollis* F.). Les faux ouvriers ou pseudergates et les mues régressives. CR Acad Sci 224:219-221
Grassé PP, Noirot C (1951) Nouvelles recherches sur la biologie de divers Termites champignonnistes (Macrotermitinae). Ann Sci Nat Zool 13:291-342
Grassé PP, Noirot C (1960a) Rôle respectif des mâles et des femelles dans la formation des sexués néoténiques chez *Calotermes flavicollis*. Insect Soc 7:109-124
Grassé PP, Noirot C (1960b) L'isolement chez le termite à cou jaune (*Calotermes flavicollis* Fab.) et ses conséquences. Insect Soc 7:323-331
Han SH, Bordereau C (1982) Origin and formation of the royal fat body of the higher termite queens. J Morphol 173:17-28
Han SH, Noirot C (1983) Développement de la jeune colonie chez *Cubitermes fungifaber* (Sjöstedt) (Isoptera, Termitidae). Ann Soc Entomol France 19:413-420
Haverty MI (1977) The proportion of soldiers in termite colonies: a list and a bibliography. Sociobiol 2:199-217
Hewitt PH, Nel JJC (1969) The influence of group size on the sarcosomal activity and the behaviour of *Hodotermes mossambicus* alate termites. J Insect Physiol 15:2169-2177
Hewitt PH, Watson JAL, Nel JJC, Schoeman I (1972) Control of the change from group to pair behaviour by *Hodotermes mossambicus* reproductives. J Insect Physiol 18:143-150
Howard RW, Haverty MI (1979) Termites and juvenile hormone analogues, a review of methodology and observed effects. Sociobiol 4:269-278
Karlson P, Butenandt A (1959) Pheromones (ectohormones) in insects. Ann Rev Entomol 4:39-58
Karlson P, Lüscher M (1959) "Pheromones": a new term for a class of biologically active substances. Nature 183:55-56
Lacy RC (1980) The evolution of eusociality in termites, a haplodiploid analogy? Am Nat 116:449-451
Lacy RC (1984) The evolution of termite eusociality: reply to Leineaas. Am Nat 123:876-879
Lanzrein B, Gentinetta V, Fehr R (1985) Titres of juvenile hormone and ecdysteroids in reproductives and eggs of *Macrotermes michaelseni*: relation to caste determination? In: Watson JAL, Okot-Kotber BM, Noirot Ch (eds) Caste differentiation in social insects. Pergamon Press, Oxford, pp 307-327
Lebrun D (1985) The role of hormones in social polymorphism and reproduction in *Kalotermes flavicollis* Fabr. In: Watson JAL, Okot-Kotber BM, Noirot Ch (eds) Caste differentiation in social insects. Pergamon Press, Oxford, pp 227-238
Lee KE, Wood TG (1971) Termites and soils. Academic Press, London and New York, 251 p
Lefeuve P, Bordereau C (1984) Soldier formation regulated by a primer pheromone from the soldier frontal gland in a higher termite. Proc Nat Acad Sci 81:7665-7668
Lepage M (1974) Les termites d'une savane sahélienne (Ferlo septentrional, Sénégal): peuplement, populations, consommation, rôle dans l'écosystème. Thesis, Univ Dijon
Leinaas HP (1983) A haplodiploid analogy in the evolution of termite eusociality? Reply to Lacy. Am Nat 121:302-304
Light SF (1944) Experimental studies on ectohormonal control of the development of supplementary reproductives in the termite genus *Zootermopsis* (formerly *Termopsis*). Univ Calif Publ Zool 43:413-454
Luamba LN (1980) Recherches sur le polymorphisme et aperçu sur l'influence de l'analogue de l'hormone juvenile sur le développement d'un termite, *Hodotermes mossambicus* (Isoptera, Hodotermitidae). Thesis, Univ Bern, 127 p
Lüscher M (1955) Zur Frage der Übertragung sozialer Wirkstoffe bei Termiten. Naturwiss 42:186
Lüscher M (1964) Die spezifische Wirkung männlicher und weiblicher Ersatzgeschlechtstiere auf die Enstehung von Ersatzgeschlechtstieren bei der Termite *Kalotermes flavicollis* (Fab.). Insect Soc 11:79-90
Lüscher M (1974) Kasten und Kastendifferenzierung bei niederen Termiten. In: Schmidt GH (ed) Sozialpolymorphismus bei Insekten. Wiss Verlagsges, Stuttgart, pp 694-739

üscher M (1975) Pheromones and polymorphism in bees and termites. In: Noirot Ch, Howse PE, Le Masne G (eds) Pheromones and defensive secretions in social insects. Symposium IUSSI, Univ Dijon, pp 123-141

üscher M (1976) Evidence for and endocrine control of caste determination in higher termites. In: Lüscher M (ed) Phase and caste determination in insects. Pergamon Press, Oxford, pp 91-103

üscher, M (1977) Queen dominance in termites. Proc Int IUSSI Congr, Wageningen, 8:238-242

uykx P (1981) A sex-linked esterase locus and translocation heterozygosity in a termite. Heredity 46:315-320

uykx P (1985) Genetic relations among castes in lower termites. In: Watson JAL, Okot-Kotber BM, Noirot Ch (eds) Caste differentiation in social insects. Pergamon Press, Oxford, pp 17-25

uykx P, Michel J, Luykx J (1986) The spatial distribution of the sexes in colonies of the termite *Incisitermes schwartzi* Banks (Isoptera: Kalotermitidae). Insect Soc 33:406-421

uykx P, Syren RM (1979) The cytogenetics of *Incisitermes schwarzi* and other Florida termites. Sociobiol 4:191-209

lac Arthur RH, Wilson EO (1967) The theory of island biogeography. Princeton University Press, Princeton, 203 p

lednikova TK (1977) Caste differentiation in the termite *Anacanthotermes ahngerianus* Jacobson (Isoptera, Hodotermitidae). Proc Int IUSSI Congr, Wageningen, 8:118-120

lensa-Bonsu A (1978) Differenciation des castes chez les termites supérieurs (Isoptera, Termitidae): determination précoce des larves neutres et séxuées, cycle saisonnier, sexués de remplacement. Thesis, Univ Dijon:115 p

leyer DR, Lanzrein B, Lüscher M, Nakanishi K (1976) Isolation and identification of a juvenile hormone (JH) in termites. Experientia 32:773

liller EM (1969) Caste differentiation in the lower termites. In: Krishna K, Weesner FM (eds) Biology of termites I. Academic Press, New York, pp 283-310

lyles TG, Nutting WL (1988) Termite eusocial evolution: a re-examination of Bartz' hypothesis and assumptions. Quart Rev Biol 63:1-23

lagin R (1972) Caste determination in *Neotermes jouteli* (Banks). Insect Soc 19:39-61

lielsen MG, Josens G (1978) Production by ants and termites. In: Brian MV (ed) Production biology of ants and termites. Cambridge Univ Press, Cambridge, pp 45-53

loirot C (1952) Les soins et l'alimentation des jeunes chez les termites. Ann Sci Nat Zool 14:405-414

loirot C (1955) Recherches sur le polymorphisme des termites supérieurs (Termitidae). Ann Sci Nat Zool 17:399-595

loirot C (1956) Les sexués de remplacement chez les termites supérieurs (Termitidae). Insect Soc 3:145-158

loirot C (1969) Formation of castes in higher termites. In: Krishna K, Weesner FM (eds) Biology of termites. Academic Press, New York, Vol I, pp 311-350

loirot C (1970) The nests of termites. In: Krishna K, Weesner FM (eds) Biology of termites. Academic Press, New York, Vol II, pp 73-125

loirot C (1974) Polymorphismus bei höheren Termiten. In: Schmidt GA (ed) Sozialpolymorphismus bei Insekten. Wiss Verlagsges, Stuttgart, pp 740-765

loirot C (1985a) Pathways of caste development in the lower termites. In: Watson JAL, Okot-Kotber BM, Noirot Ch (eds) Caste differentiation in social insects. Pergamon Press, Oxford, pp 41-57

loirot C (1985b) Differentiation of reproductives in higher termites. In: Watson JAL, Okot-Kotber BM, Noirot Ch (eds) Caste differentiation in social insects. Pergamon Press, Oxford, pp 177-186

loirot C (1989) Social structure in termite societies. Ethol Ecol Evol:in press

loirot C, Bodot P (1964) L'essaimage d'*Allognathotermes hypogeus* Silv. (Isoptera, Termitidae). CR Acad Sci 258:3357-3359

Noirot C, Bordereau C (1988) Termite polymorphism and morphogenetic hormones. In: Gupta AP (ed) Morphogenetic hormones of arthropods. Rutgers Univ Press, New Brunswick:in press

Noirot C, Noirot-Timothée C, Han SH (1986) Migration and nest building in *Cubitermes fungifaber* (Isoptera, Termitidae). Insect Soc 33:361-374

Noirot C, Pasteels JM (1987) Ontogenetic development and evolution of the worker caste in termites. Experientia 43:851-860

Noirot-Timothée C, Noirot C (1965) L'intestin moyen chez la reine des termites supérieurs. Etude au microscope électronique. Ann Sci Nat Zool 7:180-208

Nutting WL (1969) Flight and colony foundation. In: Krishna K, Weesner FM (eds) Biology of termites. Academic Press, New York, Vol I, pp 233-282

Nutting WL (1979) Termite flight periods: strategies for predator avoidance? Sociobiol 4:141-151

Okto-Kotber BM (1980) Competence of *Mastotermes michaelseni* (Isoptera: Macrotermitidae) larvae to differentiate into soldiers under the influence of juvenile hormone analogue (ZR 515, Methoprene). J Insect Physiol 26:655-659

Oster GF, Wilson EO (1978) Caste and ecology in the social insects. Monographs in Population Biology 12, Princeton University Press

Reilly LM (1987) Measurements of inbreeding and average relatedness in a termite population. Amer Nat 130:339-349

Renoux J (1976) Le polymorphisme de *Schedorhinotermes lamanianus* (Sjöstedt) (Isoptera, Rhinotermitidae). Essai d'interprétation. Insect Soc 23:281-491

Roisin Y (1987) Polygyny in *Nasutitermes* species: field data and theoretical approaches. In Pasteels JM, Deneubourg JL (eds) From individual to collective behavior in social insects. Birkhäuser, Basel, pp 379-404

Roisin Y (1988) The caste system of *Parrhinotermes browni* (Isoptera: Rhinotermitidae). Sociobiol:in press

Roisin Y, Pasteels JM (1985) Imaginal polymorphism and polygyny in the Neo-Guinean termite *Nasutitermes princeps* (Desneux). Insect Soc 32:140-157

Roisin Y, Pasteels JM (1986) Replacement of reproductives in *Nasutitermes princeps* (Desneux) (Isoptera: Termitidae). Behav Ecol Sociobiol 18:437-442

Roisin Y, Pasteels JM (1987) Caste developmental potentialities in the termites *Nasutitermes novarumhebridarum*. Entom exp appl 44:277-287

Ruppli E (1969) Die Elimination überzähliger Ersatzgeschlechtstiere bei der Termite *Kalotermes flavicollis* (Fab.). Insect Soc 16:235-248

Sands WA (1972) The soldierless termites of Africa. Bull British Mus (Nat Hist) Entomol Suppl 18:1-244

Sieber R, Darlington JPEC (1982) Replacement of the royal pair in *Macrotermes michelseni*. Insect Sci Applic 3:39-42

Sieber R, Leuthold RH (1982) Development of physogastry in the queen of the fungus-growing termite *Macrotermes michaelseni* (Isoptera: Macortermitinae). J Insect Physiol 28:979-985

Springhetti A (1970) Influence of the king and queen on the differentiation of soldiers in *Kalotermes flavicollis* (Fab.). Monitore Zool Ital (NS) 4:99-105

Stuart AM (1979) The determination and regulation of the neotenic reproductive caste in the lower termites (Isoptera) with special reference to the genus *Zootermopsis* (Hagen). Sociobiol 4:223-237

Stuart AM (1981) The role of pheromones in the initiation of foraging, recruitment and defence by the soldiers of a tropical termite, *Nasutitermes corniger* (Motschulsky). Chemical Senses 6:409-410

Syren RM, Luykx P (1977) Permanent segmental interchange complex in the termite *Incisitermes schwartzi*. Nature 266:167-168

Syren RM, Luykx P (1981) Geographic variation of sex-linked translocation heterozygosity in the termite *Kalotermes approximatus* Snyder (Insecta, Isoptera). Chromosoma 82:65-88

Thorne BL (1982) Polygyny in termites: multiple primary queens in colonies of *Nasutitermes corniger* (Motschulsky) (Isoptera, Termitidae). Insect Soc 29:102-117

Thorne BL (1983) Alate production and sex ratio in colonies of the neotropical termite *Nasutitermes corniger* (Isoptera, Termitidae). Oecologia 58:103-109

Thorne BL (1984) Polygyny in the neotropical termite *Nasutitermes corniger*: Life history consequences of queen mutualism. Behav Ecol Sociobiol 14:117-136

Thorne BL (1985) Numerical and biomass caste proportions in colonies of the termites *Nasutitermes corniger* and *N. ephratae* (Isoptera: Termitidae). Insect Soc 32:411-426

Thorne BL, Noirot C (1982) Ergatoid reproductives in *Nasutitermes corniger* (Motschulsky) (Isoptera, Termitidae). Int J Insect Morphol Embryol 11:213-226

Traniello JFA (1981) Enemy deterrence in the recruitment strategy of a termite: soldier-organized foraging in *Nasutitermes costalis*. Proc Nat Acad Sci 78:1876-1979

Truckenbrodt W (1973) Über die imaginale Ovarvergrößerung im Zusammenhang mit der Physogastrie bei *Odontotermes badius* Haviland (Insecta, Isoptera). Insect Soc 20:21-40

Truckenbrodt W (1986) Reproductive caste differentiation and endocrine gland activities in females of the higher termite, *Odontotermes stercorivorous* (Termitidae: Macrotermitinae). Abstr Int Congr IUSSI, München, 10, p 103

Vincke PP, Tilquin JP (1978) A sex-linked ring quadrivalent in Termitidae (Isoptera). Chromosoma 67:151-156

Vincke PP, Tilquin JP (1983) Nombres chromosomiques et mécanisme méiotique chez les Termitidae africains (Isoptera). Ann Soc R Zool Belg 113:69-80

Watson JAL, Abbey HM (1985) Development of neotenics in *Mastotermes darwiniensis* Froggatt: an alternative strategy? In: Watson JAL, Okot-Kotber BM, Noirot Ch (eds) Caste differentiation in social insects. Pergamon Press, Oxford, pp 107-124

Watson JAL, Metcalf EC, Sewell JJ (1975) Preliminary studies on the control of neotenic formation in *Mastotermes darwiniensis* Froggatt (Isoptera). Insect Soc 22:415-426

Watson JAL, Nel JJ, Hewitt PH (1972) Behavioural changes in founding pairs of the termite, *Hodoterms mossambicus*. J Insect Physiol 18:373-387

Watson JAL, Sewell JJ (1981) The origin and evolution of caste systems in termites. Sociobiol 6:101-118

Weesner FM (1955) The reproductive system of young primary reproductives of *Tenuirostritermes tenuirostris* (Desneux). Insect Soc 2:321-345

Weesner FM (1969) The reproductive system. In: Krishna K, Weesner FM (eds) Biology of termites. Academic Press, New York, Vol I, pp 125-160

Wilson EO (1971) The insect societies. Harvard Univ Press, Cambridge, 548 p

Wilson EO, Bossert WH (1963) Chemical communication among animals. Recent Progr Hormone Res 19:673-716

Zhuzhikov DP, Zolotarev EKh, Mednikova TK (1972) (Postembryonic development of *Anacanthotermes ahngerianus* Jacobs). (in Russian). In: Zolotarev EKR (ed) Termites (collected articles). Transact of Entomol Div N°2, Moscow Lomonosov State Univ, pp 46-62

Chapter 2

Regulation of Worker and Queen Formation in Ants with Special Reference to Reproduction and Colony Development

ALFRED BUSCHINGER[1]

CONTENTS

1 Castes in Ants

Caste formation is one of the most challenging problems in the study of social insects. In ants, induction and differentiation of female castes occur in a variety of social systems, and organisation forms,

[1] *Institut für Zoologie der Technischen Hochschule Darmstadt, BRD*

which imply various mechanisms for the ensurance of adequate production rates of workers and reproductives in the societies. Numerous investigations have been made to elucidate the different caste induction mechanisms, but they rarely evaluated their importance with respect to the social systems. At present we are not able to formulate a comprehensive theory on this subject which would comprise all known facts and which would be applicable to all ants.

In the following a survey shall be given, first of the different strategies of colony reproduction - which means formation of new colonies and their later development - and then of the various caste induction mechanisms known to occur in ants. As far as possible, I will try to indicate in a few selected examples which mechanisms prevail in what kind of social systems. However, it seems that nearly always a number of factors interfere in a given species. This review will be limited to the production of queens and workers, special problems such as soldier formation shall be deferred as well as that of male production. For a review including these problems see Brian (1979).

With a functional definition of castes (Buschinger and Winter 1976; Buschinger 1987), "queens" are mated, egg-laying individuals, irrespective of their external morphology, and "workers" are not inseminated and usually sterile. A clear distinction between the external morphology of an individual and its function in the society is necessary in many ant species, where form and function are often not corresponding. In such instances, the different morphs are termed "gynomorph" (the usual alate or dealate female), "ergatomorph" (workerlike habitus, wingless, may nevertheless sometimes have a receptacle and thus be a potential or actual queen), and "intermorph" (morphologically intermediate, often with receptacle, may functionally be a queen or a worker).

Some authors prefer an exclusively morphological definition of castes (Haskins and Whelden 1965; Crozier et al. 1984; Peeters and Crewe 1984; Crozier and Pamilo 1986; Peeters and Crozier 1988). Most present-day papers, however, use a definition in which castes are characterized both by form and function (cf. Wilson 1953; 1971), which is sufficient in many ordinary examples. For discussions of these terminological problems see Buschinger and Winter (1976), Buschinger (1978a, 1987), Francoeur et al. (1985), Winter and Buschinger (1986), Peeters (1987), Buschinger and Crozier (1987), Peeters and Crozier (1988).

The literature on caste differentiation mainly refers to the formation of "physical castes", which in the examples studied are usually morphologically distinct future, resp. potential, queens and workers. In this meaning the terms will be used when development, differentiation or production of the castes is discussed in the following.

2 Colony Reproduction Strategies among Ants

Concerning the formation of new colonies and their later development, among ants a variety of possibilities evolved. It has to be realized that all the recent ant species are social.

2.1 Colony Foundation by Single Queens

There is first the well-known foundation of colonies by single, newly mated queens. They may either bring up their first workers with food from their own body reserves, in the claustral type, such as *Camponotus ligniperda*, or in a partially claustral type with queens foraging, such as *Myrmecia* (Wheeler 1933), or even some higher *Myrmicinae*, for example *Manica* and *Myrmica* species, as was shown by Le Masne and Bonavita (1969), Buschinger (1975a) and others. In all these cases, the colonies go through a period of exponential increase of worker numbers, usually lasting for several years. Brian (1965) presents calculations of colony growth rates in this period.

Having reached a certain threshold number of workers, sexual production begins, and may continue for rather a long time, even for many years. One impressive example is represented by *Lasius niger*: Following Kutter and Stumper (1969), Appel observed a colony with one queen reaching the age of 29 years and producing female offspring until the end.

Such a system - with monogynous, long-lived colonies, founded by a single queen - looks rather simple at a first glance, and it is often regarded as representing the "normal" or phylogenetically primary strategy of colony reproduction in ants. However, this organization form involves rather sophisticated regulation mechanisms, such as behavior patterns to ensure the monogyny (Baroni Urbani 1968; Buschinger 1974b) or to guarantee adequate rates of queen and worker formation.

2.2 Polygyny, Polydomy, and Budding

A second system is represented by polygynous species with new colonies formed through budding or branching. In the respective species, e.g. of the genus *Formica*, a reduction of the ability of queens to set up colonies by solitary queens becomes evident. They either start new colonies in a temporary parasitic way with the aid of alien species, or the queens return to established colonies of their own species, thus replacing old queens who died because of old age, and enlarging the queen number in the nests. With causalities not yet completely understood, large colonies then branch and give rise to the formation of daughter colonies (Rosengren and Pamilo 1983). The single colony, inhabiting one or a group of nests, may reach a considerable age of up to 80 years - observed by Forel (1948) - because of regular queen replacement. Furthermore, in most cases, mother and daughter colonies remain in "amicable" contact, thus often forming "supercolonies" of several hundred nests, or, as was shown by Higashi and Yamauchi (1979), up to 45.000 nests in the case of *Formica yessensis* in Hokkaido.

A few species, such as *Iridomyrmex humilis*, even become "unicolonial" in the sense of Hölldobler and Wilson (1977), having no territorial or other intercolonial boundaries all over their range. Such a system is unique to ants, we do not know of any other social Hymenoptera forming polygynous supercolonies.

2.3 Facultatively Polygynous Species

A group of ants, with numerous species of different subfamilies, exhibit a "facultative polygyny" (Buschinger 1968, 1974b) which means that - in one and the same population - there exist monogynous and polygynous colonies. Both kinds of colonies are reproductive, yield sexuals. In several species - like *Leptothorax acervorum* or *Myrmica rubra* - it is well established that young queens are able, at least occasionally, to found their colonies like queens of the first group. Such haplometrotic colonies may remain monogynous, or become polygynous by later acceptance of further young queens.

Facultative pleometrotic colony foundation through aggregation of several young queens, and the development of such colonies, have been observed by Tschinkel and Howard (1983) in the fire ant, *Solenopsis invicta*. The formation of pleometrotic colonies was dependent on queen density on the ground, after swarming, and groups of 5 foundresses produced more workers than larger groups or single queens alone. Pleometrosis in certain populations of this usually monogynous species apparently develops into a functional monogyny (p. 4), whereas more recently several populations have been found exhibiting a true polygyny (coexistence of several simultaneously reproductive queens, Ross and Fletcher 1985).

Whether polygynous colonies then may reproduce by budding has yet to be staged. However, Alloway et al. (1982) have described examples of polydomy in facultatively polygynous *Leptothorax* species.

It has also not yet been determined whether all young queens in these species are able to found colonies alone, or whether there is a fraction which prefers to join existing colonies. And we do not know whether such queens are accepted only in their mother colony, or in alien colonies as well. Using electrophoretic studies, Douwes et al. (1987) provided evidence for a comparatively high relatedness among queens in polygynous nests of *Leptothorax acervorum*, suggesting that young queens mainly are adopted in their mother nests.

Nevertheless, in *Leptothorax acervorum* (Buschinger 1974a) monogynous colonies yield about five times more young queens per mother queen than comparable polygynous colonies (Tab. 1). This could indicate a regulatory mechanism for population density, in that many young queens get into existing colonies in dense populations where their reproduction remains low, and that - in a less dense population - more queens may found monogynous, highly productive colonies. We have some evidence from population studies that this interpretation is conclusive.

Comparing the 3 possibilities mentioned up to now, we may speculate that strictly monogynous species are r-selected (McArthur and Wilson 1967; Pianka 1978), which means that they are able to colonize newly emerging habitats rather quickly. Whereas polygynous species may be considered to be K-selected, utilizing the full carrying capacity of their habitat over long periods, and facultatively polygynous species have the option of doing both. Of course this differentiation can only be made when comparing social insect species with each other, since social insects in general exhibit more characters of K-selected species than solitary insects.

2.4 Functionally Monogynous Species

A seemingly more advanced strategy is represented by the so-called "functional monogyny" of the free-living ant *Leptothorax gredleri* from Europe (Buschinger 1968) , *L. sphagnicolus* (Francoeur 1986) and *L. "spec. A"* from North America (Heinze and Buschinger 1987) and of three species of guest ants, *Formicoxenus nitidulus* in Europe, and the American *Formicoxenus diversipilosus* and *F. provancheri* (Buschinger and Winter 1976; Francoeur et al. 1985). In one colony there we always find one fully fertile queen and alongside her often a small number of inseminated but sterile "replacement queens", who become fertile only after the alpha-queen has been removed. Usually, new colonies seem to be formed by single, inseminated queens. However, formation of branch colonies, in which a replacement queen then develops fertility, may also occur. In the fire ant, *Solenopsis invicta*, Tschinkel and Howard (1978) found evidence for a functional monogyny, in that certain colonies - after removal of the functional queen - some weeks later contained a replacement queen. The authors suggest that these replacements were surviving foundresses from pleometrotic colony founding. This system combines the advantages of monogyny with the "potential immortality" of polygynous colonies, if additional replacement queens are regularly adopted.

2.5 Army Ants

We have to mention here the highly sophisticated social system of army ants, as studied with admirable accuracy by Schneirla (1971). Their monogynous societies reproduce in a manner which resembles most that of the honey bees. Only very few young queens are produced in a colony, and finally one of these leaves the society together with about half of the workers, the old queen remaining with the other half (Gotwald 1987). This system again avoids the problems of polygyny, and also the difficulties of founding new colonies by single queens.

Table 1 - Comparison of gyne and worker production in polygynous and monogynous colonies of *Leptothorax acervorum* (14 colonies selected from 100 natural colonies, with respect to queen and worker numbers)

MONOGYNOUS COLONIES	POLYGYNOUS COLONIES
- colonies with similar queen number -	
10 colonies, mean n adult workers: 170, queens: 1 mean production of queens: 30 / colony 30 / queen workers: 90 / colony 90 / queen	10 colonies, mean n adult workers: 184, queens: 8,4 mean production of queens: 7 / colony 0.8 / queen workers: 59 / colony 7 / queen
- colonies with similar number of workers/queen -	
4 colonies, mean n adult workers: 96.5 production of queens: 53 = 10.7/queen = 10.7/colony production of workers: 159 = 39.7/queen = 39.7/colony	4 colonies, 2 queen each, mean n adult workers: 174.5 production of queens: 15 = 1.9/queen = 3.75/colony production of workers: 133 =16.5/queen = 33/colony

2.6 Exceptional Modes of Reproduction

Finally, there are several special cases which cannot be easily classified in the system as described above.

Thus, there is the Pharao's ant, *Monomorium pharaonis*, which is polgynous and "unicolonial". However, in contrast to other polygynous species and as yet uniquely among ants, the *Monomorium* queens produce - after 2 or 3 batches of worker brood - only one single sexual brood, and then die. According to Petersen-Braun (1975), the death of the old queens occurs almost simultaneously with the emergence of young sexuals in the societies who mate within or near the nest, and the young queens then replace their mother generation.

The reproductive cycle of the myrmicine, *Pristomyrmex pungens*, was described by Itow et al. (1984). The colonies of this species investigated in Japan do not contain functional queens, neither gynomorphic nor ergatomorphic ones. Apart from numerous small workers only a few larger ones with ocelli and abortive spermathecae were found (intermorphs, termed "ergatoid queens" by Itow et al. 1984), and males are also rare. Reproduction apparently is achieved predominantly or wholly by the small workers, and hence is thelytokous.

Ledoux (1950, 1954, 1974) has proposed a quite complicated life cycle of the weaver ant, *Oecophylla longinoda*. Apparently alate females and workers may develop both from fertilized, queen-laid eggs and from unfertilized worker-laid eggs, the latter being the more frequent source of alate females. Since the workers can also lay haploid eggs which develop into males, both thelytoky and arrhenotoky occur in this species. Presumably a bisexual generation (male/queen producing only workers) is alternating with parthenogenetical ones (workers producing further workers, males and queens).

Worker thelytoky yielding workers and queens finally has also been observed in the desert ant, *Cataglyphis cursor* (Cagniant 1979; Lenoir and Cagniant 1986). However, these results were obtained in laboratory culture, and hence it remains an open question whether or not thelytoky does play a role in field colonies.

Other special developments occur mainly among social parasitic ants to which I will not refer in detail, although in the second part of this survey I have to report certain mechanisms which have been investigated in social parasites. I will only mention the fact that among these ants, we know of a number of workerless species, inquilines, who depend on the social capacities of the host workers. Others, such as slavemakers of the genus *Epimyrma* (Winter 1979), have only a small number of workers and yield comparatively many sexuals, thus indicating that production rates of workers and sexuals, in the course of evolution, may be changed according to special requirements even in the opposite direction to the ordinary development towards larger colonies with more and more workers. The most derived species of this group, *Epimyrma corsica*, and *E. adlerzi*, have completely lost the worker caste (Buschinger and Winter 1985; Douwes et al. 1988).

3 Mechanisms of Caste Formation in Ants

In the second part of this survey we will see what mechanisms may be involved in these various colony reproduction strategies, to guarantee adequate rates of workers and queens. The mechanisms themselves can be classified as either promoting gyne formation, and thus rapid reproduction, or as decreasing gyne formation in favour of higher worker numbers and enlargement of the colonies. The different strategies of colony reproduction, thus, often reflect varying interactions of antithetical mechanisms influencing caste formation.

3.1 Loss of the Worker Caste

Caste induction in ants is essentially worker induction since phylogenetically the worker caste must be derived from the fully fertile female of the nonsocial ancestors. In certain socially parasitic ants, mainly inquilines, a worker caste is totally absent. Caste formation presumably went lost during the evolution of parasitic life. Examples are *Teleutomyrmex schneideri*, *Anergates atratulus*, *Leptothorax kutteri*, and others (cf. Buschinger 1970; Wilson 1971; Buschinger 1974c). In the genus *Epimyrma* we recently could demonstrate a graded series ranging from species with a quite normal worker caste (*E. ravouxi*, active slavemaker) to others with a reduced number of workers (*E. kraussei*, a degenerate slavemaker), and finally without workers (*E. corsica*, *E. adlerzi*) and thus without caste formation (Winter and Buschinger 1983; Buschinger and Winter 1985; Douwes et al. 1988).

3.2 Caste Differentiation in the Adult Instar

A "psychophysiological" caste differentiation is sometimes attributed to certain primitively social wasps and bees (Schmidt 1974). In such cases the females are usually more or less equal with respect to their morphology and the anatomy of the reproductive organs. Caste differentiation, the decision on their future function as queens or workers, is often achieved through insemination. In ants, who usually have a marked female di- or polymorphism, sometimes a kind of psycho-physiological caste differentaition can nevertheless still be observed. In certain *Leptothorax* species e.g. a number of well developed gynomorphs by chance may fail to be inseminated. They remain in their mother nests, dealate, and act as additional workers alongside their normal, ergatomorphic nestmates (Buschinger 1974b). In functionally monogynous species (see 2.4) psychophysiological factors may also be involved in the decision on whether an inseminated, potential queen becomes the reproductive queen of the colony, or a subordinate replacement which, at least for a certain time, does not lay eggs.

Recently Peeters and Crewe (1984, 1985) have reported an example where psychophysiological caste differentiation apparently is the only mode in an ant species. In the ponerine *Ophthalmopone berthoudi*, only one female morph exists. All the specimens examined are workerlike, ergatomorphic, but have equally well developed ovaries and spermathecae. And in each colony a certain fraction of them is inseminated and functions as ergatomorphic queens ("gamergates" in the terminology of Peeters and Crewe), whereas the remaining, unfertilized specimens represent the worker force. In another ponerine, *Pachycondyla krugeri*, monogyny with only one reproductive "gamergate" was found (Wildman and Crewe 1988). Like the gynomorphic or intermorphic queens of other ant species this specimen inhibits the reproductive activity of the not inseminated nestmates.

3.3 Larval Nutrition and Caste Formation

In most ant species the time of caste induction, or rather predisposition, is advanced into the period of larval development or even into the egg stage (see 3.6). Several factors have been found to bias larval development towards worker or gyne formation.

Thus, quantity and quality of larval food seem to play a crucial role in the female caste induction of most ant species. It is impossible, in a short review, to refer even to the most important papers in this field with any thoroughness. Thus, I must confine to only a few examples. A detailed account is given by Wilson (1971) or by Brian and several other authors in the book of Schmidt (1974), and by Brian (1979).

In general, starvation of larvae reduces the rate of gyne formation, and ample food supply enhances queen production. However, the quality of food is even more important.

Thus, Wilson (1940) could raise the percentage of queens formed in *Leptothorax* colonies by feeding an excess of protein-rich food. Frenz (1977) has found out more complex reactions in *Leptothorax acervorum* and *Harpagoxenus sublaevis*. There the larval development stops during experimental protein starvation and continues again when protein food is supplied. She could, in this way, prevent pupation of larvae to workers until hibernation, but after hibernation these larvae had gained queen potentiality, a phenomenon to which I will have to refer again in the following section.

However, not only the original food which is distributed by the workers among the brood may influence caste formation. There are many studies revealing that special food substances are provided by the workers in the form of glandular secretions. This is most obvious in all the cases where young workers must be present in a colony and care for the larvae which otherwise yield only workers. Examples are given by Bier (1958) in *Formica*, Schneirla (1971) in army ants, Brian (in Schmidt 1974) in *Myrmica*, Frenz (1977) in *Leptothorax* and *Harpagoxenus*, Passera (1980) in *Pheidole*, and several other authors.

3.4 Larval Hibernation and Climatic Cycles

Hibernation of larvae is another important factor in caste induction. While some ant species, such as *Formica*, hibernate without larvae, we find overwintering larvae in many other genera like *Lasius*, *Camponotus* and many *Myrmicinae*. Larvae of *Leptothorax acervorum* and *Harpagoxenus sublaevis* may even hibernate twice, and thus pupate after a larval period of two years or more.

In general, hibernated larvae have queen potentiality, and "rapid brood" larvae who pupate without a hibernation develop into workers. Examples were given by Brian (1974, 1977) in *Myrmica*, Poldi (1963) for *Tetramorium*, Plateaux (1971) for *Leptothorax* and Frenz (1977) for *Leptothorax* and *Harpagoxenus*. Following Brian (1977), the larvae undergo during hibernation a diapause development without simultaneous growth which may be essential for gaining queen potentiality.

Frenz (1977), using genetically marked eggs of *Harpagoxenus sublaevis* ("ee"-eggs, see 3.7) could demonstrate that eggs laid in the first half of the summer mainly develop after one hibernation into workers, whereas eggs from the second half of the summer period develop into larvae which hibernate twice, and then are irreversibly queen destined. On the contrary, the larvae of *Leptothorax acervorum* - a slave species of *Harpagoxenus sublaevis* - rarely hibernate twice, but (even then) are still able to develop into workers or gynes.

A completely different impact of climatic cycles on sexual or worker formation has been demonstrated in American army ants by Schneirla (1971). The *Ecitoninae* he observed, normally, under humid conditions, have all-worker broods. With the beginning of the dry season, however, they often yield sexual broods, with several thousand males and only a few females. Schneirla suggests a direct influence of drought on the queen.

3.5 Queen Influence on Caste Formation

Queen influence on larval development has been demonstrated in several species, for example by Gößwald and Bier (1953, 1954a,b) in *Formica*, Brian (1974) in *Myrmica*, Frenz (1977) in *Leptothorax* and *Harpagoxenus*, Petersen-Braun (1975, 1977) and others in *Monomorium* etc.. In general, queens lower the rate of gyne formation in favour of worker production. Thus, if the queens are removed from a colony, a considerably higher number of larvae develop into gynes than in the presence of a queen, and gyne formation is only possible in colonies with a higher worker/queen ratio.

Several ant species exhibit a "physiological queenlessness" during the development and differentiation of queen larvae: In *Formica polyctena* the functional queens withdraw into the cool depths of the nest after having laid the eggs predisposed to develop into sexuals (Gößwald and Bier 1954); in *Monomorium pharaonis*, the old queen generation seems to lose her inhibitory power on sexual's development at the end of a cycle (Petersen-Braun 1975), and, similarly, the monogynous *Myrmica sulcinodis* colonies only produce young queens near the end of the queen's life time, according to Elmes (1974).

A further example of queen influence is represented by the facultatively polygynous *Leptothorax acervorum*. As was already shown before (see 2.3), queen production is much lower in polygynous colonies when compared with monogynous colonies, even if the worker/queen ratio is kept constant (Tab. 1).

The mechanisms involved in queen influence may be food competition between queens and larvae, especially competition for food secretions of young workers, as was suspected by Bier (1954) and Mamsch (1965, 1967). Moreover, queens often beg food from larvae, as was shown in *Harpagoxenus* by Frenz (1977), or in *Leptothorax* by Wilson (1974). Brian (in Schmidt 1974) describes that *Myrmica* queens even cause their workers to bite the larvae who then produce droplets of a salivary secretion. *Leptanilla japonica* queens exclusively feed on hemolymph which they obtain from their own mature larvae via specialized exsudatory organs. They thus extract some energy from the brood which they use for a highly cyclic egg production (Masuko 1987). Similarly, queens of *Amblyopone* and other ant genera pierce the skin of larvae for obtaining hemolymph. Little, however, is known on a possible function of this behavior with respect to caste differentiation (Masuko 1986).

According to Brian (1975), the *Myrmica* queen switches caste development workerwards indirectly by influencing the brood care of the workers in various manners. Brian (in Hermann 1979) discusses the possible action of a pheromone, a cuticular secretion of the queen, on the workers' behavior. Formerly, Brian and Hibble (1963) in *Myrmica* and Colombel (1971, 1972) in *Odontomachus*, a ponerine, could

Fig. 1 - Alate female of *Harpagoxenus sublaevis* in sexual calling position ("Locksterzeln").

Fig. 2 - Intermorphic ("ergatoid") female of *Harpagoxenus sublaevis*. During sexual calling the cloaca is held open and the stinger extruded, releasing a sexual pheromone.

demonstrate an inhibitory effect of queen odours and secretions on the formation of gynes in the brood.

Nevertheless, there is not yet sufficient evidence for a general pheromonal control of queens over caste formation in ants.

3.6 Caste Predisposition of Eggs

By the mechanism of "blastogenic caste determination" as described by Bier (1954) in the polgynous *Formica polyctena*, the beginning of caste predisposition is advanced into the egg stage. During winter oogenesis, eggs are formed with a special pole plasm and rich in RNA. These may more easily develop into queens than summer eggs which have less RNA and polar plasm. This non-genetic mechanism may be interpreted as a trophogenetic predetermination during the period of egg formation. Other factors, however, chiefly nutrition, must be favourable during the larval development in order to maintain queen bias of the winter eggs. Larvae from summer eggs on the other hand never develop into gynes, even under most favourable conditions.

A similar mechanism has been demonstrated for *Myrmica* by Brian and Hibble (1964), and for *Pheidole* by Passera (1980). Passera and Suzzoni (1978a,b, 1979) could induce the formation of queen eggs by juvenile hormone treatment of *Pheidole* queens. We have to refer to these papers later (see 4), however it would be very interesting to know whether this hormone treatment induces similar changes in the *Pheidole* eggs as those found in winter and summer eggs of *Formica.* Egg predisposition is clearly a mechanism which promotes gyne formation in spring, and worker production throughout the summer.

3.7 Genetically mediated Caste Formation

Finally, there is the problem of a genetic predisposition of fertilized eggs to develop either into a queen or into a worker larva. This mechanism, which Kerr (1950) suggested to occur in stingless bees, hitherto could never be found in ants. In a recent investigation of isozymes, in two *Rhytidoponera* and one *Pheidole* species, Craig and Crozier (1978) could find no evidence for a role of heterozygosity in ant caste formation. However - as the authors emphasize - much more work is necessary in this field.

In the slavemaker ant, *Harpagoxenus sublaevis*, we were able to demonstrate a genetical mechanism where one pair of alleles is responsible for morph differentiation. Only larvae who are homozygous for a recessive allele "e" are capable of developing into fully alate females - gynomorphs - provided that food and other conditions are favourable. Otherwise ee-larvae can also develop into ordinary

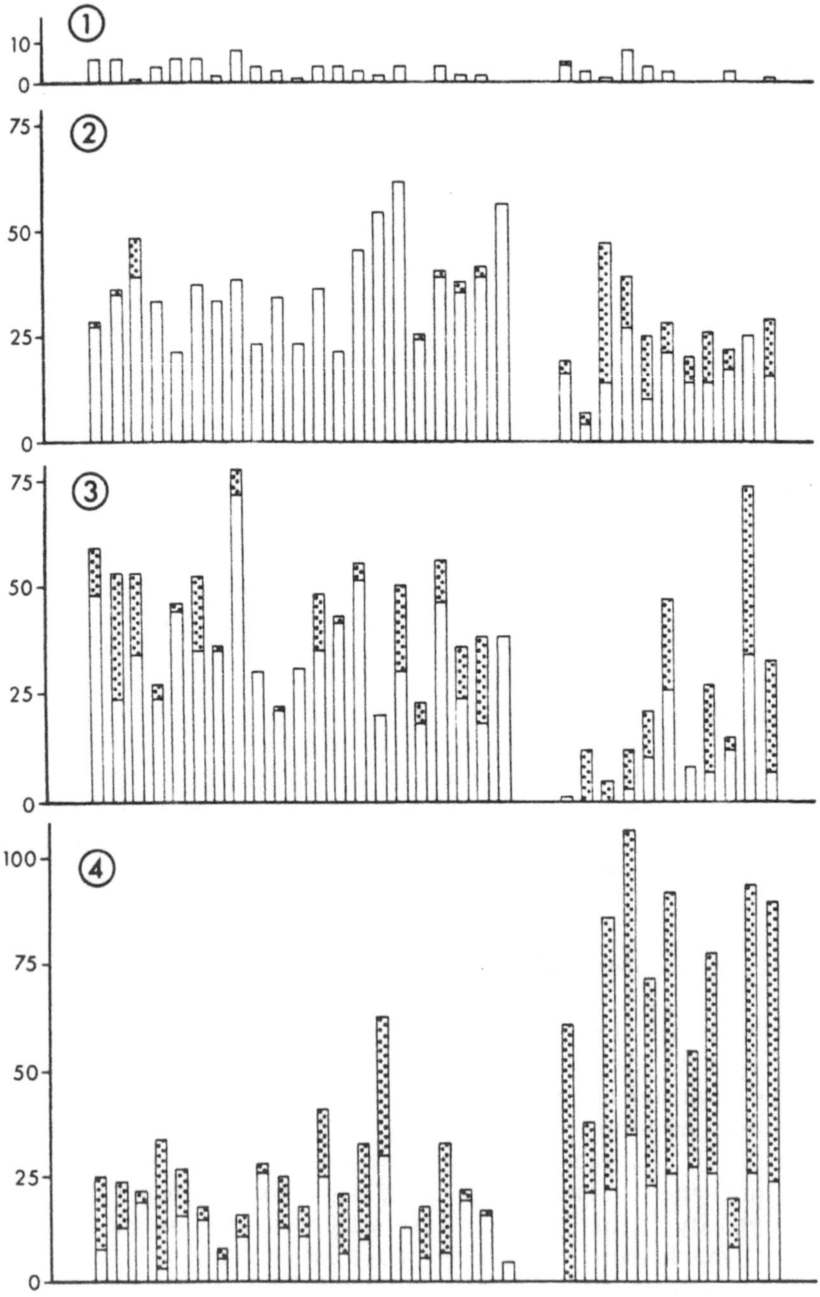

Fig. 3 - Number of young queens (dotted parts of columbs) and queens produced in homozygous EE x E or ee x e laboratory colonies of *Harpagoxenus sublavis* during their 1st to 4th breeding cycles (corresponding to the first four years after colony foundation).

workers. Larvae who are homozygous for "E", or heterozygous (eE), never develop into alates, but only into more or less workerlike "intermorphs" which may possess a receptacle and thus be potential intermorphic queens, or lack a spermatheca and then be true workers (Buschinger 1975b, 1978b; Buschinger and Winter 1978) (Fig. 1, 2).

Further investigations on this ant revealed that the two alleles e and E, besides their morphogenetic effects, have a considerable influence on true caste differentiation in that they bias the rates of formation of workers and potential queens (Winter and Buschinger 1986). In identical breeding conditions, the female offspring of homozygous strains (ee x e and EE x E) during their first 4 breeding cycles (which correspond to natural years) is equal with respect to the total numbers of specimens produced per colony (EE x E: mean 107.5, s.d. = 22.7, n = 21; ee x e: mean 123.7, s.d. = 45.5, n = 11; $p = 0.20$). The ratios of queens/workers produced, however, differ significantly ($p < 0.001$): Only 18.8% of the EE-offspring were potential queens, most of them arising in the 4th breeding cycle, whereas in ee x e-colonies the production of large numbers of gynes already began in the 2nd cycle and reached a total of 60.3% over the 4 cycles (Fig. 3).

Heterozygous colonies in all possible combinations yield intermediate results (Fig. 4), with one remarkable exception: Whereas all groups with Ee- or ee-queens already produced young queens in their 1st breeding cycle, the colonies with EE-queens did not, even in the group where heterozygous brood was present (EE x e).

With exchanging broods between EE x E- and ee x e-colonies, and with rearing such broods in host colonies without the influence of a *Harpagoxenus* queen, the mechanisms involved could be determined in more detail. The results clearly demonstrated that the allels e and E have a twofold effect. E increases the potency of queens to inhibit the formation of young queens in their brood, and it slows down the development of EE-larvae which then often hibernate twice before reaching the adult instar. Thus, the critical periods in larval growth, when queen-development can be influenced by promoting (special food, winter chilling) or suppressing (queen control) factors, are extended. These caste-biasing effects of the two allels very probably guarantee a female/worker ratio in *H. sublaevis* which is well adapted to its needs as a slavemaking ant. The more conspicuous effects on queen morphology presumably merely represent an unimportant by-product of this yet unique genetical caste-biasing mechanism.

In the North American *Leptothorax "spec. A."* a very similar genetical mechanism involved in the formation of intermorphic and gynomorphic queens has been demonstrated (Heinze and Buschinger 1987). Whether or not caste ratios in this species are also affected could not yet be staged.

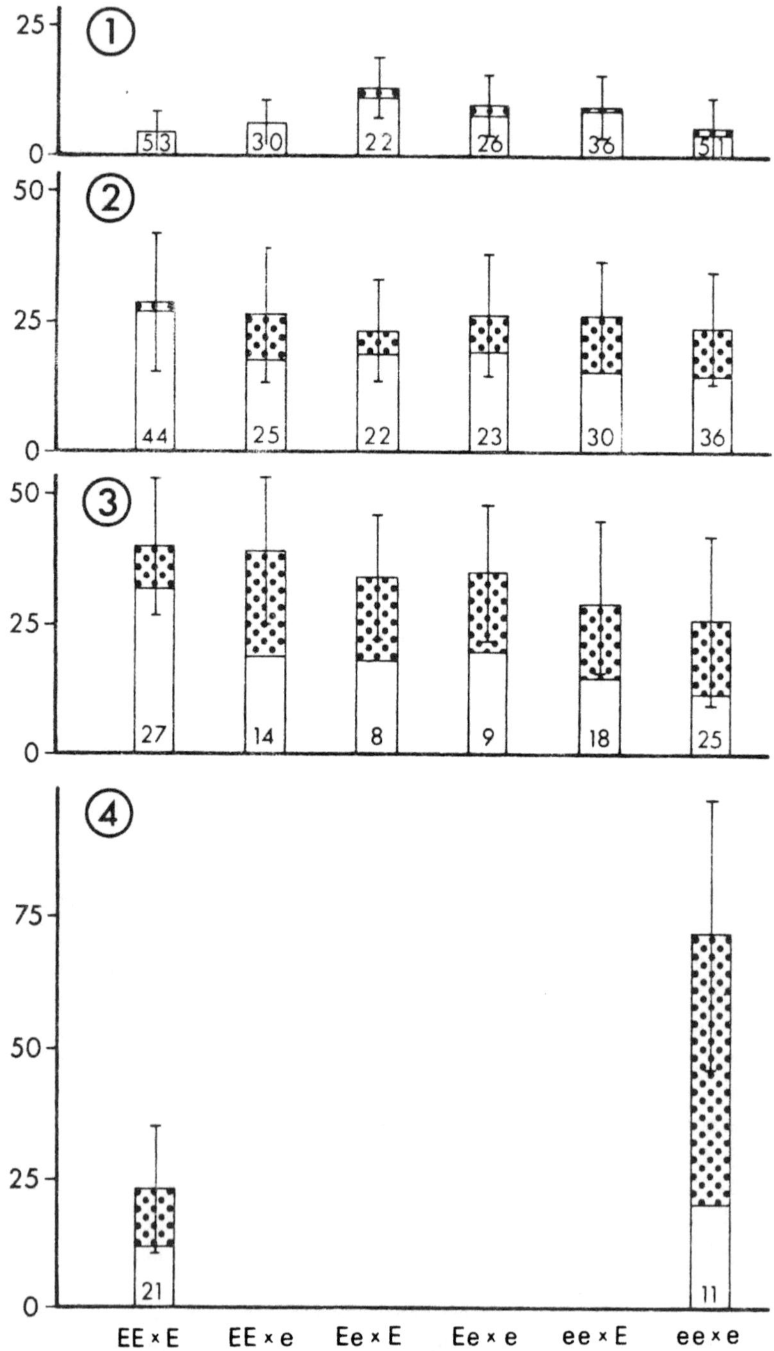

Fig. 4 - Mean numbers of young queens (dotted parts of columns) and queens produced in laboratory colonies of *Harpagoxenus sublaevis* with the 6 possible allelic combinations (numbers in columns = n colonies; bars = s.d), during the first to third resp. fourth breeding cycles.

The interesting and highly important conclusion is that genetical mechanisms may influence caste formation rates. This is supposedly achieved by shifting certain threshold values for the reactivity of larvae to caste inducing factors. Such mechanisms can be suspected to be responsible for changes in caste production rates in social parasites up to even causing complete workerlessness.

4 Hormonal Effects in Caste Development

In the final section we have to deal with hormonal effects on caste development. Only few experiments were made in order to show effects of known insect hormones on caste induction and differentiation, effects of ecdysone and juvenile hormones or juvenile hormone analogues. Thus, Brian (1974) observed that ecdysone advanced the metamorphosis of female larvae in the last instar; here they developed into small gynes. Juvenile hormone analogues applied topically or injected into larvae delayed metamorphosis and generated more and larger gynes. From his experiments with *Myrmica*, Brian (1976) concluded that worker differentiation is effected by stopping corpora allata activity and destroying all juvenile hormone at a certain stage in larval development. However, hormone titers in ant larvae hitherto have not been determined.

Robeau and Vinson (1976) fed a juvenile hormone analogue to *Solenopsis invicta* colonies and treated the queens topically with the same substance. This stimulated the production of major workers, intermorphs, and gynes. Similar results - by feeding juvenile hormone I to *Pheidole pallidula* colonies - were reported by Passera and Suzzoni (1978a). The same authors (1978b, 1979) observed the formation of queen-biased eggs by *Pheidole* queens after the topiacl application of juvenile hormone I. When the *corpora allata* in queens of this species were damaged by precocene II, the production of sexuals was diminished. Thus, juvenile hormone in general seems to promote queen formation, whereas topical application of ecdysteroid to queens of *Pheidole* lowered the production rate of young queens (Suzzoni et al. 1980). These authors could also demonstrate that in *Pheidole* queens laying worker-biased eggs, and in these eggs themselves, the ecdysteroid level was higher than in queen-biased eggs and in the queens laying them.

These findings indicate at least that the hormonal system plays a crucial role in the development of castes. However, very little is known about the interactions of this system and the various factors involved in caste formation.

5 References

Alloway TM, Buschinger A, Talbot M, Stuart R, Thomas C (1982) Polygyny and polydomy in three North American species of the ant genus *Leptothorax* Mayr (Hymenoptera: Formicidae). Psyche 89:249-274

Baroni Urbani C (1968) Monogyny in ant societies. Zool Anz 181:269-277

Bier K-H (1954) Über den Saisondimorphismus der Oogenese von *Formica rufa rufopratensis minor* Gössw. und dessen Bedeutung für die Kastendetermination. Biol Zentralbl 73:170-190

Bier K-H (1958) Die Bedeutung der Jungarbeiterinnen für die Geschlechtstieraufzucht im Ameisenstaat. Biol Zentralbl 77:257-265

Brian MV (1965) Social insect populations. Academic Press, London, New York, 135 p

Brian MV (1974) Caste differentiation in *Myrmica rubra*: the role of hormones. J Insect Physiol 20:1351-1365

Brian MV (1976) Endocrine control over caste differentiation in a Myrmicine ant. In: Lüscher M (ed) Phase and caste determination in insects. Pergamon Press, Oxford and New York, pp 63-70

Brian MV (1977) The synchronisation of colony and climatic cycles. Proc Int Congr IUSSI, Wageningen, 8:pp 202-206

Brian MV (1979) Caste differentiation and division of labor. In: Hermann HR (ed) Social insects, Vol I. Academic Press, New York, San Francisco, London, XV + 437 p

Brian MV, Hibble J (1963) Larval size and the influence of the queen on growth in *Myrmica*. Insect Soc 10:71-81

Brian MV, Hibble J (1964) Studies of caste differentiation in *Myrmica rubra* L. 7: Caste bias, queen age and influence. Insect Soc 11:223-238

Buschinger A (1968) Mono- und Polygynie bei Arten der Gattung *Leptothorax* Mayr (Hymenoptera, Formicidae). Insect Soc 15:217-226

Buschinger A (1970) Neue Vorstellungen zur Evolution des Sozialparasitismus und der Dulosis bei Ameisen (Hym., Formicidae). Biol Zentralbl 88:273-299

Buschinger A (1974a) Polymorphismus und Kastendetermination im Ameisentribus *Leptothoracini*. In: Schmidt GH (ed) Sozialpolymorphismus bei Insekten. Wiss Verlagsges, Stuttgart, pp 604-623

Buschinger A (1974b) Mono- und Polygynie in Insektensozietäten. In: Schmidt GH (ed) Sozialpolymorphismus bei Insekten. Wiss Verlagsges, Stuttgart, pp 862-896

Buschinger A (1974c) Polymorphismus und Polyethismus sozialparasitischer Hymenopteren. In: Schmidt GH (ed) Sozialpolymorphismus bei Insekten. Wiss Verlagsges, Stuttgart, pp 897-934

Buschinger A (1975a) Die Ameisenfauna des Bausenberges, der nordöstlichen Eifel und Voreifel (Hym., Formicidae) mit einer quantitativen Auswertung von Fallenfängen. Beiträge Landespflege Rhld.-Pfalz Beiheft 4:251-273

Buschinger A (1975b) Eine genetische Komponente im Polymorphismus der dulotischen Ameise *Harpagoxenus sublaevis*. Naturwiss 62:239

Buschinger A (1978a) Queen polymorphism in ants. Bull Int, Réunion Scientifique de la Section Française UIEIS, Besançon, pp 12-22

Buschinger A (1978b) Genetisch bedingte Entstehung geflügelter Weibchen bei der sklavenhaltenden Ameise *Harpagoxenus sublaevis* (Nyl.) (Hym., Form.). Insect Soc 25:163-172

Buschinger A (1979) Functional monogyny in the American guest ant *Formicoxenus hirticornis* (Emery) (=*Leptothorax hirticornis*), (Hym., Form.). Insect Soc 26:61-68

Buschinger A (1987) Polymorphism and reproductive division of labor in advanced ants. In: Eder J, Rembold H (eds) Chemistry and biology of social insects. Peperny, München, pp 257-258

Buschinger A, Crozier RH (1987) Introduction to "Towards a unified reproductive biology of the Hymenoptera". In: Eder J, Rembold H (eds) Chemistry and biology of social insects. Peperny, München, pp 257-258

Buschinger A, Winter U (1976) Funktionelle Monogynie bei der Gastameise *Formicoxenus nitidulus* (Nyl.) (Hym., Form.). Insect Soc 23:549-558

Buschinger A, Winter U (1978) Echte Arbeiterinnen, fertile Arbeiterinnen und sterile Wirtsweibchen in Völkern der dulotischen Ameise *Harpagoxenus sublaevis* (Nyl.) (Hym. Form.). Insect Soc 25:63-78

Buschinger A, Winter U (1985) Life history and male morphology of the workerless parasitic ant *Epimyrma corsica* (Hymenoptera: Formicidae). Entomol Gener 10:65-75

Cagniant H (1979) La parthenogénèse thélytoque et arrhénotoque chez la fourmi *Cataglyphis cursor* Fonsc. (Hym. Form.). Cycle biologique en élevage des colonies avec reine et des colonies sans reine. Insect Soc 26:51-60

Colombel P (1971) Etude de l'inhibition de la ponte des ouvrières d'*Odontomachus haematodes* L. (Hym. Form. Poneridae). CR Acad Sci Paris 272:970-972

Colombel P (1972) Recherches sur la biologie et l'éthologie d'*Odontomachus haematodes* L. (Hym. Form. Poneridae). Biologie des ouvrières. Insect Soc 19:171-194

Craig R, Crozier RH (1978) No evidence for role of heterozygosity in ant caste determination. Isozyme Bull 11:66

Crozier RH, Pamilo P (1986) Relatedness within and between colonies of a queenless ant species of the genus *Rhytidoponera* (Hymenoptera: Formicidae). Entomol Gener 11:113-117

Crozier RH, Pamilo P, Crozier YC (1984) Relatedness and micro-geographic genetic variation in *Rhytidoponera mayri*, an Australian arid-zone ant. Behav Ecol Sociobiol 15:143-150

Douwes P, Jessen K, Buschinger A (1988) *Epimyrma adlezi sp. n.* (Hymenoptera: Formicidae) from Greece: morphology and life history. Entomol scand (in press)

Douwes P, Sivusaari L, Niklasson M, Stille B (1987) Relatedness among queens in polygynous nests of the ant *Leptothorax acervorum*. Genetica 75:23-29

Elmes GW (1974) Colony populations of *Myrmica sulcinodis* Nyl. (Hym. Formicidae). Oecologia 15:337-343

Forel A (1948) Die Welt der Ameisen. Rotapfel, Zürich, 275 p

Francoeur A (1986) Deux nouvelles fourmis néarctiques: *Leptothorax retractus* et *L. sphagnicolus* (Formicidae, Hymenoptera). Can Entomol 118:1151-1164

Francoeur A, Loiselle R, Buschinger A (1985) Biosystématique de la tribu *Leptothoracini* (Formicidae, Hymenoptera). 1. Le genre *Formicoxenus* dans la région holarctique. Naturaliste can 112:343-403

Frenz G (1977) Untersuchungen zur Kastendetermination bei Ameisen des Tribus *Leptothoracini* (Hym., Formicidae). Doctoral thesis, Univ Darmstadt, 118 p

Gößwald K, Bier K-H (1953) Untersuchungen zur Kastendetermination in der Gattung *Formica.* 2: Die Aufzucht von Geschlechtstieren bei *F. rufa pratensis* Retz. Zool Anz 151:126-134

Gößwald K, Bier K-H (1954a) Untersuchungen zur Kastendetermination in der Gattung *Formica.* 3: Die Kastendetermination von *F. rufa rufopratensis minor* Gößw.. Insect Soc 1:229-246

Gößwald K, Bier K-H (1954b) Untersuchungen zur Kastendetermination in der Gattung *Formica.* 4: Physiologische Weisellosigkeit als Voraussetzung der Aufzucht von Geschlechtstieren im polygynen Volk. Insect Soc 1:305-318

Gotwald WH jr (1987) The relationship of form and function in army ant queens. In: Eder J, Rembold H (eds) Chemistry and biology of social insects. Peperny, München, pp 255-236

Haskins CP, Whelden RM (1965) Queenlessness, worker sibship, and colony versus population structure in the Formicid genus *Rhytidoponera.* Psyche 72:87-112

Heinze J, Buschinger A (1987) Queen polymorphism in a non-parasitic *Leptothorax* species (Hymenoptera, Formicidae). Insect Soc 34:28-43

Hermann HR (1979) Social Insets. Vol I. Academic Press, New York, San Francisco, London, XV + 437 p

Higashi S, Yamauchi K (1979) Influence of a supercolonial ant *Formica (Formica) yessensis* Forel on the distribution of other ants in Ishikari coast. Jap J Ecol 29:257-264

Hölldobler B, Wilson EO (1977) The number of queens: an important trait in ant evolution. Naturwiss 64:8-15

Itow T, Kobayashi K, Kubota M, Ogata K, Imai T, Crozier RH (1984) The reproductive cycle of the queenless ant *Pristomyrmex pungens*. Insect Soc 31:87-102

Kerr WE (1950) Genetic determination of castes in the genus Melipona. Genetics 35:143-152

Kutter H, Stumper R (1969) Hermann Appel, ein leidgeadelter Entomologe (1892-1966). Proc Int Congr IUSSI, Bern, 6:275-279

Ledoux A (1950) Etude du comportement et de la biologie de la fourmi fileuse *Oecophylla longinoda* (Latr.). Ann Sci Nat 11: Série 12, 312-461

Ledoux A (1954) Recherches sur le cycle chromosomique de la fourmi fileuse *Oecophylla longinoda* (Latr.). Insect Soc 1:149-175

Ledoux A (1974) Polymorphismus und Kastendetermination bei den Weberameisen. In: Schmidt GH (ed) Sozialpolymorphismus bei Insekten. Wiss Verlagsges, Stuttgart, pp 533-542

Le Masne G, Bonavita A (1969) La fondation des sociétés selon un type archaique par une fourmi appartenant à une sous-famille évoluée. CR Acad Sci Paris 269:2373-2376

Lenoir A, Cagniant H (1986) Role of worker thelytoky in colonies of the ant species *Cataglyphis cursor* (Hymenoptera: Formicidae). Entomol Gener 11:153-157

Mac Arthur RH, Wilson EO (1967) The theory of island biogeography. Princeton Univ Press, Princeton, NJ, XI + 203 p

Mamsch E (1965) Regulation der Fruchtbarkeit von Ameisenarbeiterinnen ohne Königin und ohne "Königinsubstanz". Naturwiss 52:168

Mamsch E (1967) Quantitative Untersuchungen zur Regulation der Fertilität im Ameisenstaat durch Arbeiterinnen, Larven und Königin. Z vergl Physiol 55:1-25

Masuko K (1986) Larval hemolymph feeding: a nondestructive parental cannibalism in the primitive ant *Amblyopone silvestrii* Wheeler (Hymenoptera: Formicidae). Behav Ecol Sociobiol 19:249-255

Masuko K (1987) *Leptanilla japonica*: The first bionomic information on the enigmatic ant subfamily Leptanillinae. In: Eder J, Rembold H (eds) Chemistry and biology in social insects. Peperny, München, pp 597-598

Passera L (1980) La ponte d'oeufs préorientés chez la fourmi *Pheidole pallidula* (Nyl.) (Hymenoptera - Formicidae). Insect Soc 27:79-95

Passera L, Suzzoni JP (1978a) Sexualisation du couvain de la fourmi *Pheidole pallidula* (Hymenoptera Formicidae) après traitement par l'hormone juvénile. CR Acad Sci Paris 286, Série D:615-618

Passera L, Suzzoni JP (1978b) Traitement des reines par l'hormone juvénile et sexualisation du couvain chez *Pheidole pallidula* (Nyl.) (Hymenoptera, Formicidae). CR Acad Sci Paris 287, Série D:1231-1233

Passera L, Suzzoni JP (1979) Le rôle de la reine de *Pheidole pallidula* (Nyl.) (Hymenoptera, Formicidae) dans la sexualisation du couvain après traitement par l'hormone juvénile. Insect Soc 26:343-353

Peeters C (1987) The diversity of reproductive systems in ponerine ants. In: Eder J, Rembold H (eds) Chemistry and biology of social insects. Peperny, München, pp 253-254

Peeters C, Crewe R (1984) Insemination controls the reproductive division of labour in a ponerine ant. Naturwiss 71:50

Peeters C, Crozier RH (1988) Caste and reproduction in ants: not all mated egg-layers are "queens". Psyche (in press)

Petersen-Braun M (1975) Untersuchungen zur sozialen Organisation der Pharaoameise *Monomorium pharaonis* (L.) (Hymenoptera, Formicidae). Insect Soc 22:269-292

Petersen-Braun M (1977) Studies on the endogenous breeding cycle in *Monomorium pharaonis* L. (Formicidae). Proc Int Congr IUSSI, Wageningen, 8:211-212

Pianka ER (1978) On r- and K-selection. In: Clutton-Brock TH, Harvey PH (eds) Readings in sociobiology. Freeman and Company, Reading and San Francisco, 393 p

Plateaux L (1971) Sur le polymorphisme social de la fourmi *Leptothorax nylanderi* (Förster). II: Activité des ouvrières et déterminisme des castes. Ann Sci Nat Zool 13:1-90

Poldi B (1963) Studi sulla fondazione dei nidi nei Formicidi. I: *Tetramorium caespitum* L.. Atti IV Congresso UIEIS Pavia 1961, Symp Gen Biol Ital 12:132-199

Robeau RM, Vinson SB (1976) Effects of juvenile hormone analogues on caste differentiation in the imported fire ant, *Solenopsis invicta*. J Georgia Entomol Soc 11:198-203

Rosengren R, Pamilo P (1983) The evolution of polygyny and polydomy in mound-building *Formica* ants. Acta Entomol Fennica 42:65-77

Ross KG, Fletcher DJC (1985) Comparative study of genetic and social structure in two forms of the fire ant *Solenopsis invicta* (Hymenoptera: Formicidae). Behav Ecol Sociobiol 17:349-356

Schmidt GH (1974) Sozialpolymorphismus bei Insekten. Wiss Verlagsges, Stuttgart, XXIV + 974 p

Schneirla TC (1971) Army ants. In: Topoff HR (ed), Freeman and Company, San Francisco, X + 349 p

Suzzoni JP, Passera L, Strambi A (1980) Ecdysteroid titre and caste determination in the ant, *Pheidole pallidula* (Nyl.) (Hymenoptera: Formicidae). Experientia 36:1228-1229

Tschinkel WR, Howard DF (1978) Queen replacement in orphaned colonies of the fire ant, *Solenopsis invicta*. Behav Ecol Sociobiol 3:297-310

Tschinkel WR, Howard DF (1983) Colony founding by pleometrosis in the fire ant *Solenopsis invicta*. Behav Ecol Sociobiol 12:103-113

Wesson LG jr. (1940) An experimental study on caste determination in ants. Psyche 47:106-111

Wheeler WM (1933) Colony-founding among ants, with an account of some primitive Australian species. Harvard University Press, Cambridge, Mass., X + 179 p

Wildman MH, Crewe RM (1988) Gamergate number and control over reproduction in *Pachycondyla krugeri* (Hymenoptera: Formicidae). Insect Soc (in press)

Wilson EO (1953) The origin and evolution of polymorphism in ants. Quart Rev Biol 28:136-156

Wilson EO (1971) The insect societies. Belknap Press of Harvard University Press, Cambridge, X + 548 p

Wilson EO (1974) Aversive behavior and competition within colonies of the ant *Leptothorax curvispinosus*. Ann Ent Soc America 67:777-780

Winter U (1979) *Epimyrma goesswaldi Menozzi*, eine sklavenhaltende Ameise. Naturwiss 66:581

Winter U, Buschinger A (1983) The reproductive biology of a slavemaker ant, *Epimyrma ravouxi*, and a degenerate slavemaker, *E. kraussei* (Hymenoptera: Formicidae). Entomol Gener 9:1-15

Winter U, Buschinger A (1986) Genetically mediated queen polymorphism and caste determination in the slave-making ant, *Harpagoxenus sublaevis* (Hymenoptera: Formicidae). Entomol Gener 11:125-137

Chapter 3

Physiology and Reproduction in Social Wasps

ALAIN STRAMBI[1]

CONTENTS

1 Introduction

Of all social insects, probably the least well known are social wasps. This is especially true of the mechanisms of caste induction and the regulation of reproductory functions in the wasp colony.

This lack of knowledge is undoubtedly due, for the most part, to difficulties involved in making experimental observations. Except in the least evolved sub-families, all wasps nests are wrapped in an envelope which impedes direct observation and renders access inconvenient. Until now, very few VESPINAE nest foundations have been able to be obtained in a laboratory environment.

Nevertheless, articles or books enumerating the main problems have been published in the last decades (Richards 1971; Wilson 1971; Spradbery 1973; West-Eberhard 1978; Brian 1979, 1980; Jeannė 1980; Edwards 1980; de Wilde and Beetsma 1982; Akre 1982).

---

[1] *CNRS, Laboratoire de Neurobiologie, Marseille, France*

The vespid family is made up of three sub-families of social wasps according to the classification of Richards: STENOGASTRINAE, POLISTINAE and VESPINAE. The knowledge of STENOGASTRINAE has been greatly increased by Turillazzi (1985, 1987).

The most varied group is POLISTINAE which is subdivided into POLISTINI, ROPALIDIINI and POLYBIINI. The latter two dwell in tropical and subtropical climates while POLISTINI (genus *Polistes* and *Sulcopolistes*) live also in temperate zones. VESPINAE are widely found in temperate regions.

2 Caste Differences

Vespid colonies go through characteristic yearly cycles. Even in tropical conditions with a more or less stable climate, the colony changes in size and above all in composition. These cycles are all the more visible in temperate zones.

In these regions, the colony disappears at the beginning of winter when the nest is abandoned. Only future foundresses are able to survive the winter and found a new colony the following spring.

When warmer weather returns, the nest is founded by a single female in VESPINAE as opposed to several as is often the case for POLISTINAE. In the tropical POLIBIINI family, new nests are founded by a swarm of workers accompanying a one or more founding females (Jeanne 1980).

The foundation of a nest is therefore the most significant event in the definition of caste.

The terms "queen" and "worker" used to describe castes in bee societies are not readily applicable to wasp societies in general and in particular to POLISTINAE which do not often exhibit any external morphological feature on which to base such a distinction.

2.1 Vespinae

It is in VESPINAE that morphological differences between queens and workers are the most pronounced. The queen is always visibly larger than the workers even though workers do vary in size (Archer 1972; Spradbery 1972). In addition to this difference in size, there are often slight variations in color. The species exhibiting the most pronounced polymorphism is *Vespula squamosa* (Drury): in this type of wasp, the queen is deep orange and nearly is two times longer than the worker which has black and yellow strips.

In temperate zones wasps, the fat body of the young adult queen stores reserves of fat and other materials (Spradbery 1973).

Egg laying is not exclusively limited to founding females but unmated workers can only lay eggs which hatch into males.

2.2 Polistinae

In the POLISTINAE family, morphological differences between castes are, however, very small and even inexistent. The queens have a tendency to be larger in size but size distribution overlaps in most species. Instead moderate seasonal variations in size have been reported for *Polistes exclamans*, *Polistes fuscatus*, *Polistes gallicus* (Eickwort 1969; West-Eberhard 1969; Turillazzi 1980). Only in *Polistes fuscatus* a statistical slight difference between workers and foundresses has been pointed out.

There is, however, one notable exception: queen *Polybia dimidiata* wasps (Richards 1971) are smaller than workers of the species.

The only reliable criteria by which castes may be defined are biological ore ethological in nature. For those species living in temperate zones only founding females are able to hibernate. It is widely believed that although workers are endowed with a complete set of sexual organs including a spermatheca and attract males, they refuse to copulate and are wholly responsible for finding food and wood fiber, constructing the nest and feeding the brood. However, once the nest is firmly established, the queen remains in the nest and lays eggs while the workers assume the outside work and feed the brood and often the queen itself.

In the tropical wasp *Polistes canadensis* (West-Eberhard 1969), coupling is in some cases postponed and nest building may begin before the female is inseminated.

Of all the POLISTINAE, the POLIBIINI wasp *Stelopolybia areata* appears to be the species with the greatest polymorphism (Jeanne and Fagen 1974). The gaster, especially the first segment, is ostensibly larger in queen wasps to make room for its more voluminous ovaries. The petiole is larger in *S. flavipennis* queens. Jeanne and Fagen report that caste differences are most conspicuous in species living in large colonies.

3 Social Parasitism

It should be mentioned that some species of social Vespids live as social parasites; the females are not divided into castes. They are called cuckoo wasps. Among VESPINAE, *Vespula austriaca* lives on the nests of *Vespula rufa*. The female parasite enters an already established nest containing brood and takes the place of the host queen. Its descendants are raised by workers born from the brood existing before its arrival. Only males and females of the parasitic species are hatched (Archer 1978). *Dolichovespula integra* and *Dolichovespula adulterina* are the two other European parasite species. Greene et al. (1978) report on the behavior of an american species, *D. arctica*, which invades the nests of *D. arenaria* and replaces the queen after several days. The morphology of the social parasites of VESPINAE wasps is very close to that of their hosts. In POLISTINAE wasps, however, the difference is more apparent and parasites have been classed into a different genus: *Sulcopolistes*. The parasite is able to impose itself on a polygynous colony of *Polistes* by means of its dominant behavior, but the eggs which it deposits in the nests are sometimes destroyed even by females from the lowest ranks of the social hierarchy (Demolin and Martin 1980).

Finally, we should mention some intermediary cases of parasitic activity. Certain species are known to be capable of either building a nest of its own ore invading a nest already established by another species. In case of invasion the worker caste continues to exist. Examples of this duality are *Vespa dybowskii* (Sakagami and Fukoshima 1957) in Japan and *Vespula squamosa* in North America (Mac Donald and Matthiews 1975).

4 Physiological Correlates of Caste Differentiation

We have seen above that wasp societies are subject to regulation from ecological, biological and social conditions which together bring about differentiation of females into worker and foundress castes. These interactions have often been studied; we have listed here a few examples. Pheromones, food or perhaps other sensorial information affect ovarian physiology in larvae or young adults and thereby induce their caste by means of endocrine factors. These questions have received less attention and knowledge is fragmentary.

4.1 Vespinae

In VESPINAE wasps, caste dimorphism is clear-cut. The queen alone can found a nest and than give birth to a very large population (one queen can have several thousand descendants in *Vespa crabro*, *Paravespula germanica* or *P. vulgaris* (Spradbery 1973)). Females are endowed with twelve ovarioles instead of six as in the other vespid females. This gives them a much greater fertility.

The VESPINAE are also the only wasp group in which the queen hatches from larger cells located in the lower part of the nest, i.e. in the last combs to be constructed. The queen induces these cells to be built just before the end of the colony life cycle. As there is no transition, they clash with the appearance of the rest of the nest. Once they have started the construction of large cells, it is exceptional to see workers resume building small ones. In *Vespa orientalis* (Ishay 1975) these larger cells are not built if the queen is removed beforehand. In *Vespula vulgaris*, Potter (1965) proved that a queen taken from an old colony producing future foundresses and placed in a newer nest, immediately, induces the workers to construct large cells. Conversely, a young queen transfered to an old colony in the process of constructing large cells, induces the workers to resume construction of small cells.

In *Vespa orientalis*, Ishay (1975) observed that if a queen is transfered from a new nest in which worker cells are being built into one in which construction of the larger cells is in progress, the workers will reduce the size of the opening of the larger cells to approximately the same diameter as worker cells. The larvae fed in these downsized cells hatch into workers.

Ishay (1975) also succeeded in displacing eggs inside the nest. When the eggs extracted from queen-producing cells are glued into smaller cells they develop into workers. Conversely eggs from small cells transported into queen cells at the end of the season turn into queens. The same sort of transfer can be made with larvae. The same results are obtained as long as the larvae are not older than the third larval instar. In other words larvae from worker-cells transfered into large cells at the end of the cycle become queens. Queen larvae become workers if transfered to small cells during the first to the third instar. In contrast, at the fourth and fifth instar, it is too late and larvae become workers and not queens inspite of transfer to larger cells. The same results have been recorded for *Paravespula germanica.*

These experiments bring out two basic factors in the understanding of female caste induction in VESPIDAE.

1. Castes are induced before the end of the third larval instar. Larvae which will become larger-sized queens, able to found new nests and lay hundreds and thousands of eggs, are raised in larger cells where they are fed more and richer food.

Montagner and Courtois (1963) have shown that worker larvae are fed less than queen larvae. Montagner (1966) further adds that queen larvae receive a much greater quantity of the salivary gland secretions which impregnate the regurgitations of exogenous origin. The size of the cells containing the larvae appears to signal to the workers the quality and quantity of food to be supplied. Fischl and Ishay (1971) have recorded differences in carbohydrates metabolism between the two castes larvae.

2. The construction of small or large cells by the workers according to the degree of completion of the nest seems to be under the control of the queen. Ishay reports that in absence of a queen,

oriental hornet workers do not build any queen cells at the end of the season. Potter has shown in *Vespula vulgaris* that the age of the queen is an important factor in the type of cell constructed. In their experiments Ishay et al. (1965) and Ikan et al. (1969) have been able to isolate a pheromone which, when placed in a queenless nest on a piece of cotton, is very attractive to workers and triggers the construction of large cells. This substance which is obtained from cephalic glands was identified as a lactone with 16 carbon atoms (delta-n hexadecalactone).

Fertility, however, is not the absolute characteristic of the queen. Under certain conditions, workers are capable of laying eggs. This is the case when the queen is removed from the nest (Montagner 1963; Ishay 1964; Spradbery 1973).

In a study of *Paravespula vulgaris* and *P. germanica*, Montagner (1966) showed that if a group of workers is deprived of their queen, their ovaries begin to develop after ten days if they come from a young colony and after only two if they are taken from an old nest in which male larvae have already appeared. The group is an important stimulating factor for the growth of ovaries in workers without a queen. An isolated worker exhibits no such development (Motro et al. 1979).

By splitting colonies of various *Vespula* species into queenless groups and queenright groups, Akre and Reed (1983) showed that the queen inhibits the ovaries of the workers of its group. The results of switching the groups to different sides of a vertically devided nest box, strongly suggested that the queen may deposit an inhibitory pheromone on the comb. This pheromone not only reduces ovarian development but also affects the foraging behavior which was reduced in queenless workers.

In larger nests, the queen occupies the large cell-combs and the workers sometimes lay eggs in the upper levels. A queen who has used all her spermatozoa can no longer lay diploid eggs. In this case, the responsibility for egg laying falls on the workers (Montagner 1963, 1966). Workers can therefore become the main source of males in aging colonies.

We can thus assume that in VESPINAE, at least in some species, the queen maintains the coherence of the colony by preventing workers from reproducing. Towards the end of the cycle, a pheromone is produced and serves as a signal to begin the construction of large cells. Workers, probably guided by the size of the opening, supply the larvae inside with more and richer food which by trophogenic induction releases the development of the further queens.

Such a scheme is not so easy to establish for POLISTINAE colonies.

4.2 Polistinae

Certain genera of POLISTINAE, although they are considered as social, do not exhibit caste differentiations. One such example is *Belonogaster griseus*. In an important study of this species, Marino-Piccioli and Pardi (1970), Pardi and Marino-Piccioli (1970) performed measurements on the population of the nest with mated and unmated females and uncovered certain statistical differences. One the one hand, inseminated females tended to have larger thoraces as well as more fully developed ovaries. They were more dominant and oophagic, undertook more building and stayed most often close to the brood. Uninseminated females, on the other hand, were less dominant and stayed away from the nest longer in search of food. However, they were capable of great fecundity and could rise to the highest ranks of the hierarchy and even found a nest on their own.

The question of whether the caste of a fertile or unfertile female is already determined at the preimaginal instars is left unanswered by these authors.

It is equally difficult to uncover the existence of castes and assign them to POLISTINAE in general and *Polistes* in particular, for which there are two mechanisms affecting female reproduction: a real caste separation between workers and foundresses and a phenomenon of social hierarchy which separates foundresses into queen and auxiliaries.

4.2.1 Workers and Foundresses

In view of the shortage and, in many cases, absence of morphological differences, authors have generally relied on behavioral differences as a basis for assigning castes. In summer, progeniture of *Polistes* living in temperate zones where only future foundresses survive the winter, normally have underdeveloped ovaries. Furthermore it is widely accepted that workers are more active and while they attract males, can not be inseminated. Conversely future foundresses work less and can be inseminated (Pardi 1946; Deleurance 1946, 1952; West-Eberhard 1969). West-Eberhard (1969) suggests that in *P. fuscatus* and *P. canadensis* only those females with mature ovaries actually construct new cells whereas workers with underdeveloped ovaries simply extend the cell.

Under certain conditions, workers can however develop their ovaries (Pardi 1946, 1951; Haggard and Gamboa 1980). In foundresses, the developmental cycle of the ovaries is of course under the control of the complex neuroendocrine system.

Juvenile hormone injection or *corpora allata* implantation have been seen to stimulate ovarian growth in *P. metricus* and *P. gallicus* (Bohm

1972; Girardie and Strambi 1973). Ecdysone can be found in large amounts in the ovaries and also circulating in the haemolymph (Strambi, Strambi and de Reggi 1977). If the ovaries of a mature queen are removed, the *corpora allata* first increase, then decrease in size (Strambi 1967a, 1969) while neurosecretory material accumulates in the cisternae of rough reticulum of medial neurosecretory cells. This material is probably a precursor of neurosecretion (Strambi 1967a, Strambi and Strambi 1973a).

Castration is caused naturally by the Strepsipteron *Xenos vesparum* Rossi, the presence of which affects also the neuroendocrine system (Strambi 1965, 1966, 1967b, 1969). In new born parasitized females which have smaller than normal ovaries, the *corpora allata* greatly decrease in size while a rapid discharge from the neurosecretory cells of the *pars intercerebralis* is observed (Strambi and Strambi 1973b).

In new born females, the haemolymphatic protein level is low and no depletion is detected in parasitized females up to the tenth day of imaginal life. After that, in some normal females, the circulating proteins increase. Such an increase is never observed in stylopized wasps for which protein level remains unchanged (Strambi et al. 1982).

In spite of castration, the qualitative composition of haemolymphatic proteins is the same in both normal and parasitized wasps, even for the female specific fraction (Roux 1973).

The implantation of *corpora allata* induces the growth of the ovaries even in such stypolized females (Girardie and Strambi 1973). Probably endocrine factors are responsible for both ovarian physiology and reproductive behavior; an ovariectomized queen of *P. gallicus* is able to show the whole sequence of egg-laying behavior (egg deposition being of course included) but the ventral nerve cord had to be severed in order to exclude communication between the last abdominal nervous ganglion and the brain (Deleurance 1963).

Several authors have used physiological criteria to distinguish between female castes. Eickwort (1969) approached the caste problem in *P. exclamans* on the basis of internal structure. She reports that the parietal fat body is highly developed in certain females and in her opinion, this feature enables them to survive the winter. She therefore classifies such females as foundresses. Fat body appearance is, according to Strassmann et al. (1984), a reliable indicator of caste in *P. annularis* and *P. exclamans*. Survival following exposure to cold was better for future queens (separated on the basis of the appearance of their fat bodies) which had elevated cryoprotectant levels consisting mainly of fructose in the hemolymph. On the other hand, *P. gallicus* females parasitized by *Xenos*, have very thin yellowish fat bodies and nevertheless overwinter as well as normal females. Therefore the highly developed fat body found in future foundresses is probably necessary for egg development the next spring but not for survival during the winter.

Pratte et al. (1981) used physiological criteria (fat body, ovary size, haemolymphatic levels of ecdysone and proteins) and ethological criteria (presence on the nest, brood care) to divide a population of *P. gallicus* females into two groups which may correspond to castes. Females with small fat bodies seem to spend more time on nest maintenance while females with large fat bodies can remain on a nest without brood. In spite of an obvious analogy, these two groups seem not to coincide perfectly with the two female castes.

The photoconductivity of the cuticle has also been utilized as a criteria for caste distinction by Perna et al. (1978). Bohm (1972) used environmental factors for *P. metricus*. The ovaries of new born wasps raised in June-like conditions (26 °C, 16 hours day light) grew in size while females raised in July-like conditions (22 °C, 14 hours day light) had smaller ovaries. The latter females are the future foundresses with diapausing ovaries. The "June group" consists of workers, the ovaries of which developed normally in the absence of queen, but which would have been stunted in a queen right colony. Turillazzi and Conte (1980) do not agree with the results of Deleurance (1952) who reported that chilling the nurse-workers at 5 °C during the night induced the larvae they fed during the day to hatch into workers.

West-Eberhard (1969) suggests that the induction of castes takes place at the larval stage and depends on the amount of food the larvae receive according to the season and their location in the nest.

4.2.2 Social Hierarchy

In POLISTINAE, at least in most species, there is another mechanism of regulation of reproduction which takes place at the time of polygynous foundation.

Generally a nest is started by a foundress which is joined by another foundress which works in the same nest.

A hierarchy is set up in this small society or before the beginning of nest building and leads progressively to a functional monogynous arrangement (Pardi 1940, 1941, 1947, 1951; Deleurance 1946, 1952; Gervet 1964a, b, 1965, 1968; West-Eberhard 1969). Social rank is dependent on behavior which has been described in detail for many species but which can be outlined as follows. When two females meet, the dominated one assumes a motionless posture with its antennae folded down and the dominant one holding itself high on its legs moves around and strikes the dominated one with its antennae.

Usually, the foundresses which are attracted to a nest under construction have mature, functional ovaries and can lay eggs. However, Gervet (1962) reports that the dominant female is the most prolific. He further notes that dominance in a nest is related to fecundity, i.e. the number of eggs laid a day. Eggs laid by lower-ranked females are detected and destroyed by the dominant female. In this way, althoug the nest is polygynous, the only eggs which hatch are those of the dominant female known as the queen. The others are called auxiliaries.

Gamboa (1980) reports as an advantage of association that in such multiple-foundress colonies of *P. metricus*, the first workers hatch earlier than in single-foundress colonies.

Along with this behavior-related regulation, there is also a physiological phenomenon at work since the number of eggs deposited by auxiliaries gradually decreases and their ovaries regress (Pardi and Cavalcanti 1951).

The dominant female is able to increase its egg production (Gervet 1968; Hermann and Dirks 1975). At normal temperatures, the queen's ovaries produce more ovocytes than she actually lays (Strambi 1962, 1963, 1965; Gervet and Strambi 1965). This fact explains how the quantity of eggs laid can be quickly and easily increased or decreased according to the needs of the colony.

Juvenile hormone would also appear to play a role in the phenomenon of dominance. In a worker population of *P. annularis*, Barth et al. (1975) revealed that experimentally administrated juvenile hormone not only stimulated ovarian growth but also strenghtened dominant behavior in interrelations to such a point that the whole social order was perturbed. In multiple foundresses associations of *P. gallicus*, queens are larger than their auxiliaries (Turillazzi and Pardi 1977).

Foundresses of *P. gallicus* (Röseler et al. 1980) were submitted to measurements to determine the volume of the *corpora allata* and the amount of juvenile hormone synthesized by them. During the activation period the amount of juvenile hormone synthesized was seen to be correlated to the volume of the *corpora allata.* Vitellogenesis was also observed to be linked to an increase in the volume of *corpora*

Fig. 1 - *Corpora allata* (CA) volume in foundress associations in relation to social rank (ordinates). The number of females investigated is indicated above the columns. - One day after hibernation, the social hierarchy was determined, the subordinated females were dissected and the volume of their CA was measured. The three most dominant (upper diagram, trigynous association) or the two most dominant (lower diagram, digynous association) females were left together for six additional days. They were then dissected and the sizes of their CA compared to those of subordinate females investigated before. The CA of the most dominated females (gamma in trigynous associations, beta in digynous associations) were found to be smaller than those of the lower ranked females which they had dominated 6 days before; their sizes have therefore decreased. (After the data of Röseler et al. 1984).

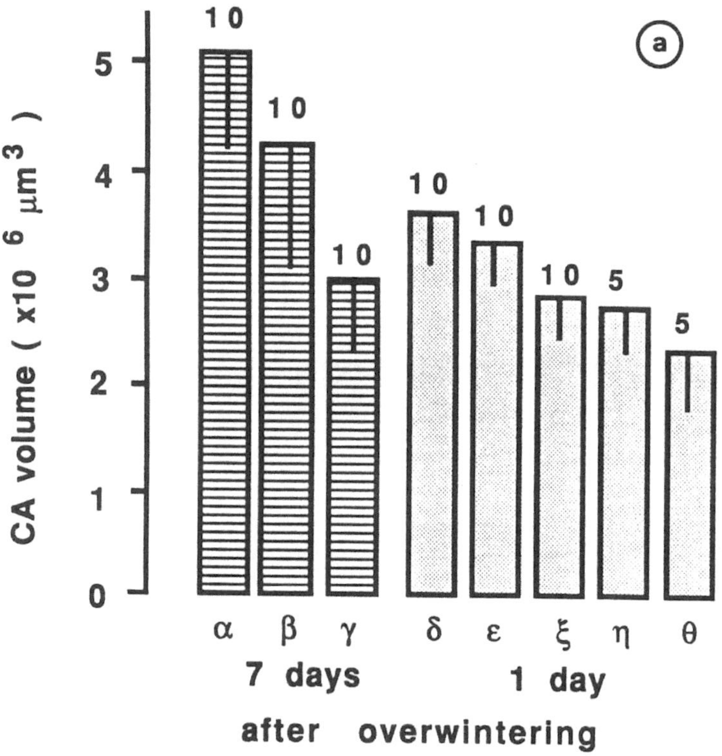
a
CA volume (x10 6 μm 3)
0
1
2
3
4
5
10
10
10
10
10
10
5
5
α β γ δ ε ξ η θ
7 days
1 day
after overwintering

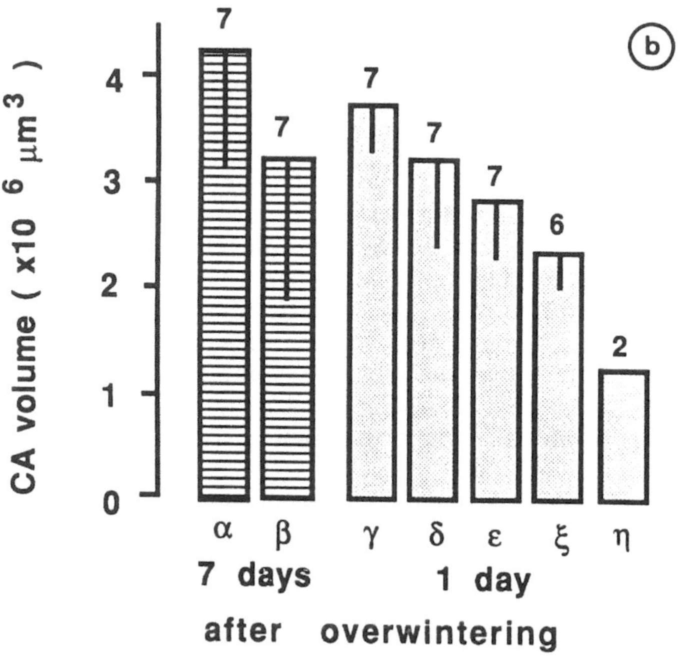
b
CA volume (x10 6 μm 3)
0
1
2
3
4
7
7
7
7
7
6
2
α β γ δ ε ξ η
7 days
1 day
after overwintering

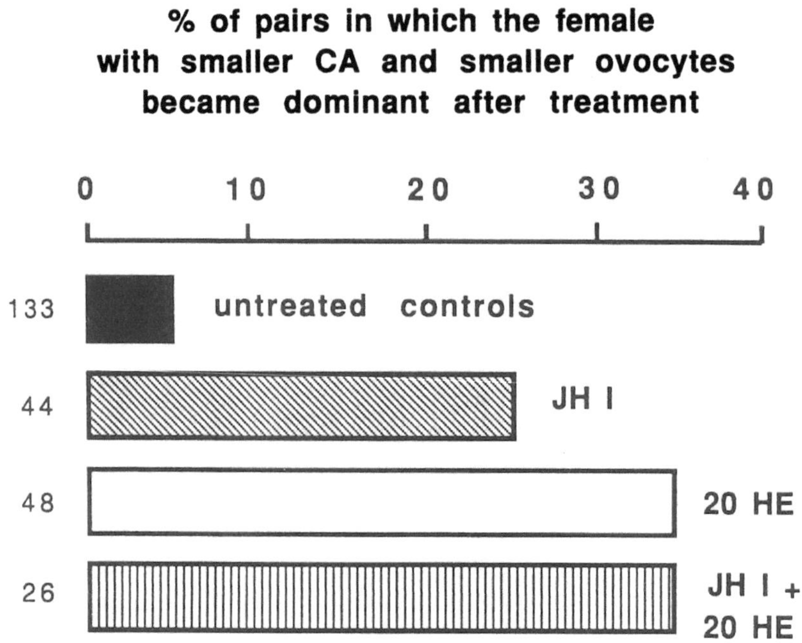

Fig. 2 - Influence of hormone injections on dominance in females which had smaller ovocytes and smaller *corpora allata* (CA) than sham operated females of the pair. Untreated control pairs are also shown. The number of pairs of females investigated is indicated on the left of the bars. JH I: juvenile hormone I; 20 HE: 20-hydroxyecdysone. (After the data of Röseler et al. 1984).

allata. Moreover, the *corpora allata* of subordinated females were smaller and less active than the dominant ones. It was therefore interesting to know if the *corpora allata* of the subordinated females shrank as a result of the dominant behavior of the queen or if, on the other hand, subordination was the result of the hormonal context to which the *corpora allata* belong. In order to answer this question, foundresses just emerging from hibernation were divided into small groups of two and three. As soon as the social hierarchy was established in each group, the insects were dissected. In all cases this revealed a connection between the volume of the *corpora allata* and the social hierarchy. The dominant ones always had the largest *corpora allata* while the most dominated female always had the smallest.

Moreover, it could be shown experimentally that in fact the dominant female in such small societies causes a decrease in the *corpora allata* volume of the most dominated ones. The *corpora allata* of these most dominated females were found to be smaller than the glands dissected from the females they had dominated one week before.

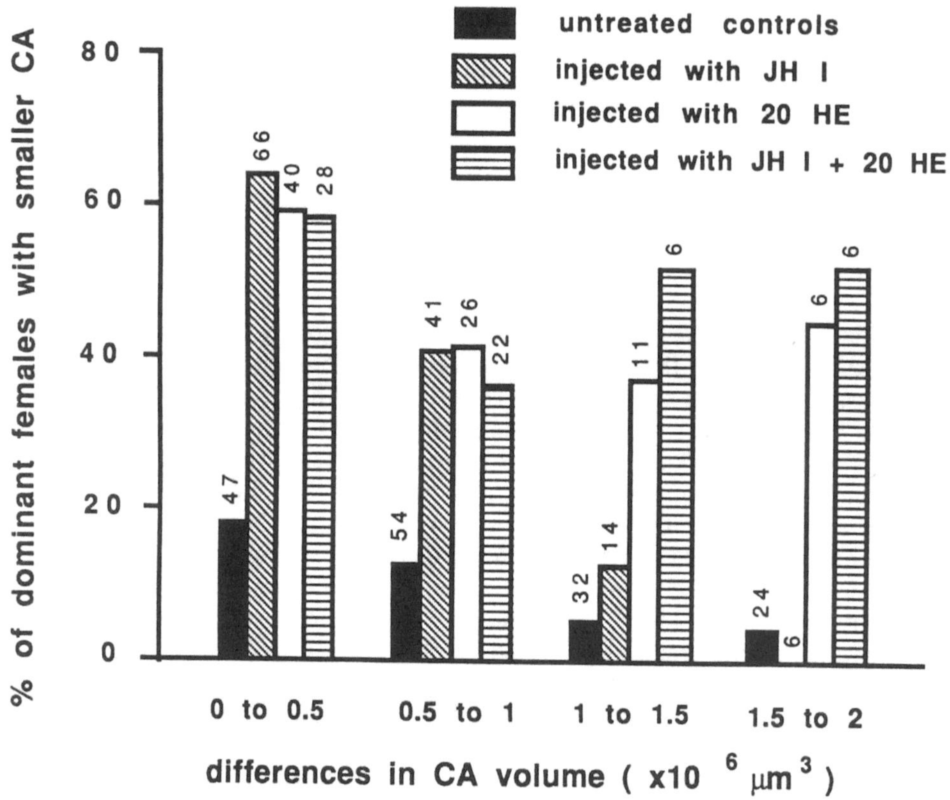

Fig. 3 - Influence of hormone injections on dominance in females which had smaller *corpora allata* (CA) but nevertheless became dominant over sham operated females of the pair. - Ordinates: differences between CA volumes of sham injected subordinate female and treated female. Untreated control pairs are also shown. The number of pairs of females investigated is indicated above the columns. JH I: juvenile hormone I; 20 HE: 20-hydroxyecdysone. (After the data of Röseler et al. 1984).

Therefore, the *corpora allata* of the newly subordinated must have diminished. In contrast, the glands of the dominant wasps continued to increase during that time (Röseler et al. 1984), (Fig. 1).

Environmental conditions (light and temperatute) could be responsible for the inequality which exists prior to the social hierarchy (Röseler 1985). This inequality is accentuated by the dominant/subordinate behavior (Röseler et al. 1986). Finally the social hierachy is established that leads to a functional monogyny in which only the dominant reproduces while the subordinates assume most of the social tasks (foraging, building...).

The influence of the *corpora allata* could be mimicked by injection of juvenile hormone or of 20-hydroxyecdysone (Röseler et al. 1984). One of these hormones or a mixture of both was injected into foundresses prior to their first meeting. They were dissected after the hierarchy had been established. A hormone injection into females having smaller *corpora allata* provided an advantage for obtaining a dominant position (Fig. 2). The smaller was the difference in *corpora allata* volume between the females, the more effective was the hormone injection (Fig. 3).

5 Conclusions[2]

Vespids are divided into two zoological categories. This distinction is underlined by their rather different biological characteristics. VESPINAE exhibit a monogynous foundation. In the progeny, the two female castes are well distinguishable. The queen inhibits the ovaries of the workers and is able to provoke caste differentiation by means of pheromones which she secretes. So she maintains the yearly regulation of the colony.

In contrast POLISTINAE, at least in certain species, are borderline examples of social insects. Castes are often morphologically indistinguishable. The polygynous foundation of their nests necessitates social regulation even within the foundress caste. It is not yet known if any pheromones are involved in the regulation of their reproductive activity. Their physiology and especially endocrinology is, however, a little better known and we are beginning to understand the endocrine regulation of the society.

6 References

Akre RD (1982) Social wasps. In: Hermann HR (ed) Social insects, vol 6. Academic Press, pp 1-105

Akre RD, Reed HC (1983) Evidence for a queen pheromone in *Vespula* (Hymenoptera: Vespidae). Can Entomol 115:371-377

Archer ME (1972) The significance of worker size in the seasonal development of the wasps *Vespula vulgaris* (L.) and *Vespula germanica* (F.). J Entomol 46:175-183

Archer ME (1978) The cuckoo wasp, *Vespula austriaca* (Panzer) (Hymen. Vespidae) in Yorkshire. Naturalist Hall 103:133-134

Barth RH, Lester LJ, Sroka P, Kessler T, Hearn R (1975) Juvenile hormone promotes dominance behavior and ovarian development in social wasps (*Polistes annularis*). Experientia 31:691-692

Bohm M (1972) Effects of environment and juvenile hormone on ovaries of the wasp *Polistes metricus*. J Insect Physiol 18:1875-1883

Brian MV (1979) Caste differentiation and division of labour. In: Hermann HR (ed) Social insects, vol 1. Academic Press, pp 121-222

---

[2] This chapter is a modified and updated version of "*Physiological aspects of caste differentiation in social wasps*", Strambi (1985).

Brian MV (1980) Social control over sex and castes in bees, wasps and ants. Biol Rev 55:379-415

Deleurance EP (1946) Une régulation sociale à base sensorielle périphérique; l'inhibition de la ponte des ouvrières par la présence de la fondatrice chez les *Polistes* (Hyménoptères Vespidae). CR Acad Sci 223:871-872

Deleurance EP (1952) Le polymorphisme social et son déterminisme chez les guêpes. Coll Int CNRS 34:141-155

Deleurance EP (1963) Analyse segmentaire du comportement de ponte chez les *Polistes* (*P. gallicus* L., *P. nimpha* Christ, Hyménoptères Vespides). CR Acad Sci 257:2175-2177

Demolin G, Martin JC (1980) Biologie de *Sulcopolistes semenowi* (Morawitz) parasite de *Polistes nimpha* (Christ), Hyménoptères Vespidae). Biol Ecol Medit 7:181-182

De Wilde J, Beetsma J (1982) The physiology of caste development in social insects. Adv Insect Physiol 17:167-246

Edwards R (1980) Social wasps. Rentokil, East-Grinstead, 398 p

Eickwort K (1969) Separation of the castes of *Polistes exclamans* and notes on its biology (Hym. Vespidae). Insect Soc 15:67-72

Fischl J, Ishay J (1971) The glucose levels and carbohydrate autolysis in *Vespa orientalis* haemolymph. Insect Soc 18:203-314

Gamboa GJ (1980) Comparative timing of brood development between multiple and single-foundress colonies of the paper wasp, *Polistes metricus*. Ecol Entomol 5:221-225

Gervet J (1962) Etude de l'effet de groupe sur la ponte dans la société polygyne de *Polistes gallicus* L. (Hymen. Vesp.). Insect Soc 9:231-263

Gervet J (1964) La ponte et sa régulation dans la société polygyne de *Polistes gallicus* L. Ann Sci Nat Zool, 12e série 6:601-778

Gervet J (1965) La ponte et sa régulation dans la société polygyne de *Polistes gallicus* L. Behaviour 25:221-233

Gervet J (1968) L'effet de groupe dans la société polygyne de Polistes (Hymen. Vesp.). Coll Int CNRS, L'effet de groupe chez les animaux, 173:77-103

Gervet J, Strambi A (1965) Dynamique de la fonction ovarienne chez les *Polistes* (Hymen. Vesp.). Cas de l'ouvrière. CR Acad Sci 260:4599-4601

Girardie A, Strambi A (1973) Effet de l'implantation de *corpora allata* actifs de *Locusta migratoria* (Orthoptère) dans les femelles de *Polistes gallicus* L (Hyménoptère) saines et parasitées par *Xenos vesparum* Rossi (Insecte, Strepsiptère). CR Acad Sci 276:3319-3322

Greene A, Akre RD, Landolt PJ (1978) Behaviour of the yellow jacket social parasite *Dolichovespula arctica* (Rohwer) (Hymenoptera, Vespidae). Melanderia 29:1-28

Haggard M, Gamboa GJ (1980) Seasonal variation in body size and reproductive condition of a paper wasp, *Polistes metricus* (Hymenoptera, Vespidae). Can Entomol 122:238-248

Hermann HR, Dirks TF (1975) Biology of *Polistes annularis* (Hymenoptera: Vespidae). I. Spring behavior. Psyche 82:97-108

Ikan R, Gottlieb R, Bergmann ED, Ishay J (1969) The pheromone of the queen of the oriental hornet, *Vespa orientalis*. J Insect Physiol 15:1709-1712

Ishay J (1964) Observations sur la biologie de la guêpe orientale *Vespa orientalis* F. Insect Soc 11:193-206

Ishay J (1975) Caste determination by social wasps; cell size and building behavior. Anim Behav 23:425-431

Ishay J, Ikan R, Bergmann ED (1965) The presence of pheromones in the oriental hornet, *Vespa orientales* F. J Insect Physiol 11:1307-1309

Jeanne RL (1980) Evolution of social behavior in the Vespidae. Ann Rev Entomol 25:371-396

Jeanne RL, Fagen R (1974) Polymorphism in *Stelopolybia areata* (Hymenoptera, Vespidae). Psyche 81:155-166

Mac Donald JF, Matthews RW (1975) *Vespula squamosa* a yellow jacket wasp evolving towards parasitism. Science 189:1003-1004

Marino-Piccoli MT, Pardi L (1970) Studi sulla biologia di *Belonogaster* (Hymenoptera, Vespidae). 1. Sull' etogramma di *Belonogaster griseus* (Fab.). Monit Zool Ital NS Suppl 3:197-225

Montagner H (1963) Contribution à l'étude du déterminisme des castes chez les guêpes du genre *Vespa*. CR Soc Biol 157:147-150
Montagner H (1966) Sur le déterminisme des castes femelles chez les guêpes du genre *Vespa*. CR Acad Sci 263:547-549
Montagner H, Courtois G (1963) Données nouvelles sur le comportement alimentaire et les échanges trophallactiques chez les guêpes sociales. CR Acad Sci 256:4092-4094
Motro M, Motro U, Ishay JS, Kugler J (1979) Some social and dietary prerequisites of oocyte development in *Vespa orientalis* L. workers. Insect Soc 26:155-164
Pardi L (1940) Poliginia vera ed apparente in *Polistes gallicus* L. Atti Soc Tosc Sci Nat 49:3-9
Pardi L (1941) Ricerche sui Polistini. 3. Ancora sulla poliginia iniziale di *Polistes gallicus* L. e sul compartamento delle femmine associate fino alla schiusa delle prime operaie. Atti Soc Tosc Sci Nat 50:3-15
Pardi L (1946) Ricerche sui Polistini. 7. La dominazione e il ciclo ovarico annuale in *Pollistes gallicus* L. Boll Ist Entomol Univ Bologna 15:25-84
Pardi L (1947) Beobachtungen über das interindividuelle Verhalten bei *Polistes gallicus* L. Behaviour 1:138-172
Pardi L (1951) Ricerche sui Polistini. 12. Studio della attività e della divizione di lavoro in una società di *Polistes gallicus* L. dopo la comparsa delle operaie. Arch Zool Ital 36:365-431
Pardi L, Cavalcanti M (1951) Esperienze sul meccanismo della monoginia funzionale in *Polistes gallicus* (L.) (Hymenopt., Vesp.). Boll Zool 18:247-252
Pardi L, Marino-Piccioli MT (1970) Studi sulla biologia di *Belanogaster* (Hymenoptera, Vespidae) 2. Differenziamento castale incipiente in *B. griseus* (Fab.). Monit Zool Ital NS Suppl 3:235-265
Perna B, Croitoru N, Ishay JS (1978) Dominance and hierarchy in *Polistes gallicus* colonies attained through photoelectric properties. Experientia 34:1022-1023
Potter NB (19??) A study of the biology of the common wasp, *Vespula vulgaris* L., with special reference to the foraging behaviour. Ph D Thesis Bistrol Univ
Pratte M, Strambi C, Gervet J, Strambi A (1981) Paramètres physiologiques et éthologiques dans un guêpier de *Polistes gallicus* L. Insect Soc 29:383-401
Richards OW (1971) The biology of the social wasps (Hymenoptera, Vespidae). Biol Rev 45:483-528
Röseler PF (1985) Endocrine basis of dominance and reproduction in polistine paper wasps. In: Hölldobler B, Lindauer M (eds) Fortschr Zool 31. Fischer, Stuttgart, New York, pp 259-272
Röseler PF, Röseler I, Strambi A (1980) The activity of *corpora allata* in dominant and subordianted females of the wasp *Polistes gallicus*. Insect Soc 27:97-107
Röseler PF, Röseler I, Strambi A, Augier R (1984) Influence of insect hormones on the establishment of dominance hierarchies among foundresses of the paper wasp *Polistes gallicus*. Behav Ecol Sociobiol 15:133-142
Röseler PF, Röseler I, Strambi A (1985) Role of ovaries and ecdysteroids in dominance hierarchy establishment among foundresses of the primitively social wasp *Polistes gallicus*. Behav Ecol Sociobiol 18:9-13
Röseler PF, Röseler I, Strambi A (1986) Studies on the dominance hierarchy in the paper awsp *Polistes gallicus* L. (Hymenoptera, Vespidae). Monitore Zool Ital 20:283-290
Roux C (1973) Etude comparée de la protéinémie chez les imagos de *Polistes* (Hyménoptères Vespides) sains ou parasités par le Strepsiptère *Xenos vesparum* Rossi. CR Acad Sci 276:3159-3162
Sakagami SF, Fukushima K (1957) *Vespa dybowskii* André as a facultative temporary social parasite. Insect Soc 4:1-12
Spradbery JP (1973) Wasps. Sidgwick and Jackson, London, 408 p
Strambi A (1962) Sur le fonctionnement de l'ovaire chez la guêpe Poliste. Détermination du temps de croissance ovocytaire (*Polistes gallicus* L. et *P. nimpha* Christ). CR Acad Sci 255:2012-2014

Strambi A (1963) Action différentielle de la température sur le débit ovarien et la ponte chez la guêpe Poliste (*Polistes gallicus* L. et *Pollistes nimpha* Christ., Hyménoptères Vespides). CR Acad Sci 256:5642-5643

Strambi A (1965a) Essai d'analyse de la dynamique ovarienne chez les *Polistes*. (*P. gallicus* et *P. nimpha* Christ) en activité reproductrice. Insect Soc 12:1-17

Strambi A (1965b) Influence du parasite *Xenos vesparum* Rossi (Strepsiptère) sur la neurosécrétion des individus du sexe femelle de *Polistes gallicus* L. (Hyménoptères Vespides). CR Acad Sci 260:3768-3769

Strambi A (1966) Action de *Xenos vesparum* Rossi (Strepsiptère) sur la neurosécrétion des fondatrices-filles de *Polistes gallicus* L. (Hyménoptère, Vespide) en diapause. CR Acad Sci 263:533-535

Strambi A (1967a) Quelques effets de la castration sur la neurosécrétion protocérébrale des femelles de *Polistes* (Hyménoptères, Vespides). CR Acad Sci 264:2031-2034

Strambi A (1967b) Effets de la disparition du parasite *Xenos* (Strepsiptère) sur la neurosécrétion protocérébrale de son hôte *Polistes* (Hymen. Vespides). CR Acad Sci 264:2646-2648

Strambi A (1969) La fonction gonadotrope des organes neuroendocrines des guêpes femelles du genre *Polistes* (Hyménoptères). Influence du parasite *Xenos vesparum* Rossi (Strepsiptères). Thèse Doct Sci Nat Paris

Strambi A (1985) Physiological aspects of caste differentiation in social wasps. In: Watson JAL, Okot-Kotber BM, Noirot C (eds) Caste differentiation in social insects. Pergamon Press, Oxford, New York, pp 371-384

Strambi A, Strambi C (1973a) Etude histochimique et ultrastructurale des sécrétions élaborées par les péricaryones neurosécréteurs de la *pars intercerebralis* chez la guêpe *Polistes*. Acta histochem 46:101-119

Strambi A, Strambi C (1973b) Influence de développement du parasite *Xenos vesparum* Rossi (Insecte, Strepsiptère) sur le système neuroendocrinien des femelles de *Polistes* (Hyméoptère, Vespide) au début de leur vie imaginale. Arch Anat Micr Morphol exp 62:39-54

Strambi A, Strambi C, De Reggi M (1977) Ecdysones and ovarian physiology in the adult wasp *Polistes gallicus*. Proc Int IUSSI Congr, Wageningen, 8:19-20

Strambi C, Strambi A, Augier R (1982) Protein level in the hemolymph of the wasp *Polistes gallicus* at the beginning of the imaginal life and during overwintering. Action of the strepsipterian parasite *Xenos vesparum* Rossi. Experientia 38:1189-1191

Strassmann JE, Lee RE, Jr, Rojas RR, Baust GJ (1984) Caste and sex differences in cold-hardiness in the social wasps, *Polistes annularis* and *P. exclamans* (Hymenoptera: Vespidae). Insect Soc 31:291-301

Turillazzi S (1980) Seasonal variation in the size and anatomy of *Polistes gallicus* L. (Hymenoptera, Vespidae). Monit Zool Ital 14:63-75

Turillazzi S (1985) Colonial cycle in *Parischnogaster nigricans serrei* (Du Buysson) in West Java (Hymenoptera, Stenogastrinae). Insect Soc 32:43-60

Turillazzi S (1987) Social biology of the Stenogastrinae: temporary dynamic reproductive strategy in the west tropics. In: Eder J, Rembold H (eds) Chemistry and biology of social insects. Peperny, München, pp 381-382

Turillazzi S, Conte A (1980) Influence of temperature on caste differentiation in *Polistes foederatus* (Kohl) (Hymenoptera Vespidae). Monit Zool Ital 14:116

Turillazzi S, Pardi L (1977) Body size and hierarchy in polygynic nests of *Polistes gallicus* L. (Hymenoptera, Vespidae). Monit Zool Ital 11:101-112

West-Eberhard MJ (1969) The social biology of Polistine wasps. Miscellaneous publications of the Museum of Zoology, Univ Michigan 140:1-101

West-Eberhard MJ (1978) Polygyny and the evolution of social behavior in Wasps. J Kansas Entomol Soc 51:832-836

Wilson EO (1971) The Insect Societies. Harvard Univ Press, Cambridge, Mass, 548 p

Chapter 4

Reproduction and Castes in Social Halictine Bees*

CHARLES D. MICHENER[1]

CONTENTS

1 Introduction

Parasocial and primitively eusocial bees are found in various families of the Apoidea, but the majority of such forms are in the HALICTINAE (Family HALICTIDAE, the sweat bees). The HALICTINAE

[1] *Departments of Entomology and of Systematics & Ecology and Snow Entomological Museum, University of Kansas, Lawrence, Kansas, USA*
* Contribution number 1779 from the Department of Entomology, University of Kansas, Lawrence

are an enormous and abundant group, worldwide in distribution, arctic to tropical, and every continent has forms whose social biologies remain unknown. Although a few species nest in rotting wood, most make burrows in the soil. New and interesting types of social organization probably remain to be discovered in this subfamily, for only a tiny fraction of the species have been studied behaviorally.

One recent finding concerns food transfer. In 1974 I emphasized that mouth to mouth food transfer was unknown in HALICTINAE, although well known in the other major group of primitively social bees, the XYLOCOPINAE (see chapter 5), as well as in the APIDAE and other social insects. It is therefore of special interest that Kukuk and Schwarz (1987, 1988) reported what appeared to be such trophallaxis in communal colonies of *Lasioglossum* (*Chilalictus*) *erythrurum* in Australia. This is an entirely new finding for halictid bees; if nectar or other materials are transferred, it considerably expands the diversity of social behaviors known among HALICTINAE.

The present chapter and the next are largely limited to findings made since 1972. The HALICTINAE were treated by Sakagami (1974) and by Michener (1974), the material in the latter relevant to the present paper being largely in Chapters 8-11, with background natural history information in Chapters 22-26. In the following pages, although the 1974 book is specifically referred to on various occasions, unattributed statements are also based upon it. A more recent review, largely limited to biology of the genus *Halictus*, is by Knerer (1980a).

The two following paragraphs provide some classificatory and nomenclatural background:

The HALICTINAE are divisible into three tribes, the HALICTINI, AUGOCHLORINI, and NOMIOIDINI (Michener 1978). Nearly all of this chapter concerns the HALICTINI; except for *Pereirapis* the occasional references to AUGOCHLORINI will be recognized by the generic names based on *Augochlora*.

Within the HALICTINI there is some nomenclatorial confusion at the generic and subgeneric levels. Some authorities, especially in the past, placed all the species referred to below in the genus *Halictus* and included *Lasioglossum* under the generic name *Halictus*. Others recognize not only *Halictus* and *Lasioglossum* but divide the latter and use the names *Dialictus* (Fig. 1) and *Evylaeus* at the genus level. Along with A. W. Ebmer and others I prefer to consider *Dialictus* and *Evylaeus* as subgenera (or as a subgenus) of *Lasioglossum*. To save space in the following account, I make extensive use of the abbreviations *H.* for *Halictus* and *L.* (*D.*) and *L.* (*E.*) for *Lasioglossum* (subgenus *Dialictus*) and *Lasioglossum* (subgenus *Evylaeus*). The subgenera *Evylaeus* and *Dialictus*, as commonly recognized, intergrade completely. I find it convenient, nevertheless, to use the names in the following way: The small

Fig. 1 - Overwintered female of *Lasioglossum* (*Dialictus*) sp.? on flower of *Isopyrum biternatum*, Lawrence, Kansas (photograph by R. W. Brooks).

greenish species whose biologies are well known are in *Dialictus*. The larger, usually banded, non-greenish species, usually with a carinate propodeum, are in *Evylaeus*. Many small non-greenish species, while technically included in *Evylaeus*, are in fact black *Dialictus*. Little is known of their behavior and they are scarcely mentioned in the rest of this review. The name *Chloralictus*, which sometimes appears in the literature, is a synonym of *Dialictus*.

While there are many solitary HALICTINI, often in the same genera and subgenera that contain social ones, this chapter concerns almost exclusively species such as *L.* (*D.*) *zephyrum* that are regularly social during at least part of the year or life cycle.

Yanega's (1988, 1989) studies emphasize that in social halictines there are commonly two sorts of principal egg layers, (1) those derived from foundresses that never functioned as workers, and (2) those derived from among daughters (often workers) of queens that disappeared. Following Eickwort and Kukuk (1987), I call these foundress queens and replacement queens, respectively ("replacement" because such bees replace foundress queens or other replacement queens). In temperate climates foundress queens have overwintered and replacement queens have not. A colony consisting of a foundress queen and her daughter workers is eusocial. A colony of a replacement queen and her sister workers is semisocial, with the potential of becoming eusocial if the queen's sisters die and are replaced as workers by her daughters. While from the viewpoint of function (egg laying, dominance over workers, etc.) a replacement queen may be as much a queen as is a foundress queen, from the viewpoints of origin, pre-reproductive physiology and behavior (lack of diapause, failure to establish a new nest), and mean size she is more like a worker. It can be argued, therefore, that a replacement queen should be thought of as a reproductive worker rather than as a queen. Nonetheless caste differences are much the same regardless of the type of queen in a colony (see Packer 1986d).

Particularly in the tropics, the distinction between foundress and replacement queens may not be clear. There is no winter and there may be no season of inactivity and no diapause of foundresses. The most fully studied tropical species is *Pereirapis semiaurata* investigated in the state of Sao Paulo, Brazil, by Campos (1980). The species is active throughout the year. Nests are started by lone foundresses and eusocial colonies develop when there are workers. In the larger colonies new female reproductives are produced but no clear distinction is evident in Campos' work between those that will be foundress queens and those that will be replacements. Unlike workers they do not forage for pollen; in this respect they are like young foundresses of *H. rubicundus* (Yanega 1988).

Elaborating further upon the caste terminology of Eickwort and Kukuk (1987), **foundresses** initiate nests and if working alone, become **foundress queens** when they have workers and thus a eusocial colony. If several cofoundresses jointly initiate a colony, and if there is reproductive division of labor among them, a **foundress queen** and **foundress workers** (or **auxiliaries**) are recognizable in the semisocial colony. [Eickwort and Kukuk use instead the terms dominant cofoundress and subordinate cofoundress, but I prefer to avoid the implications that one must have data on dominance interactions and that there are multiple dominant and subordinant individuals; there is ordinarily one queen and may be only one worker (subordinant) in a colony.] If a foundress worker replaces a dead foundress queen, she becomes a **secondary foundress queen**. Daughters that are of reduced reproductivity and that remain in their mother's nest are **workers** (**daughter workers** if the distinction is needed). If a daughter worker (even if young and little differentiated as a worker) replaces a dead foundress queen, she becomes a **replacement queen**. For convenience, I use the terms foundresses or **gynes** for females that will found nests, as well as those that have done so, but the word "queens" is applied only to individuals that have worker associates.

Yanega's (1988, 1989) findings suggest that the caste system among most social HALICTINI is hierarchial, as follows:

1. **foundresses** (diapause, at least in temperate species; establish new nests or colonies)

   a. **foundress queens**
   b. **foundress workers** or auxiliaries

2. **non-foundresses** (no diapause; in existing nests or colonies)

   a. **workers** (= daughter workers)
   b. **replacement queens**.

Packer (1986b) has described an unfamiliar cuckoo caste of intraspecific cleptoparasites in a population of *Halictus ligatus*. His data are inconclusive but such bees do not fall into any of the castes listed above.

Because of extensive field and laboratory studies of *Lasioglossum* (*Dialictus*) *zephyrum* made at the University of Kansas, work on that species constitutes the backbone of this review. A vast array of information on nest architecture, life history, foraging behavior, individual odor differences, defense behavior by guards, and the like is not treated here because, while certainly not irrelevant to reproduction and caste, it is not central to those issues. There are various other species with life cycles and social organizations similar to those of *L* (*D.*) *zephyrum*. Such a species is *L.* (*D.*) *lineatulum*, which differs in that most years about 50% of the spring nests contain colonies of two or more foundresses having a semisocial organization, a foundress queen and foundress workers of reduced reproductivity (Eickwort 1986). In *L.* (*D.*) *zephyrum* such colonies exist, but less commonly than in *L.* (*D.*) *lineatulum*. The latter is not discussed further in this review. In *L.* (*D.*) *zephyrum*, as in most studied species of *Dialictus*, foundresses provision cells for a first brood of young and then await emergence of workers and construction and provisioning of cells before resuming egg laying. After this first discrete brood, reproduction seems to be more or less continuous until late summer. However, after the first brood in *L.* (*D.*) *laevissimum* (Batra 1987), and in most *Evylaeus* and some *Halictus*, there is a second (rarely more) discrete brood rather than protracted laying.

2 Caste Differences

In forms like many halictines, in which there is often intergradation between castes (at least between workers and replacement queens), and in which caste changes can occur during the life of an adult female, application of the terms queen and worker suggests a sharper distinction than actually exists. However, these terms are useful and are used here with the admonition that the reader should remember the usually great similarity of the castes.

2.1 Behavior

Frequency of mating is a common difference between castes. Essentially all foundress queens are mated. In most species, workers are largely unmated, although small numbers of mated but otherwise normal workers often exist; their number increases toward the end of colony life probably because of the increasing frequency of males. In *H. ligatus* mating of workers is common, attaining well over 50% toward the end of the season in a Canadian population (Packer 1986c). Even higher figures are known for *H. farinosus* in California (Eickwort 1985) and for *Augochlorella* (Ordway 1965). Such findings have suggested that mating is often not relevant to caste induction, but see below.

A study of behavioral differences among members of colonies of *L. (D.) zephyrum* was made by Brothers and Michener (1974). The colonies used were made in the laboratory by assembling three to six female pupae or sometimes teneral adults of about the same age and often not related to one another. Subsequent work on eusocial (mother-daughters) laboratory colonies has not demonstrated different caste characteristics. In both cases, since no diapause was involved, the queens should have been more like replacement queens than foundress queens. The colonies lived in observation nests made in soil between two glass plates, like that shown in Figure 2. The bees were individually color marked for recognition.

The castes in *L. (D.) zephyrum* are externally indistinguishable and intergrade with one another. The behavioral attributes that characterize the castes are mostly relative frequencies; few attributes are characteristic of one caste and never seen in the other. Some workers even lay eggs (probably true of most social halictines). Earlier students (Batra, Michener, Brothers, and Kamm; references in Michener 1974) hesitated to use words like queen and

→

Fig. 2 - Observation nest of *Lasioglossum (Dialictus) zephyrum* in earth layer between two pieces of glass separated by the diameter of the plastic tube. The bee nearest the entrance, the guard, is hidden by earth. The bees marked Q are in locations where the queen is commonly found; this does not imply the presence of two queens in this colony. The burrows are round in section and of rather uniform diameter but appear irregular here because of the earth pushed against the glass to round out the burrows. Only a "window" opens into each cell for the same reason. Heavily stippled areas represent earth packed into place by the bees (from Michener 1974, courtesy of Harvard University Press).

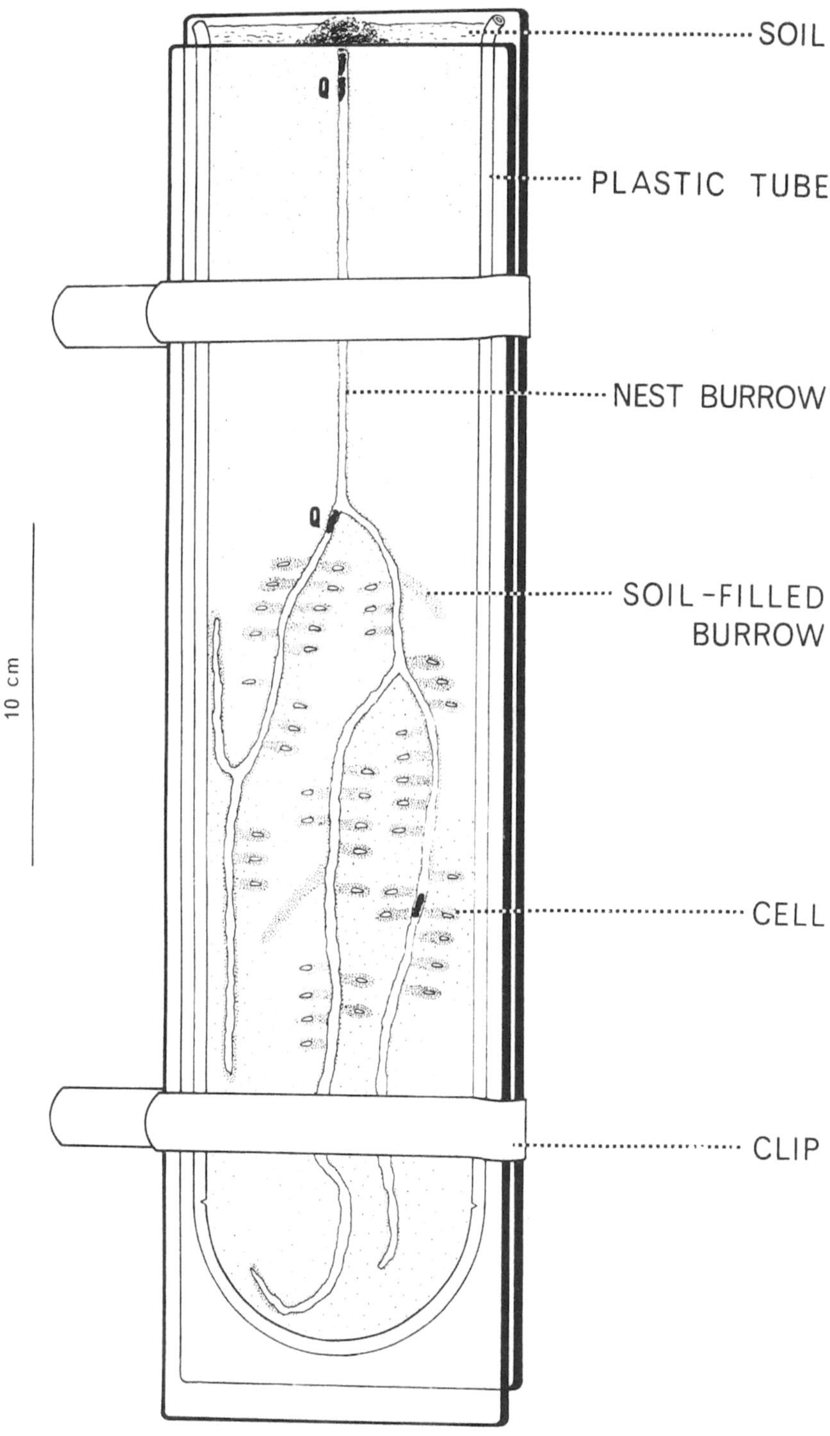
SOIL
PLASTIC TUBE
NEST BURROW
SOIL-FILLED BURROW
CELL
CLIP
10 cm
Q
Q

worker in this species. However, the bee with the largest or most active ovaries in each colony shows behavioral specializations that justify calling her a queen (if she recently laid, her ovaries may not be the largest). Indeed there is also division of labor among the workers. Table 1 summarizes the frequencies for 10 kinds of behavior recorded by Brothers and Michener, all observed in the laboratory, and except for pollen collecting, inside of observation nests. The behaviors listed are self explanatory except as follows: "Nudging" occurs when a bee moving along the nest burrow hesitates as she approaches another bee, then moves forward in a short, sharp movement that brings her face into contact with the other bee. "Backing"

Table 1 - Activity levels of queens, guards, and foragers of *L.* (*D.*) *zephyrum* for ten behavioral categories. (*0.05 > p > 0.01; **p < 0.01). Sample sizes (numbers of colonies) are in parentheses. For explanation, see text. (From Brothers and Michener 1974.)

	Queen	Guard	Forager
Nudging (1a)	max** (25)	var (25)	low[c] (9)
Being nudged (1b)	min** (25)	int[c] (25)	var (9)
Backing (2a)	max[a]	min* (24)	var (8)
Following (2b)	min** (24)	var (24)	var (8)
Passing (3a)	min** (25)	var (25)	high* (9)
Being passed (3b)	var (25)	var (25)	var (9)
Guarding (4)	min** (25)	max[a]	int[c] (9)
Pollen collecting (5)	min*[b] (9)	var (9)	max[a]
Working on burrows (6)	max[c] (23)	var (23)	var (9)
Working on cells (7)	var (17)	min* (17)	var (9)

a. Position by definition.
b. Tested by grouping the two highest ranks *vs*. the remainder.
c. 0.1 > p > 0.05.

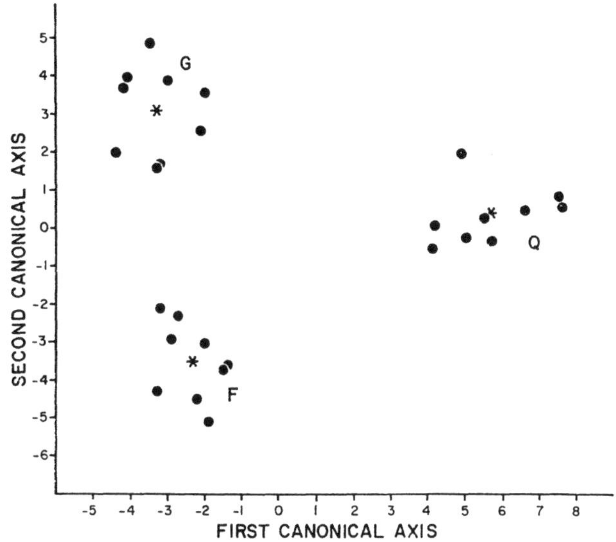

Fig. 3 - Members of colonies of three females of *Lasioglossum* (*Dialictus*) *zephyrum* allocated to three groups (queens, guards, foragers) and plotted on the first and second canonical axes for nine behavioral variables. *Group mean. (From Brothers and Michener 1974.)

occurs when a bee rapidly backs down the burrow for several centimeters. As she backs, she is often followed by another bee; this is "following." The data were examined as ranks, thus "max" indicates the highest rank, bees showing the behavior maximally; "min," the lowest; "int," intermediate. For each test involving foragers the distribution of ranks was also divided into two equal groups, bees with "high" activity and those with "low" activity, in an attempt to minimize the effects of small sample size. The notation "var" signifies no significant deviation from uniform distribution among the three columns of the table.

Canonical analysis of the data on the 3-bee colonies (based on 9 behavioral attributes, pollen collecting being omitted) produced the clusters shown in Figure 3, one bee from each nest falling in each cluster. In short, there was a queen, a principal guard, and a principal forager in each colony of three bees. (The proportion of each activity contributed by each bee in its colony, rather than the rank of each bee, was used in the multivariate studies.) When occupants of 4-, 5-, and 6-bee nests were similarly treated and plotted onto Figure 3, Figure 4 was obtained. The distinct worker clusters shown in Figure 3 merge in Figure 4, but the result is an elongate cluster with guards above, foragers below. Queens remain rather distinct. Thus even in this species whose eusocial attributes are poorly developed, castes are clearly discernible upon multivariate analysis.

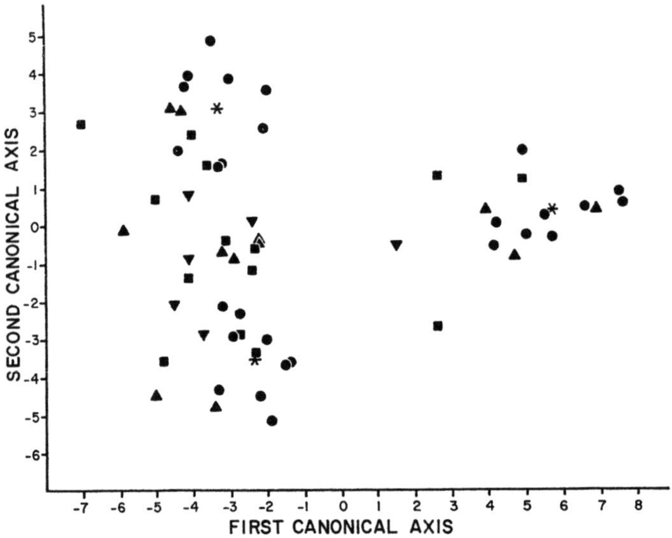

Fig. 4 - Members of colonies of four (▲), five (■) and six (▼) *L.* (*D.*) *zephyrum* plotted on the canonical axes derived from colonies of three (●) bees (see Fig. 3). *Group means for colonies of three bees. (From Brothers and Michener 1974.)

An outstanding feature of queens is their high total activity (Breed and Gamboa 1977). The widespread idea that hymenopterous queens are inactive egg producers is not applicable to this species and probably is equally inapplicable to many primitively eusocial HYMENOPTERA. Preeminent among queen behavioral categories is backing; the queen in every colony backed away from other bees more frequently than any other bee in the colony. Queens are usually also maximal nudgers; they have low ranks for pollen collecting and guarding. Both backing and nudging, as one would expect, are positively related to ovarian size. Following and guarding are usually minimal for queens and progressively more common for workers with smallest to largest ovaries; thus the ovarially most queen-like workers are behaviorally least queen-like in these respects. The principal guard, which usually has ovaries larger than does the average worker and often the largest, except for the queen, in the colony, is usually the least active bee. The maximal pollen collector commonly is the minimal nudger and frequently passes other bees, and usually has the smallest ovaries in the colony. The workers that are neither maximal guards nor maximal foragers often do some of both, but in contrast to the queen, tend to be inactive. Kukuk (1980) found that over one third of the bees in colonies of five or six individuals are inactive at any given time.

Since the Brothers and Michener (1974) study, further work by Buckle (1982a) has demonstrated other caste differences in *L.* (*D.*) *zephyrum*. He worked with colonies of young bees that were added day by day as they eclosed. They were kept in observation nests like that shown in Figure 2 except that bees of each colony foraged in a small closed container instead of in a common flight room. Most colonies consisted of six bees but some were smaller. As expected, one of the bees in each colony became the queen and the others workers. Perhaps because the colonies (and bees) were young (< 17 days), Buckle did not find a recognizable principal guard; various workers guarded the nests. (Brothers and Michener had noted that although nests had principal guards, other and perhaps all individuals sometimes functioned as guards.) Buckle corroborated many of the Brothers and Michener findings, e.g., the prevalence of backing by the queens. In fact this behavior was the quickest way of identifying the queens.

One of Buckle's additional findings was avoidance of the queen by workers. In Table 1 queens are listed as being variable with respect to frequency of being passed. Buckle (1982a), however, showed that if one looks not at frequencies but at the percentage of occasions that a bee, when approached by another bee, is passed by that bee, then one finds that queens always have the lowest rank. That is, they are less frequently passed than are guards and foragers. A higher proportion of approaches by workers to queens terminated by the workers turning around (37% compared to 11% for nonqueens), even when the queen responded by backing away. Workers passed other (stationary) workers much more readily (75% of approaches) than queens (23%). Avoidance of dominant individuals is well known in other social insects; even if it were initially an aggression-avoidance tactic, it could serve to reduce the possible effects of dominant individuals on the reproductive success of subordinates.

In the burrow or tube behind the nest entrance and guard, other bees tend to line up in a queue (Fig. 2). The queen is commonly the last (innermost) bee in the queue. This may be partly because the queen often backs when approached, for example by a returning forager, and then returns to the queue; also with no obvious stimulus she frequently goes deep into the nest and then returns to the queue. Perhaps an aspect of the same behavior pattern is that unlike workers, the queen often rests in the outer two thirds of the nest, sometimes at burrow junctions, instead of returning all the way to the queue. Queens also make more visits to cells than do workers. Such brief inspections may make the queen aware of cell conditions and may influence her control (if any) of traffic by workers from front to back of the nest. Her frequent location at the back of the queue puts her in a position to have such an influence and also to escape to the deepest nest burrows (which she does) in case of any disturbance at the nest entrance.

Smith (1987) indicated that the willingness of workers to pass the queen is influenced by genealogical relationship. Passing the queen was more frequent in artificial colonies of related bees than in such colonies of unrelated bees. Both levels of aggression displayed by queens toward workers, and the workers' ovarian development, are greater in colonies of unrelated bees than in colonies of related bees. Thus development of social relationships and caste must depend in part on kin recognition. Such findings suggest that, had the studies of Brothers and Michener (1974) and Buckle (1982a) been made with normal colonies of related bees, the division of labor among females might have been even more striking.

Different behaviors among subcastes of workers are not limited to *L.* (*D.*) *zephyrum*. In *L.* (*E.*) *calceatum*, Bonelli (see Michener 1974) reports that an occasional female of the first brood (otherwise workers) becomes a major guard instead of foraging, survives through the summer, and becomes the queen of a colony the next year. Curiously, he says that such individuals are the size of their sisters (workers), but the following spring he says they are the size of queens. In a more recent study Plateaux-Quénu (1978) notes similar diversity of behavior among first brood individuals of the same species; she says that occasional females are larger than their sisters, that they seldom leave the nest, and that they replace the queen in case she dies or is removed. Plateaux-Quénu does not mention Bonelli's work but it seems likely that both authors are talking about the same class of bees. Each was working with very few individuals and, while Bonelli's conclusion that they survive the winter is equivocal in view of his statements about their size, Plateaux-Quénu's bees were dissected and had no opportunity to survive the winter. Further work is obviously needed. Plateaux-Quénu and Plateaux (1980a) found that the second female produced was often the heaviest in the first brood of 2 to 7 workers, and suggest that this may be the one most likely to become a replacement queen. The observation that in the closely related *L.* (*E.*) *duplex* some old queens pass the winter (Sakagami 1977) may be related to the findings on *L.* (*E.*) *calceatum*.

Plateaux-Quénu aptly compared her "replacement queens" of *L.* (*E.*) *calceatum* with the apparently unspecialized workers of *L.* (*D.*) *zephyrum* that sometimes become replacement queens. Her "replacement queens" may be a subgroup of workers that, along with the foundress queen in *L. calceatum*, specialize in nest guarding as noted by Bonelli, but that also are biased toward becoming replacement queens; if they do not do so, perhaps they can overwinter and become foundress queens the next year. But it is equally logical to regard them as a subgroup of foundresses, produced early in the season instead of late, active in guarding like other queens of this species, and able to become reproductive either during the summer or in the following spring. This illustrates the difficulty of trying to draw a sharp distinction between castes applicable to every species, especially when information is sparse.

Yanega (1988) found that in *H. rubicundus* in New York about half of the females in the first brood of young, instead of becoming workers, mate and disappear into estivation-hibernation. They return unworn and establish new nests the following spring and commonly survive to become the foundress queens of eusocial colonies in the summer. They are thus foundresses produced in the first brood rather than later. (They may be the equivalent for this species of the major guards or replacement queens reported for *L.* (*E.*) *calceatum* by Bonelli and Plateaux-Quénu.) If each year all females disappeared in this way, *H. rubicundus* would be nonsocial and univoltine. Other females in the same nests, however, remain there as workers, forming with their mothers eusocial colonies that produce a second brood, the females of which mate and go into hibernation. For the first brood of young, the earlier a female matures, the more likely it is that she will remain in the nest as a worker. Females of the first brood that mate while young depart within a few days and do not engage in worker-like activities like pollen collecting before leaving.

These findings, of course, further obfuscate the differences between castes and especially between social levels (solitary and eusocial) as applied to species or even to the inhabitants of a single nest. Moreover, without marked individuals and observations of their return the following year, disappearance of females of the first brood could be misinterpreted as worker mortality. Probably this has often happened; Yanega cites suggestive findings in various earlier studies. Life cycles of other halictids will have to be re-examined in the light of Yanega's work, for worker "mortality" is often presumed. It may be that such a mixture of solitary and eusocial life cycles is common (Yanega, 1989). Yet for species that produce few or no males in the first brood of offspring, there is little possibility that females of that brood could mate and disappear into diapause.

Caste- and subcaste-specific recognition odors or behaviors exist in *L.* (*D.*) *zephyrum*. Individual odor differences are well known in this species (Barrows 1975a, b; Bell 1974; Barrows, Bell, and Michener 1975; Greenberg 1979; Hölldobler and Michener 1980; Greenberg 1982a; Michener 1982; Smith 1983; Wcislo 1987; Smith and Wenzel 1988; summarized by Michener and Smith 1987) and occur also in a solitary species, *Augochlora pura* (Barrows 1975a). These odors are used by females to distinguish nestmates from other bees and by males to distinguish familiar from unfamiliar (possibly receptive) females. Females of *L.* (*D.*) *zephyrum* also respond differently to individuals of different castes and subcastes (foragers, guards). Breed, Silverman, and Bell (1978) showed that in circular plastic tubes about the diameter of nest burrows, guards from different field nests never pass one another even after many encounters while foragers and queens often do so. Young adults (1 day old) pass, but bees five days

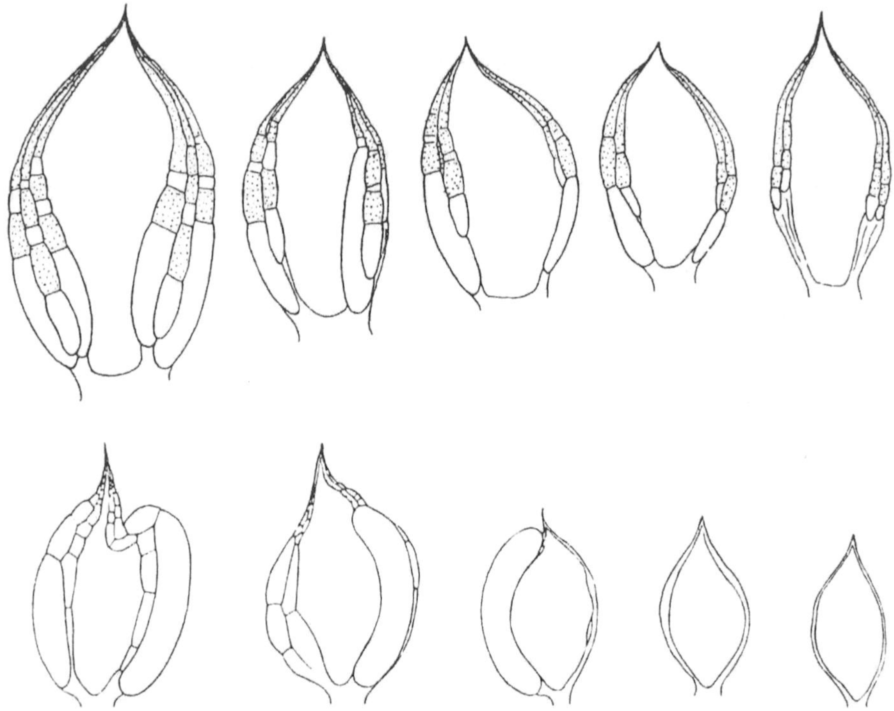

Fig. 5 - Ovaries of *L.* (*D.*) *zephyrum* (above, from Michener, Brothers, and Kamm 1971) and *L.* (*D.*) *imitatum* (below, from Michener and Wille 1961). The first four pairs of ovaries of *L.* (*D.*) *zephyrum* were from bees in a single nest.

old that had been kept in isolation and had never encountered other bees were not observed to pass. These authors suggest that they must learn to associate with nestmates while young if they are to develop as colony members.

2.2 Physiology

The physiological caste differences are not well known. The relationship that probably exists between juvenile hormone titer and aggression, ovarian development, etc., has not been clarified for any halictid. Of course ordinarily queens have enlarged ovaries and lay eggs while workers have slender ovaries and lay few if any eggs (Fig. 5). There are intermediates, however; as indicated elsewhere, workers with enlarged ovaries are likely to be guards in *L.* (*D.*) *zephyrum*. An unusually large number of ovarially developed workers is found in *H. ligatus*. In a population in Ontario, 49.5% had one or two developed ovarioles and 10.5% had three or four (Packer 1986c). Bell (1973) examined the initiation of vitellogenesis in *L.* (*D.*) *zephyrum*.

Queens usually live much longer than the average worker, but there is intraspecific variation. In 1974 I reported that overwintered (foundress) queens of *L.* (*D.*) *zephyrum* die in June and are replaced; this is probably usually true but at least in the laboratory some foundress queens survive until August (L. Greenberg, personal communication). Workers, however, survive for only about three weeks. Yanega (1989) notes that in *H. rubicundus* foundresses may survive for 12 to 14 months, while workers and replacement queens survive as adults for only about three weeks.

In *L.* (*D.*) *zephyrum* in the laboratory, so long as temperature remains high and light cycle long, there is no obligate diapause and activity and reproduction can continue throughout the year (Greenberg 1982b). There seems to be a series of replacement queens, although mortality of some females in autumn suggests the existence of some obligate diapausing foundresses. According to Plateaux-Quénu (1978), however, in *L.* (*E.*) *calceatum*, foundresses must diapause before they can lay eggs while replacement queens and laying workers can lay without diapause. This is in agreement with Yanega's (1988, 1989) findings concerning *H. rubicundus.*

2.3 Size

Size is commonly related to caste, the queens being larger than workers. Species exist in which the queens found in the field do not, or scarcely, overlap the size range of workers (Michener 1974); such species include: *L.* (*E.*) *cinctipes*, *malachurum*, *pauxillum*, and *L.* (*D.*) *umbripenne* (Damitas population but not that from Turrialba, both in Costa Rica). New additions to this list are *H. hesperus* (in which the castes not only differ in size but in the much larger hypostomal tooth of the queen, Brooks and Roubik 1983; Packer 1985), *L.* (*E.*) *lineare* (Knerer 1983), and perhaps *L.* (*D.*) *exiguum* (Packer 1985). In all such cases, the queens concerned are foundress queens.

In the majority of eusocial halictines, however, while the mean sizes of the castes are significantly different, the size ranges overlap so broadly that frequency distributions of size classes of all females are not bimodal. Recently studied examples of such species are *L.* (*E.*) *glabriusculum* (Knerer 1981), *L.* (*E.*) *affine* (Sakagami et al. 1982), *L.* (*E.*) *laticeps* (Packer 1983), *L.* (*D.*) *rohweri* (Kansas population but not that of Ontario) (Breed 1975a), *L.* (*D.*) *laevissimum* (Batra 1986), *H. ligatus* (Michener and Bennett 1977; Knerer 1980a; Packer and Knerer 1986a; Packer 1986c), *H. maculatus* (Knerer 1980a), and *H. rubicundus* (Yanega 1988, 1989). The much studied *L.* (*D.*) *zephyrum* and *L.* (*E.*) *calceatum* are also members of this group. *Pereirapis semiaurata* females have a clearly bimodal size distribution but there are many intermediates (Campos 1980).

Species also exist in which there are no significant mean size differences between castes, the size range for each caste being about the same. The best known example is *L.* (*E.*) *marginatum* but Michener, Breed, and Bell (1979) have shown the same to be true of *L.* (*D.*) *breedi* and perhaps *seabrai*.

The behavioral classes or subcastes of workers of *L.* (*D.*) *zephyrum* do not have the same mean size. Breed, Silverman, and Bell (1978) show that guards (individuals captured as guards in field nests) are significantly larger than foragers (captured returning to nests with pollen loads) although smaller than foundress queens. Mean wing lengths of foragers, guards, and queens were 4.14, 4.24, and 4.61 mm, respectively. Thus the difference between the two worker groups is less than the difference between guards and queens. This is paralleled by ovarian size, which is also smallest in foragers, largest in queens, with guards intermediate but nearer foragers. (There must be an error in Table III of the paper cited, which shows ovarian width of guards less than that of foragers; this does not agree with the text, nor with Brothers and Michener 1974.) Thus considering major functional groups (castes and subcastes), there is a positive correlation between body size and ovarian development; within a group, there is no such correlation according to Breed, Silverman, and Bell.

As noted above, subcastes of *L.* (*E.*) *calceatum* also differ not only in behavior but in size. It seems likely that such differences occur in most or all species in which queens and workers differ in mean size.

The magnitude of the difference in mean size between foundresses and workers is 13 to 18% for the species or populations with castes more or less completely different from one another in size, whereas it is less in most forms that have broad overlap of worker with queen size. In *H. ligatus*, however, foundress queens are 12.7 to 16% larger than their workers although there is complete size intergradation and no bimodality in sizes of all females (Packer 1986c; Packer and Knerer 1986a). Among species of the subgenus *Dialictus* the differences between caste size means are as follows: *L.* (*D.*) *breedi*, 2.1% (not statistically significant; Michener, Breed, and Bell 1979); *rhytidophorum*, 6.0%; *laevissimum*, 9.0% (Batra 1986); *zephyrum*, 9.1%; *imitatum*, 9.9%; *umbripenne* (Turrialba population), 10.0%; *rohweri* (Kansas population), 10.0%; *versatum*, 11.9%; *umbripenne* (Damitas population), 16.9%; *exiguum*, 19% (Packer 1985) (data from Breed 1975b, 1976, except as otherwise noted).

Marked variation in the amount of difference in mean size between castes sometimes occurs in different populations of the same species. Note the two populations of *L.* (*D.*) *umbripenne* listed above. *L.* (*D.*) *rohweri* is said to have a population in Ontario with few intermediates in size between workers and queens (see Michener 1974) while in Kansas there is a continuum (Breed 1975a).

In summary, while species exist in which there is no significant mean size difference between foundress queens and daughter workers, foundress queens usually average larger than their workers and are sometimes discretely larger. While replacement queens fall within the size range of daughter workers, they nonetheless average larger than the workers although smaller than foundress queens. The caste size differences are usually evident in population means but are more consistent on a nest by nest basis. Nests are sometimes found, however, with one or more workers larger than the queen, their presumed mother.

3 Caste Induction

It seems clear that differentiation of replacement queens and of workers occurs in an environment of competition among females, all of whom aspire to become queens because of individual fitness advantages. In the evolutionary past, i.e., among solitary bees, all females developed into foundresses, capable of initiating nests, making and provisioning cells, and laying eggs in them. Probably, therefore, all females have a tendency to develop into foundresses, and if they are to become workers, must be inhibited from following that course by general environmental factors and/or by environmental factors shaped by nestmates. The foundress queen *vs.* foundress or daughter worker dichotomy must be studied by investigating these inhibitory factors and learning how they function. Replacement queens, however, are derived from bees that at least started to develop as workers. Here the question is, do replacement queens arise because of relaxation of the same inhibitory factors, or are there other, possibly stimulatory, factors that promote differentiation from associated workers? So far no such stimulatory factors are known. Some workers, however, fail to become replacement queens even in the absence of a queen; the colony remains orphaned.

Most of the experimental and observational work described below concerned the differentiation of replacement queens *vs.* associated workers, or caste induction in laboratory semisocial colonies where, because there was no diapause, the processes are probably similar to those involved in induction of replacement queens. When formation of foundress queens is involved, this is specifically stated.

Caste differentiation (at least worker *vs.* replacement queen) occurs during adult life in halictids. There is often a pre-adult caste bias, for example due to size or season, but at least in many species the bias can be overridden by conditions that the adult experiences. Caste differentiation may not be the best expression for the phenomena discussed below, for we know that specialization as one caste or the other is, for some (I suspect most) species, not necessarily permanent. A worker can become a replacement queen and a foundress can become a worker (auxiliary). For monogynous species with marked size differences between castes, such breakdown of differentiation is not usual in nature.

Unlike highly social bees such as *Apis*, there seems to be no regular ontogenetic sequence of worker age "castes" in halictids. In *L.* (*D.*) *zephyrum* once individual bees have taken up their caste activity, and so long as colony members remain the same, their caste behaviors remain about the same. As shown below, however, loss of a nestmate (e.g., the queen) can influence the behavior of others and can result in caste change.

It seems likely that caste-specific behaviors are not only the principal caste differences but some of them are themselves the factors that determine caste. Buckle (1982b), in a study of interactions among young adult females of *L.* (*D.*) *zephyrum* assembled into his colonies as described above, investigated the development of caste differences. The individual becoming the queen sometimes showed queen-like behavior on the first day that two or more bees were together in a nest. The longest time between the artificial establishment of a colony and the first observed queen-like behavior was four days. Within two days before or after the first observed backing episode, the differentiating queen showed a sharp increase in frequency of nudging, which dropped again to a level somewhat higher than the average worker rate after a few days. Frequent nudging may be the means by which queens establish dominance, but once they have done so, they may not need to nudge so often to maintain it. Sometimes one bee started to become queen, but after much nudging another became the queen. The main point, however, is that nudging and other dominance behaviors are major characteristics of queens and at the same time the factors that inhibit queen-like activity of workers and thus determine their caste.

Workers are not aggressive toward queens. Of 4471 times that Buckle recorded queens (or developing queens, individuals that backed) nudging workers, the workers nudged back only 44 times. The rarity of conflicts suggests that without fighting, bees can assess one another's potential to become queens.

3.1 Age

The life cycles of eusocial halictines are such that the foundress queen is commonly the oldest bee in a colony. This is because in most species a queen starts a colony and her first daughters become her workers. In artificial colonies of young *L.* (*D.*) *zephyrum* of different ages, Michener, Brothers, and Kamm (1971) found that the oldest bee was more likely to become the queen than younger ones, and that age commonly was more important than size, so that if the oldest bee was small, it nonetheless often became the queen. Bees in these semisocial colonies, while of different ages, were much more similar in age than the bees of a typical eusocial colony; the age difference between youngest and oldest bees in the artificial colonies was never over 10 days.

The importance of age has also been shown for *L.* (*E.*) *calceatum*. Plateaux-Quénu (1979) placed four pupae together in a nest, the smallest being the oldest. After emergence this oldest of four became the queen of the artificial colony, in the absence of the large foundress queen normally present in summer nests of this species. Similar conclusions resulted from a study of a colony from which the queen was removed (Plateaux-Quénu 1985). The replacement queen was the oldest, not the largest, worker, and later laid eggs that produced large overwintering foundresses.

It is likely that the first (i.e., oldest) female in a queenless nest promptly undergoes ovarian development so that by the time another is added (as in the Buckle, 1982, study; or in the field, by the time another emerges), the first usually has a decisive reproductive advantage. This would explain the observation that the queen is usually the oldest bee in a colony. There are rare exceptions. For example, Greenberg and Buckle (1981) report two artificial colonies in which the youngest bee became the queen. Yanega (1989) reported for *H. rubicundus* that the only orphaned nests in which the oldest or next oldest female did not become the replacement queen were those in which the foundress queen died in midseason, after her older progeny had become regular foragers. This finding suggests that in this species, as in *L.* (*E.*) *calceatum* (Plateaux-Quénu 1974), inhibition by the foundress can permanently prevent workers from becoming replacement queens.

3.2 Size, Larval Food, and Cell Size

Castes do not differ in size in all halictine species. Yet larger absolute or average size of queens compared to workers is so prevalent that the causes of the size differences seem worth examining in connection with caste differentiation. For many species, in spite of larger average size of queens, some colonies have queens smaller than some or even all of their workers. This makes one suspect that the generally large size of queens is a product of seasonally controlled size variation and might have no causal relation to caste. On the other hand, in *L.* (*D.*) *zephyrum* in artificial colonies of presumably unrelated bees of approximately the same age (all eclosed within a 24 hour period), the average size of 51 queens was larger than that of 105 workers (Michener, Brothers, and Kamm 1971). Thus when age, season, and the like are excluded, larger bees do have an advantage in the competition to become a replacement queen.

In *L.* (*D.*) *zephyrum* and various other species, offspring produced in mid-summer average smaller than those produced in autumn; for some species including *L.* (*D.*) *zephyrum*, those produced in spring have an intermediate mean size. These statements apply to male offspring as well as females.

The size of a bee like *L.* (*D.*) *zephyrum* depends on the amount (or quality) of food stored since each larva eats all of the food provided by mass-provisioning for it. The quantity of food, in turn, is influenced by the size of the cell constructed before food storage begins. Kamm (1974) found a high correlation between cell size and size of the bee emerging from it. (In HALICTINI cells are not re-used; a new one is constructed for each bee to be produced.) Packer and Knerer (1986b) verified the relation of cell size to the size (and weight) of its pollen mass, and of course to the size of the resulting bee, in *H. ligatus*.

The interplay of at least four factors (temperature, day length, colony size, and size of bees in the colony) influences sizes of cells and of progeny reared in them in *L.* (*D.*) *zephyrum*, and therefore may bias the caste into which the female progeny will develop. Kamm first noted that in a cool corner of a bee room the cells and bees were larger than elsewhere. With a later contribution from Kumar (1975), he examined the influences of various factors, each with the others held constant, on cell and resultant bee sizes. Temperatures of 21°, 26°, and 30° were used, as were day lengths of 11, 13, and 15 hours. Cell and resultant bee sizes of both sexes were inversely related to both temperature and day length. Increasing the numbers of adults in a nest also increased sizes of the cells that they produced. Furthermore, colonies consisting of bees of larger mean size tend to produce larger cells than colonies of smaller bees. Knerer (1987) attributes size (and number of worker broods) to photoperiod in various *Evylaeus* species, but does not mention temperature which is often correlated with day length; a complex of factors could have been responsible as in *L.* (*D.*) *zephyrum*.

Cells of *L.* (*D.*) *zephyrum* that produced males had a smaller mean size than those that produced females. Perhaps cell size influences not only the amount of food put in the cell, but also, either directly or through the amount of food, whether or not the egg placed in the cell is fertilized. This response, however, must be standardized by the environmental conditions; as an example, female-producing cells at one temperature would have the same mean size as male-producing cells at a lower temperature. By artificially adding pollen, Plateaux-Quénu (1983) showed that the sizes of pollen masses of *L.* (*E.*) *calceatum* influenced the sexes of the eggs laid upon them, increasing the frequency of females.

In summary, the cell construction (which at least in colonies of *L.* (*D.*) *zephyrum* is often done by several bees working successively), i.e., its size, is evidently to a major extent under environmental influence and biases the sex and caste of the progeny reared.

That some of the environmental factors listed above influence size of progeny in other halictids is likely. Experimental data, however, are available for only one species other than *L.* (*D.*) *zephyrum*. It is *L.* (*E.*) *calceatum*, for which Plateaux-Quénu and Plateaux (1979) found that workers reared in the warm laboratory were smaller than those reared in the field, and that those reared in the field in a cold spring were larger than usual. The same authors (1980b) found size differences in first brood workers and also males between those reared at 14-17° and those reared at 20-22°, the latter producing significantly smaller individuals. These results, generally corroborating those of Kamm, are of special interest because *L.* (*E.*) *calceatum* is not closely related to *L.* (*D.*) *zephyrum*; probably the results are indicative of trends that are widespread in HALICTINAE.

The probable relation of the laboratory findings to observed field conditions can be summarized for *L.* (*D.*) *zephyrum* as follows: Interactions among the environmental factors lead to the seasonal size variation, high temperatures and long days producing small midsummer bees, lower temperatures and shorter days producing the larger, earlier and late summer bees, the latter being particularly large because of the large colonies. These largest (on the average) females overwinter and are the foundresses the following spring and early summer. It may be that other factors are also involved in foundress determination. The production of relatively large, early summer bees has not been found in other species.

Field observations (without laboratory verification) of some other species suggest that size is associated with caste largely through the circumstance that large females produced late in the active season (late summer in temperate climates) are the bees that hibernate or estivate and become foundresses the following year. This view is supported by the following: (1) The one species whose workers and queens are produced at exactly the same season of the year (for any one colony, in different years) has castes of the same size; this is *L.* (*E.*) *marginatum* (see review in Michener 1974). (2) *L.* (*D.*) *breedi* which inhabits an equable, equatorial montane climate and either produces both castes throughout the year or if not, must rear them under similar conditions since there is little seasonal variation in weather, also shows little or no caste difference in size. The same is probably true of *L.* (*D.*) *seabrai* from the same habitat although the data are less adequate (Michener, Breed, and

Bell 1979). (3) Of the species of *Dialictus* considered by Breed (1976), the one with minimal caste size difference, *L.* (*D.*) *rhytidophorum*, inhabits a relatively equable climate on the southern Brazilian plateau.

In *H. rubicundus* (Yanega, 1989) foundresses are produced throughout the active season, overlapping the two to three week period when workers are produced. Foundresses and workers which emerge simultaneously do not differ in mean size, but there is a temporal bias such that the seasonal increase in female mean size coincides with increasing proportions of females that will become foundresses. Thus, even within the first brood, future foundresses average significantly larger than workers, because foundresses tend to emerge later; size does not appear causally related to caste. Second brood females of this species average even larger, but are destined to be foundresses regardless of size.

In *H. ligatus* also (in the Florida keys, Packer and Knerer 1986a) young foundresses and workers (distinctly different in size) are produced simultaneously in the same nest. Perhaps other species in tropical or subtropical areas are similar in this respect. Such cases need further study to determine why a given cell is made large and, appropriately provisioned, produces a foundress while at the same time another cell is small and produces a worker.

This section would be incomplete without noting that in at least two species the cells that produce the largest females (queens) are not only markedly larger than other cells, but also the food masses placed in them are differently shaped, rather rectangular or ovoid rather than flattened spherical. These species are *L.* (*D.*) *umbripenne* (Damitas population) (summary in Michener 1974) and *H. ligatus* (some but perhaps not all populations) (Michener and Bennett 1977). In tropical populations of the latter species, large females appear to be produced in small numbers at any time of year; the factors that stimulate their production are unlikely to have anything to do with day length or temperature. Yanega (1989) reports for *H. rubicundus* large oval pollen masses only in nests with many workers and a laying queen. Such nests musts therefore produce the largest of the overwintering foundresses (as well as the largest males). In other nests and in other species, even those with maximum differences in size between castes like *L.* (*E.*) *malachurum*, the food masses in large, foundress-producing cells are not reported to differ in shape from those in smaller cells (see Michener 1974).

Finally, there is an indication of pre-adult foundress *vs.* worker differentiation in *L.* (*E.*) *malachurum*, a species with a relatively great caste size difference. Bohrer (1987) found that pollen balls destined to produce foundresses (which are about twice as heavy as worker pollen balls) contain about 0.5 mg/g protein and 6% sugar while for those destined to produce workers equivalent figures are

3.4 mg/g and 16%. Thus quality as well as quantity of larval food may be involved in caste differentiation at least in some species. However, since in *L.* (*E.*) *malachurum* worker and foundress broods are produced at different seasons, the qualitative differences in larval food may reflect different floral sources rather than caste-inducing factors. Bohrer suggests that quantity is the major factor and quality of minor importance.

3.3 Mating

In *L.* (*E.*) *marginatum* females that mate become foundresses, those that do not, become workers (for details, see summary by Michener 1974). In *H. rubicundus*, Yanega (1988, 1989) showed that all foundresses are mated, and that their mating occurs within a very few days after emergence of a female from her cell. The first brood is slightly protogynous. Some females mate and enter next year's foundress pool throughout the period of first brood emergence. In a nesting aggregation the percentage of the females produced per day that become foundresses was closely correlated ($r = .96$) with the cumulative number of males in the population at that time. While correlation does not necessarily indicate causation, it seems likely that Yanega is correct in inferring that mating by recently emerged females leads to their departure and diapause, i.e., to production of foundresses. Females that do not mate remain in the nest as daughter workers, sometimes becoming replacement queens and mating later. In the first brood, worker production, of course, is highest in the first days of emergence when there are few males in the area, and therefore little opportunity for early mating. The females of the second brood all mate and become foundresses. This brood is protandrous and every female seems to have an opportunity to mate when young. Yanega's view, then, is that there is a brief period after emergence when foundress induction occurs. If a female mates at this stage she diapauses and becomes a foundress. Females that do not mate at this time do not diapause but become workers or replacement queens. Later mating does not strongly influence their behavior.

Other studies have involved production of replacement queens, of queens in artificial colonies where conditions suggest those of queen replacement, or differentiation of a foundress queen from foundress workers. It is increasingly apparent that caste differentiation in these circumstances is different from that involving foundress production. Laboratory studies have repeatedly verified the observation that unmated females of *L.* (*D.*) *zephyrum* can be replacement (or equivalent) queens. There is no evidence that such queens do not function normally; they back and nudge at high rates like other queens. Of course they produce only male offspring.

It is also evident that in the same species, mated workers function normally and are not known to show any more tendency toward ovarian development than unmated workers. In a recent study colonies were synthesized with unmated queens and with all the workers mated (see below under Male Production). Behavior seemed normal in such colonies. Thus for *L.* (*D.*) *zephyrum*, mating, while correlated with caste, is not causally related to replacement queen production.

For *L.* (*D.*) *zephyrum* Greenberg and Buckle (1981) have investigated the control of mating by equivalents of replacement queens. In the field essentially all queens but only about 18% (varying seasonally) of workers have sperm cells in the spermathecae. Mating occurs outside the nest, yet it is under social control by colony members. In the laboratory 73% of the two day old females, isolated since removal from their cells, mated, while 66% of three day old females did so. Thus most young females, in the absence of a queen, will mate. Artificial colonies of unmated young bees were established by adding a newly emerged female more or less daily until a population of six (or less) was attained. When a queen was behaviorally recognized (by backing), each bee in the colony was separately removed to a small cage with males and the results of mating attempts monitored. In 29 such colonies, all 29 queens mated but only 7.7% of the 104 workers did so; the rest repulsed the males in spite of numerous mating attempts.

In some unknown way the presence of a queen in the nest controls mating by her workers outside the nest. Thus when queens were removed from the above described colonies, a worker in each soon showed queen behavior, and would thereafter mate, although only 9.4% of the remaining workers would do so. In every case the new queen had previously been a worker that refused to mate. Removal of the new queen resulted in the same phenomena; in one colony four bees successively became queens and mated.

3.4 Queen Removal Experiments

If the queen of a colony of *L.* (*D.*) *zephyrum* is removed, one of her nestmates (daughter or not, depending on the make-up of the colony) will commonly become a replacement queen; if she is removed, she will in turn be replaced (Breed 1977; Michener and Brothers 1974; Brothers and Michener 1974; Greenberg and Buckle 1981). This process can often be continued until only two or three bees are left in the colony. Commonly queen behavior (backing, frequent nudging) appears in a replacement within a few hours or a day after removal of the first queen; sometimes it does not appear for several days. Rarely a replacement fails to appear.

From these experiments one can conclude that most females have the potential to become replacement queens. Seemingly the caste biases due to age and size, described above, have effects only in the context of competition with other bees. As competitors are removed, most or all females show the ability to attain queen status. Another point that these experiments make clear is that while caste biases may be preimaginal, depending on the amount of food that the larva receives and presumably other factors such as season, caste induction occurs in the adult stage, but even then, it is sometimes reversible. Workers of *L.* (*D.*) *zephyrum* at least a month old are able to become replacement queens.

As noted in the section on Caste Differences, the queen's nudging in *L.* (*D.*) *zephyrum* is not directed uniformly toward all workers; she nudges especially the worker with the largest ovaries (commonly the principal guard). It is not usually this perhaps traumatized individual that replaces a removed queen. Most often it is the worker with the second largest ovaries that becomes the replacement, i.e., neither the top ranking guard nor the top ranking forager (who has slender ovaries) but an unspecialized intermediate (Michener and Brothers 1974; Brothers and Michener 1974).

In a study designed to see if workers that later become replacement queens are more queen-like in behavior than are worker nestmates, Buckle (1981) ranked workers of *L.* (*D.*) *zephyrum* according to the percentage of times they were passed after being approached by other workers. He then removed the queen. In nine of eleven nests, as he had predicted, it was the worker that was passed least often by other workers that became the replacement queen. He then removed nine of these replacement queens and again predicted which workers would become the new replacement queens on the basis of frequency of being passed. Six of the nine predictions were correct. Once again he removed the replacement queens from eight nests and observed which bees became the third replacement queens. Five of the eight replacements were those that had been the lowest ranking workers remaining in the nests. Overall, his success rate at predicting which bees would become the replacement queens was 71%. Since queens are minimally passed, it is evident that workers that replace lost queens often have this recognizable queen-like attribute beforehand.

3.5 Inhibition of Workers by Queens

The queen removal experiments indicate that in *L.* (*D.*) *zephyrum* the queen seems able to inhibit the ovarian development and queen-like behavior of her nestmates, who therefore become workers (Michener and Brothers 1974; Brothers and Michener 1974; Breed 1977; Michener 1977). There is no evidence of pheromonal caste control in *L.* (*D.*)

zephyrum. There is no proof that it does not occur, but I suspect that it is the repeated nudges by the queen, directed particularly at the worker with largest ovaries (often the guard) or with queen-like behavior, that traumatize that worker so that its ovaries do not develop further. Experiments described by Michener and Brothers (1974) and Michener (1977), in which young bees in nests were artificially disturbed repeatedly with a small iron ball moved through burrows with a magnet in an effort to simulate nudging, did inhibit production of queens whereas queens developed in undisturbed control nests. This shows that artificial disturbance of bees in nests can prevent ovarian development; nudging may well be the way the queen does it.

Because most observations on *L.* (*D.*) *zephyrum* described above (especially in the section on Caste Differences) were based on colonies of six bees or less, Buckle (1985) compared behavior in large colonies (10-17 workers) with that observed in small colonies (1-8 workers). The hypothesis was that suppression by the queen of queen-like behavior of workers should be less effective in large than in small colonies if suppression involves contacts. Certain workers in large colonies elicited more turn-arounds (i.e., fewer passes occurred) and more backing episodes than did workers in small colonies. Moreover, a higher proportion of workers in large colonies mated than in small colonies. Thus some workers in large colonies act and are treated as more queen-like than workers in small colonies. Thus the hypothesis appears to be supported although there are other possible explanations for the findings, such as a direct effect of colony size (perhaps through frequency of worker-worker interactions) on worker behavior, rather than an influence by the queen.

Plateaux-Quénu (1974) also showed inhibition of worker activities and ovarian development by the queen in *L.* (*E.*) *calceatum*. In this case the inhibition persists, even in the absence of the queen. If a foundress queen is removed from a colony soon after the eggs that would normally produce workers are laid, then the resultant bees, who have never seen their mother, form a functioning semisocial colony. One becomes a replacement queen in spite of being a small, summer bee; the others become workers. If a colony is de-queened after the summer workers have emerged and spent some time with their mother, they seem to be permanently inhibited in ovarian development and a new queen does not develop.

Inhibition occurs whether the queen is the mother, as in normal eusocial colonies, or a sister of or unrelated to her associates, as in many experimental colonies of *L.* (*D.*) *zephyrum* and in spring assemblages of foundresses such as occur sometimes in various species. Plateaux-Quénu (1984) has described probable inhibition in one such semisocial assemblage of *L.* (*E.*) *calceatum*.

3.6 Stimulation of Worker Activity

Although workers are in some ways inhibited by queens, a worker appears to require stimulation by the queen to perform certain activities characteristic of workers. Backing (often after nudging) by queens, the queen being followed by a worker, often seems to draw workers of *L.* (*D.*) *zephyrum* down into the lower parts of the nest where perhaps the stygmergic stimuli of incomplete cells and the like lead to constructive activity (Michener and Brothers 1974; Michener 1977). Breed and Gamboa (1977) indicated that the queen often appears to lead returned foragers with pollen loads down to the cells where the pollen is to be deposited. Subsequent studies (Buckle 1984) found such behavior less common than in Breed and Gamboa's nests, and clearly show that worker foragers can provision cells without guidance from queens. It is possible that the advance of a worker elicits backing by the queen, who thus appears to lead the worker, especially because she stops and waits for further advance by the worker. In reality the queen might be merely keeping ahead of the worker, as part of her dominance behavior.

The influence of queens of *L.* (*D.*) *zephyrum* upon many nest activities is evident, however. The constant nudging of other bees by the queen and her high total activity seem to promote worker activity. Removal of the queen results in greatly reduced activity by other bees (Breed and Gamboa 1977). Plateaux-Quénu (1974), working with *L.* (*E.*) *calceatum*, also judged that the queen plays a stimulatory as well as an inhibitory role in shaping worker activity. Workers that had been associated with a queen became permanently inactive in her absence.

4 Colony Size

Colonies of eusocial halictids range in population from a few bees to several hundred. *L.* (*E.*) *marginatum* in Europe regularly has colonies of hundreds (see summary in Michener 1974), one nest of *Halictus* (*Seladonia*) *lutescens* in Costa Rica contained 342 females (Wille and Michener 1971) and a nest of the same species in Guatemala contained over 589 females (Sakagami and Okazawa 1985). Colonies of *H.* (*S.*) *hesperus* in Panama consisted of 97 to 149 workers near the end of the nesting cycle (Brooks and Roubik 1983). In contrast, very small colonies have been reported in the past and study of two tropical American species at different seasons of the year show that mature colonies can be singularly small (Michener, Breed, and Bell 1979). The data for *L.* (*D.*) *seabrai* and *breedi* suggest similar colony sizes and are not segregated here. Offspring are apparently produced at rather long intervals so that no large number of workers is found in a nest at any one time. In both the wet season and the dry season, many nests are occupied by only one bee so that the number of bees per nest averaged about 1.5 (96 nests); in nests with cells this number was about 2.0; in nests containing two or more adults the number was 2.3 for only six nests of *seabrai*, nearly 3.0 for *breedi*. Even in these species where the mature colony size is so small, the organization is eusocial, the queen having enlarged ovaries, the one, two, or sometimes three workers having less developed ovaries and usually being unfertilized.

Colony size must be determined by the egg-laying rate of queens, egg-laying by mated workers, brood mortality, and the longevity of workers. Interestingly, colony size is not always highly correlated with other conventional measures of social level. Thus the species with the largest colonies known among halictine bees, *L.* (*E.*) *marginatum*, has no caste differences in size. However, in general, colony size seems related to other measures of social level, and such a relationship was clear for the species of the subgenus *Dialictus* studied by Breed (1976).

5 Male Production

In social HALICTINAE the early progeny of a queen are sometimes all or nearly all females (workers) but even in some rather highly social species some males are produced in the first brood. Males are typically produced in greater numbers later in the history of a colony. Among eusocial forms of the subgenus *Dialictus*, overwintered foundress queens of *L.* (*D.*) *zephyrum* produce males (less than 15%) in the first brood while other species produce almost no males in the first broods (Breed 1975a). This variation makes sense in terms of queen longevity. In *L.* (*D.*) *zephyrum* queens usually are replaced in early or mid-summer; new queens need to mate. Later in the summer male production often approaches 50%, at the time when foundresses that will mate and overwinter are being produced. Even in the subtropical Florida Keys where colony development of *H. ligatus* is not synchronized (new nests can be started at various times of year), the male production in the first brood averages 11%, but rises later in the colony cycle to 56% (Packer and Knerer 1986a).

As in most HYMENOPTERA, most male halictines come from unfertilized eggs, females from fertilized eggs. So far as I know in all laboratory studies where unmated females have produced offspring, they were all males. Males can result from eggs laid either by unmated workers or by mated individuals such as queens that lay some unfertilized eggs. Michener (1974) summarized the indecisive evidence for worker origin of males of *L.* (*E.*) *malachurum*. To shed more light on the matter for *L.* (*D.*) *zephyrum*, artificial colonies consisting of an unmated queen and mated workers were established by keeping the oldest individual of each colony unmated and mating the others individually when young, before they were added to the colonies (B. H. Smith and L. Greenberg unpublished). The offspring were mostly males, but a few females, no doubt from worker-laid eggs, also resulted. If laying and egg survival in the experimental colonies followed normal patterns, then most males in this species are probably from queen-laid eggs but a few are of worker origin.

6 Mating Systems

The number of times that queens mate is of interest, especially in view of Hamiltonian thinking and the close relationships among sisters if the mother has mated only once. Unfortunately there is little information on this subject. In *L.* (*D.*) *zephyrum* females seem reluctant to mate a second time but at least one did so in the laboratory several days after the first mating. Unfortunately no study of the number of matings has been made. However, electrophoretic studies by Kukuk, Eickwort, and May (1987) have shown that among apparently monogynous colonies of *L.* (*D.*) *zephyrum*, 23% contained progeny either of more than one female or of a single, multiply mated female. I have seen *L.* (*D.*) *rohweri* females mate two and three times in the laboratory within an hour. The above behavioral observations were based on laboratory-reared bees probably similar to replacement queens, whereas the electrophoretic study was based on progeny of foundresses. *L.* (*E.*) *malachurum* foundresses were observed to mate "many times" in the course of one day by Packer and Knerer (1985). Such observations in primitively social species diminish the probable importance of haplodiploid sex determination in explaining the origin of social behavior in Hymenoptera.

Mating behavior of halictines has been described by Barrows (1975a, 1976a, b) and reviewed by Alcock et al. (1978). Some species mate at the nest site, some elsewhere (e.g., on flowers), and some at both. In addition, mating sometimes occurs within nests, probably regularly for *L.* (*E.*) *marginatum* in Europe and for the large-headed, nonflying males of *L.* (*Chilalictus*) *erythrurum* (Kukuk and Schwarz 1987), and probably occasionally for many species. Some reports of males entering nest burrows may concern males merely looking for resting places; they often rest in other burrows as well. Other such reports, however, may concern males entering the burrow to mate, as suggested by Knerer and Plateaux-Quénu (1970) for *L.* (*E.*) *nigripes*. The following paragraphs concern mating outside of nests, which for most species is clearly usual.

L. (*D.*) *zephyrum* as well as the *Habralictus* and *Caenohalictus* species studied by Michener, Breed, and Bell (1979) seem to mate

Fig. 6 - Nest entrances of *Halictus hesperus* in Panama, made conspicuous by the earthen tumulus around each. Each nest contains a eusocial colony; the group of nests is called an aggregation. In many places and seasons, wind and rain destroy the tumuli so that nests are inconspicuous (photograph by R. W. Brooks).

primarily at the nest site; *L.* (*D.*) *rohweri*, at the nest site and on flowers; and *L.* (*D.*) *imitatum* and *seabrai*, probably largely away from nest sites. The species that mate primarily at nest sites tend to nest in aggregations (Fig. 6); those that mate away from nest sites usually have dispersed nests. In a subtropical population of *H. ligatus*, males for the most part patrol flower patches but also fly over nesting areas (Packer 1986a). Such divided attention is associated with the weak aggregating of nests of this species in subtropical and tropical areas.

Except in part for *L.* (*D.*) *rohweri* (Barrows 1976b), males of the species that form nest aggregations patrol over the nesting area, pouncing on small dark objects (including female bees) beneath. Pouncing occurs in the presence of a female sex pheromone which for *L.* (*D.*) *zephyrum* is in part a mixture of macrocyclic lactones from Dufour's gland (Smith, Carlson, and Frazier 1985). At least in *L.* (*D.*) *rohweri* and *zephyrum* and *Caenohalictus eberhardorum*, males captured, marked, and released returned to the same small area (e.g., 30 x 30 cm) for patrolling day after day. These areas are not mutually exclusive territories; the ranges of different males may broadly overlap. Lyn Garling (personal communication) trapped and marked males as they emerged in the mornings from small groups of nests of *L.* (*D.*) *zephyrum* in an aggregation 30 m long in the bank of the Kansas River; presumably many were newly emerging young males. Of 41 later recaptured, all but two were in the same areas where they were first caught. These are preliminary results but probably mean that a male's flight area commonly includes the entrance of the nest where he was reared.

Because of the likelihood of nesting success in the home vicinity, I believe it probable that a female starting a new nest usually does so near the nest where she was reared. Such phylopatry was shown by marked foundresses of *H. rubicundus* by Yanega (1988, 1989). For *L.* (*D.*) *zephyrum* phylopatry is supported by the responses of guards of at nest entrances (Kukuk and Decelles 1986). Guards usually lunge at bees introduced from a distant (several meters to kilometers) source but less commonly lunge at individuals from nearby nests. Given laboratory findings (Greenberg 1979; Buckle and Greenberg 1981) that guards admit nestmates on the basis of learned odors, but confuse close relatives of nestmates with the nestmates themselves presumably because of similar odors (for reviews, see Hölldobler and Michener 1980; Michener 1982; and Michener and Smith 1987), the data on introduction of bees to nests suggest that nearby nests are commonly inhabited by relatives. Of course species that diapause in or near their natal nest and start new nests by burrowing up from the hibernacula are automatically phylopatric.

These findings indicate that a male's flight area is likely to be over a group of nests containing not only his sisters from his own natal nest but also other relatives. Greenberg (1982a) noted that young males of *L.* (*D.*) *zephyrum* in the laboratory, given a choice of black spots behind which were female nestmates (his sisters) or non-nestmates, showed a slight initial pouncing preference for nestmates. After as little as an hour with nestmate odor alone, however, they thenceforth showed a preference for non-nestmates. In the field this could mean that, by the time they leave their natal nest, they prefer to pounce in the presence of non-nestmate odors. Field observations of a slight preference of males for females from foreign nesting aggregations tend to support these results (Smith and Ayasse 1987). Given the flight patterns of males and the presumed nest site selection behavior of females, Greenberg's findings appear to mean that there would be strong inbreeding but with discrimination in favor of matings between cousins, etc., rather than siblings.

Discrimination by males against females having familiar odors has been more fully documented in other laboratory studies of *L.* (*D.*) *zephyrum* (e.g., Smith 1983) and in field studies (Wcislo 1987). Thus a male learns the odors of his sisters, of females that have repulsed mating attempts, and probably of females with whom he already mated (although there are no data on this), and tends to avoid them in subsequent mating attempts (Michener and Smith 1987). Fuller documentation of comparable findings for *L.* (*E.*) *malachurum* was provided by Smith and Ayasse (1987). They point out that experience with a female reduces the response of a male to that female and her nestmates (? sisters) in later encounters, but does not alter the male's response to other females. A chemical label (antiaphrodisiac) deposited on the female at the time of copulation is another possible explanation for some of these findings (Kukuk 1985; Smith and Ayasse 1987).

In *H. ligatus* Packer (1986a) found that males avoid mating with the smallest available virgin females. The cues used by the males (size or odor) are not known, but as in *L.* (*D.*) *zephyrum*, males are choosy. As suggested by Michener and Smith (1987), males may not be able to mate many times, and may therefore avoid wasting time and semen on females likely to become non-reproductive workers.

That inbreeding may be common in *L.* (*D.*) *zephyrum* is suggested by the lack of inbreeding depression or unusual sex ratios after four generations of brother-sister mating (Greenberg 1982b) and by the low level of allozymic variation in the species (Snyder 1975; Kukuk and May 1985). However, in other bees, as in passerine birds, electrophoretically detected genetic variation is less than in most organisms. This is true for *Apis mellifera* in which behavioral outbreeding mechanisms are obvious, as are deleterious results of inbreeding. Recent results using the limited allozymic polymorphism of *L.* (*D.*) *zephyrum* indicate that nestmates are highly related but probably not all full sisters (Crozier, Smith, and Crozier 1987; Kukuk, Eickwort, and May 1987). Variation among colonies in the sex pheromone mixture parallels the genetic variation (Smith and Wenzel 1988).

Obviously the number of sperm cells produced by males and transferred to spermathecae of females might increase with the number of female offspring produced per female, and hence with eusocial colony size. Unfortunately, the recent study of this topic (Garófalo 1980), while giving data on ten species of HALICTINAE, includes no eusocial ones; all are solitary or parasocial.

7 Evolution of Social Features

The HALICTINAE includes many solitary species in which each female excavates her nest and makes provisions, and oviposits in her own cells. In addition, the subfamily includes species that live during part of the seasonal cycle in **parasocial** colonies (in which two or more females of the same generation work together in a single nest) and **eusocial** colonies (in which females of successive generations, mother and daughters, live together). The parasocial species are commonly facultatively so; they may also nest solitarily. Parasocial colonies have been divided into the following three categories: (a)

Communal, i.e., each female making, provisioning, and laying in her own cells. (There are many such forms, perhaps none of them obligately communal. Examples are *Pseudagapostemon*, see Michener 1974; *Agapostemon virescens*, Abrams and Eickwort 1980, 1981; *Habralictus bimaculatus* and *Caenohalictus eberhardorum*, Michener, Breed, and Bell 1979. Some species of *Nomioides* are communal or semisocial, Batra 1966.) (b) **Quasisocial**, i.e., females cooperatively making and provisioning cells; all females have enlarged ovaries and presumably lay; not known as a regular or principal social pattern in any halictine, although a nest may temporarily contain a quasisocial colony. (c) **Semisocial**, i.e., females live cooperatively with division of labor, one or more being the principal egg layers (queens), the others being workers or auxiliaries. (Examples are certain species of *Augochloropsis*, see Michener 1974. As noted by Eickwort and Sakagami 1979, the social AUGOCHLORINI tend to be semisocial.) For (a) and (b) there are no castes; for regularly semisocial species (c) there is little new information. Many of the experimental colonies of *L.* (*D.*) *zephyrum* discussed above were semisocial, and semisocial colonies occur in some nests in nature as temporary stages in this and probably most eusocial HALICTINAE (Eickwort 1981). Thus a group of foundresses in a single nest in spring, or a group of workers whose queen has died in summer, constitute semisocial colonies. Communal or semisocial initial stages of colonies are common in species that hibernate in the old nest, rare in species that disperse before hibernation. Regularly communal species are nearly all in different genera or subgenera than semisocial or eusocial species. This observation suggests that the casteless communal way of life is an evolutionary alternative to societies with castes, not a stage in the development of such societies.

The majority of well-studied species of colonial HALICTINAE are eusocial. These forms may start their nests as solitary bees but in most cases obligately become colonial, with the mother (foundress queen) the principal egg layer and daughters (workers) having reduced reproductivity. Unlike the highly eusocial bees, castes of eusocial halictines are alike or nearly so in external morphology. The principal caste differences, as noted above, are behavioral and physiological. Colonies of such bees are called primitively eusocial. Packer and Knerer (1985) compare eight such eusocial species of the subgenus *Evylaeus* in a multivariate study similar to that of Breed (1976) for *Dialictus*.

The verbal explanation of the classification of social levels outlined above suggests major steps in social evolution but Michener (1974, pp. 237-241), in a principal components approach to social levels, showed that all the halictines are behaviorally remarkably similar relative to the total spectrum of social behavior found among bees. This implies that relatively little evolution is required to shift from one halictine level to another.

The ease with which such shifts occur is supported by both ontogenetic and phylogenetic evidence. Some nests of various species, at the beginning of the active season, contain two or more foundresses, these being a foundress queen and one or more overwintered foundress workers. Such semisocial colonies of overwintered bees occur occasionally, for example, in *L.* (*E.*) *calceatum* (Plateaux-Quénu 1973), various species of the subgenus *Dialictus*, and in *H. ligatus* (Litte 1977). They occur regularly in *L.* (*D.*) *versatum* and *lineatulum* (Eickwort 1986). As the progeny of the foundress queen mature, such colonies become eusocial (or in *H. ligatus*, because of long survival of the foundress workers, mixed eusocial and semisocial). Thus in the course of its ontogeny, such a colony changes from semisocial to eusocial. In a single nesting site at the same season, one may sometimes find nests of a given species inhabited by solitary females as well as by colonies at diverse social levels. A *H. rubicundus* population in New York includes individuals that go into diapause after summer emergence, as would solitary bees, as well as individuals from the same nests that remain with their mothers as workers and form eusocial or semisocial colonies (Yanega 1988). The same species is interesting in that it is holarctic in distribution. Presumably in the short northern summer there is no time for the eusocial phase and northern populations, therefore, are presumably wholly solitary. Such populations probably crossed Berengia from Asia, becoming partially or wholly eusocial at more southern latitudes in America, like the populations in Europe. Old suggestions that the northernmost populations of *L.* (*E.*) *malachurum* are solitary, in contrast to the well known social behavior of more southern populations, are equivocal (Michener 1974, but see Knerer 1987) but Sakagami and Munakata (1972) showed that in Hokkaido, Japan, a high mountain population of *L.* (*E.*) *calceatum* is solitary while lowland populations are eusocial much like those that have been studied in Europe.

Phylogenetic evidence of shifts in social level is provided by closely related bees that have different types of social organization. Thus in the subgenus *Dialictus*, there are solitary species as well as eusocial, and *L.* (*D.*) *problematicum* is principally semisocial (Sakagami *in* Eickwort and Sakagami, 1979; Sakagami, Hoshikawa, and Fukuda 1984). The subgenus *Halictus* s. str. (in the sense of Michener 1978) contains both solitary and social species, as does the subgenus *Evylaeus* even if narrowly defined. For the subgenus *Halictus*, based upon a phylogeny by Y. Pesenko, Packer (1986e) suggests that if solitary behavior were primitive, social behavior must have arisen on seven of the presumed phyletic lines. If social behavior were primitive for the genus, then solitary behavior arose three times and reverted to social once.

In the tribe AUGOCHLORINI, a group of related taxa (*Pereirapis*, *Augochlorella*, and *Augochlora* subgenus *Oxystoglossella*) consists of eusocial species that, like nearly all other HALICTINAE, nest in the ground. The subgenus *Augochlora* s. str., however, closely related to and either derived from or the sister group of *Oxystoglossella*, consists of solitary species that nest in rotten wood. It seems likely that *Augochlora* s. str. reverted to solitary behavior as a result of nesting in rotten wood in forests where natural enemies are seemingly less common and less successful than on the open ground, and where the disadvantages of social behavior therefore outweighed the advantages. Such observations support the view that in the HALICTINAE, in spite of its apparent complexity, eusocial behavior has arisen repeatedly, dozens or hundreds of times, and that reversion to solitary behavior is also easy. The frequently asked question about the numbers of origins of eusocial behavior cannot be answered for halictines. As noted by Michener (1985) and in the following chapter on XYLOCOPINAE, direct development of eusociality from solitary forms is probable, without intervening species having lower social levels.

It is not surprising that additional ways of life are being found. Michener and Bennett (1977) found that in tropical and subtropical populations of *H. ligatus* about half the bees in mature colonies are workers, the other half are mated, egg-laying foragers, probably best considered a class of workers. A colony containing the latter alone might be considered quasisocial. Brood of all stages occurs at all times of year. It does not seem appropriate to call such colonies eusocial with multiple queens, because the egg-layers, unlike ordinary queens, are the major foragers and thus worker-like in this aspect of behavior (53% of foragers in Trinidad had enlarged ovaries). The condition is clearly derived evolutionarily from the temperate eusocial populations of the same species, and probably it is also derived ontogenetically from a eusocial state. It is an adaptation that, in a continuously warm climate, permits colonies to persist and grow for long periods of time after the death of the foundress and without a replacement queen. Presumably such a colony consists of daughters, granddaughters, or more distant descendants of the foundress. Packer and Knerer (1986a) found the population of *H. ligatus* in the Florida Keys behaving similarly, although each nest commonly had a foundress queen or a replacement queen. They also found that young foundresses are produced and seek nest sites through most of the year; this is probably also true in the tropics. The transition from seasonal, eusocial life to almost nonseasonal life occurs for this species in the southern half of the Florida peninsula (Packer and Knerer, 1987). There are hints of possibly similar modifications of colonial life of *H. farinosus* in California (Eickwort 1985) compared to ordinary colonies in Utah (Nye 1980).

Breed (1976) has analyzed eleven characteristics thought to be indicative of social level for six species of *Dialictus*. The five characteristics that appear to be related to one another and important in indicating a trend in social evolution in this subgenus are colony size, longevity of the queen, seasonal size variation among workers, size difference between queens and workers, and number of queens per colony. The last of these characteristics is of dubious value, considering the difficulty of distinguishing castes in most species of this subgenus and the small total range in mean number of queens per colony, only 1 to 3. The other characteristics, taken together, appear to provide a good guide to social level, not only in the subgenus *Dialictus* but in other eusocial halictine bees. As noted above, however, intraspecific variation within and among populations is so great that one must be cautious in interpreting these socially important features as specific characters; they may often be attributes of particular populations.

Knerer (1980b) has emphasized an inverse relation of size difference between castes and frequency of males in the first brood; abundance of such males is associated with little caste size difference in the same species. In *L. (E.) marginatum*, however, there is no size difference between castes and there are no males in the first brood; at least there are exceptions to the inverse relationship. Moreover, as noted above under Caste Differences, there is great intraspecific variation in caste size difference and probably also in frequency of first brood males. Knerer proposes, nonetheless, that caste differences in size are related to suppression during halictine phylogeny of first brood males. The hypothesis is that the ancestral sexual size-differentiating mechanism (cell size, amount of food provided) became a caste size-differentiating mechanism during the evolution of eusociality.

In the presumably solitary, bivoltine *L. (E.) villosulum*, however, females of the first (summer) generation average smaller than those of the second (overwintering) generation (Plateaux-Quénu and Plateaux 1985; Plateaux-Quénu, Plateaux, and Packer 1987). Moreover the sex ratio in the first generation is three females to one male. These findings indicate that seasonal rather than (or in addition to) sexual size differentiating mechanisms antedated caste differences. An alternative explanation is that *L. (E.) villosulum* evolved from a eusocial ancestor and retains some life history features of that ancestor.

Various efforts to relate halictid social level to nest architecture are not supported by the facts. Species that make cell clusters and that do not, are found among both solitary and eusocial forms. Kukuk and Eickwort (1987), however, have shown that because of burrow branch lengths, individuals in communal colonies are not obligated to meet one another as often as are individuals in eusocial colonies. Thus in communal colonies individuals are better able to escape domination by potential dominators (queens) than in eusocial colonies. This may explain the lack of eusocial groups related to or derived from the predominantly communal taxa such as the *Agapostemon* complex and *Lasioglossum* (*Chilalictus*).

8 Discussion and Conclusions

An outstanding observation on *L.* (*D.*) *zephyrum*, probably applicable to other halictines also, is the enormous behavioral variability among individuals and colonies, even those in the same population. Conclusions based on one or a few colonies or on data gathered at a particular time and place are likely to differ from those based on other nests or data taken at another time or place. Large series of observations are needed to determine what the population as a whole is doing. This could be in part due to inbreeding and the development of groups of related colonies, differing from neighboring groups. A major factor, however, is the lack of sharp caste differences, at least between workers and replacement queens. Unlike highly social insects, the caste functions overlap broadly and caste intermediates doubtless can play useful roles. There must be little selection for a major switch mechanism that causes every female to develop into either a well differentiated queen or a well differentiated worker. The result is great variability in behavior and physiology of females. A multivariate study of behavioral attributes distinguishes rather clearly between the castes, but examination of any one feature is likely to show complete intergradation.

Given the obviously extensive intraspecific variation in size difference between castes, and even in whether or not there is a worker caste (see information on *L.* (*E.*) *calceatum* in section on Evolution of Social Features), it seems important to note that evolutionarily and probably also environmentally induced adaptability is enormous for these features, that each population is adapted to its local environment, and that far-reaching conclusions on the social status of related species often depend on the particular populations studied. I believe that these problems exist for the variables considered by Breed (1976), Knerer (1980b), and others.

Caste differentiation (replacement queen *vs.* worker) in *L.* (*D.*) *zephyrum* is in the adult stage and I postulate that it is largely a result of the high activity of the queen, who mechanically disturbs her nestmates and inhibits their tendencies to develop as major reproductives. It seems to me that such mechanical control of caste by the queen is likely in a very primitively social species because I suspect that it could evolve more easily than a pheromonal control involving glands, synthesis of compounds, sensory organs to perceive them, and nervous equipment for an appropriate response. So long as the colony is small, the queen could have her effect through her behavior, perhaps using responses already effective in solitary species. In large colonies, of course, direct mechanical control by the queen would be impossible, and chemical control (as in *Apis*) has evolved. It is in such colonies that the traditional account of the queen as an inactive egg-layer (see Oster and Wilson 1978) is accurate.

It will be important to investigate control of caste by queens in large halictine colonies. It may be significant that in a halictine known to have colonies of several hundred, *L.* (*E.*) *marginatum*, there is indeed a switch mechanism that induces caste differences. As noted above, females that mate become queens; those that do not become workers. It is unlikely that queens' mechanical activity (nudging, etc.) has anything to do with caste differentiation in this case. This, however, has to do with differentaition of workers from foundress queens; we do not know whether replacement queens occur in this species.

Caste in *L.* (*D.*) *zephyrum* is biased by factors other than queen action. Thus age and size (the latter influenced by cell and food mass size, which are in turn influenced by temperature, day length, colony size, etc.) influence caste. In some halictines, such factors are the principal determinants of caste. Thus in *L.* (*E.*) *malachurum* and *calceatum* the life history is such that, in summer, the larger, older, overwintered foundresses are the mothers (queens) of the smaller, younger workers. It is not certain that the latter are reproductively inhibited by queen activity, although they are inhibited by queen presence at least in *L.* (*E.*) *calceatum*. Only the early death of a queen would produce a situation in which a worker might compete with its worker nestmates to become queen.

Considering further the intermediates between castes that occur in *L.* (*D.*) *zephyrum* and many other halictines, an interesting topic for physiological study would be the distinction between two types of workers with partially enlarged ovaries. There are those with slight to moderate enlargement of several or of all six ovarioles, and there are those with slender ovaries except for enlargement of a single oocyte in one or rarely two ovarioles. Dissections suggest that both types can reach the stage of laying an occasional egg. The first type is more common in *L.* (*D.*) *zephyrum*, the second, in *L.* (*D.*) *imitatum* (Fig. 5).

I believe that the interactions among females in a colony are basically competitive. The dominant female (queen, usually mother of other colony members) can control the tendency of other females (workers) to become reproductive. Her individual fitness is enhanced, the more effective this control, which even extends to inhibiting mating by her subordinates. At the same time a subordinate may be able to reduce the influence of a queen by avoiding her; this is illustrated by reluctance of workers to pass queens and queen-like individuals. Possibly the construction of branch burrows is related to this topic, offering places to avoid the queen.

A loser in competition with the queen may improve her individual fitness by laying what eggs she can and waiting for the death of the queen, perhaps to win in the competition among workers to replace her. Meanwhile, her inclusive fitness will be improved by helping her mother, the queen. If, because of the season or the difficulties of establishing a new nest, the increase in inclusive fitness plus the above mentioned potential increments to her individual fitness exceed the probable fitness advantage of leaving the mother and starting a new colony, then the daughter should stay with her mother as a worker.

As indicated by Michener and Brothers (1974), a major factor in the origin of a worker caste must be selection at the individual level for queens that more effectively control associated adult females (workers). Evidence that workers have evolved to help their queens is weak in primitively social bees; rather they seem to do so as a last resort. Escape of some daughters from maternal control is suggested by the departure of some "worker brood" females for diapause and foundress activities the next year.

Nonetheless, females that mature at the "wrong" season or that are for other reasons disadvantaged may have a higher fitness if they remain with their mother instead of trying to establish a new nest, e.g., at a season when it would likely be unsuccessful. It is mutually advantageous for both mother and such daughters if they remain together and form a eusocial unit. Much social evolution must be based on this fact, even though interactions within a colony are competitive. The origin of a worker caste must depend on advantages to workers. There may be good reason for any unmated female to remain with and help her mother. This situation would constitute the setting in which parental manipulation (of mating and other things) could evolve.

Although mutualism probably influences much social evolution, a major puzzle concerns the resistance of workers to mating attempts. Workers of social halictids are repeatedly pounced upon by males; I suspect that they produce the sex pheromone because it is also the pheromone by which they are recognized by guards at nest entrances. However, they are not usually receptive. They are evidently controlled in this respect by the queen, for they mate promptly after she is removed from the nest.

Since mating takes only a few moments, sometimes less time than fighting off a male, and involves no obvious increased dangers, one would think that mating would be advantageous to the workers. A worker with stored sperm cells would be in a position to immediately become a replacement queen on the death of the old queen, and could promptly lay female-producing eggs. One would expect workers outside the nest to escape the queen's control and mate, yet this is not usual in any social insect group. One is left with an impression of strong maternal control.

9 Acknowledgements

Above all I wish to acknowledge my numerous students and former students cited as authors here and in Michener (1974), especially for work on *L.* (*D.*) *zephyrum*. These studies have been possible largely because of grants from the National Science Foundation, of which the most recent is BNS-8200651.

Especially I thank P. F. Kukuk, L. Packer, and D. Yanega for carefully reading the manuscript of this paper and making many valuable suggestions.

I must also thank Prof. Wolf Engels whose invitation to contribute to a symposium at the XVI International Congress of Entomology in Kyoto in August, 1980, led to preparation of this manuscript.

10 References

Abrams J, Eickwort GC (1980) Biology of the communal sweat bee *Agapostemon virescens* in New York State. Search: Agriculture [Cornell Univ] 1980(1):1-20

Abrams J, Eickwort GC (1981) Nest switching and guarding by the communal sweat bee *Agapostemon virescens*. Insect Soc 28:105-116

Alcock J, Barrows EM, Gordh G, Hubbard LJ, Kirkendall L, Pyle DW, Ponder TL, Zalom FG (1978) The ecology and evolution of male reproductive behaviour in bees and wasps. Zool J Linnean Soc London 64:293-326

Barrows EM (1975a) Mating behavior in halictine bees: III: Copulatory Behavior and olfactory communication. Insect Soc 22:307-332

Barrows EM (1975b) Individually distinctive odors in an invertebrate. Behav Biol 15:57-64

Barrows EM (1976a) Mating behavior in halictine bees: I: Patrolling and age-specific behavior in males. J Kansas Entomol Soc 49:105-119

Barrows EM (1976b) Mating behavior in halictine bees: II: Microterritorial and patrolling behavior in males of *Lasioglossum rohweri*. Z Tierpsychol 40: 377-389

Barrows EM, Bell WJ, Michener CD (1975) Individual odor differences and their social functions in insects. Proc Nat Acad Sci USA 72:2824-2828

Batra SWT (1966) Nests and social behavior of halictine bees of India. Indian J Entomol 28:375-393

Batra SWT (1987) Ethology of the vernal eusocial bee, *Dialictus laevissimus*. J Kansas Entomol Soc 60:100-108

Bell WJ (1973) Factors controlling initiation of vitellogenesis in a primitively social bee, *Lasioglossum zephyrum*. Insect Soc 20:253-260

Bell WJ (1974) Recognition of resident and nonresident individuals in intraspecific nest defense of a primitively eusocial halictine bee. J Comp Physiol 93:173-181

Bohrer K (1987) Larvenernährung und Kastendifferenzierung bei *Lasioglossum malachurum*. Diplomarbeit, Fac Biol Univ Tübingen, [7]+87 pp.
Breed MD (1975a) Life cycle and behavior of a primitively social bee, *Lasioglossum rohweri*. J Kansas Entomol Soc 48:64-80
Breed MD (1975b) Sociality and seasonal size variation in halictine bees. Insect Soc 22:375-380
Breed MD (1976) The evolution of social behavior in primitively social bees: a multivariate analysis. Evolution 30:234-240
Breed MD (1977) Interactions among individuals and queen replacement in a eusocial halictine bee. Proc Int Congr IUSSI, Wageningen, 8:228-231
Breed MD, Gamboa GJ (1977) Behavioral control of workers by queens in primitively eusocial bees. Science 195:694-696
Breed MD, Silverman JM, Bell WJ (1978) Agonistic behavior, social interactions, and behavioral specialization in a primitively eusocial bee. Insect Soc 25:351-364
Brooks RW, Roubik DW (1983) A halictine bee with distinct castes: *Halictus hesperus* (Hymenoptera: Halictidae) and its bionomics in central Panama. Sociobiol 7:263-282
Brothers DJ, Michener CD (1974) Interactions in colonies of primitively social bees. III: Ethometry of division of labor in *Lasioglossum zephyrum*. J Comp Physiol 90:129-168
Buckle GR (1981) Caste differences, nestmate interactions, and nestmate recognition in *Lasioglossum zephyrum*. Ph D dissertation, Univ of Kansas, Lawrence, [iii] + 133 pp.
Buckle GR (1982a) Queen-worker behavior and nesmate (sic) interactions in young colonies of *Lasioglossum zephyrum*. Insect Soc 29:125-137
Buckle GR (1982b) Differentiation of queens and nestmate interactions in newly established colonies of *Lasioglossum zephyrum*. Sociobiol 7:8-20
Buckle GR (1984) A second look at queen-forager interactions in the primitively eusocial halictid, *Lasioglossum zephyrum*. J Kansas Entomol Soc 57:1-6
Buckle GR (1985) Increased queen-like behaviour of workers in large colonies of the sweat bee *Lasioglossum zephyrum*. Anim Behav 33:1275-1280
Buckle GR, Greenberg L (1981) Nestmate recognition in sweat bees (*Lasioglossum zephyrum*): Does an individual recognize its own odour or only odours of its nestmates? Anim Behav 29:802-809
Campos MJ de O (1980) Aspectos da sociologia e fenologia de *Pereirapis semiauratus*. Dissertaçao (Mestre) da Univ Federal de Sao Carlos, SP, Brazil, [10] + 188 pp.
Crozier RH, Smith BH, Crozier YC (1987) Relatedness and population structure in the primitively eusocial bee *Lasioglossum zephyrum* (Hymenoptera: Halictidae) in Kansas. Evolution 41:902-910
Eickwort GC (1981) Presocial insects. In: Hermann HR (ed) Social insects. Vol. 2, Academic Press, New York: pp 199-200
Eickwort GC (1985) The nesting biology of the sweat bee *Halictus farinosus* in California, with notes on *H. ligatus*. Pan-Pac Entomol 61:122-137
Eickwort GC (1986) First steps into eusociality: the sweat bee *Dialictus lineatulus*. Florida Entomol 69:742-754
Eickwort GC, Kukuk PF (1987) Reproductive castes in primitively eusocial halictid bees. In: Eder J, Rembold H (eds) Chemistry and biology of social insects. Peperny, München: pp 261-262
Eickwort GC, Sakagami SF (1979) A classification of nest architecture of bees in the tribe Augochlorini, with description of a Brazilian nest of *Rhinocorynura inflaticeps*. Biotropica 11:28-37
Garófalo CA (1980) Reproductive aspects and evolution of social behavior in bees. Rev Brasil Gen 3:139-152
Greenberg L (1979) Genetic component of bee odor in kin recognition. Science 206:1095-1097
Greenberg L (1982a) Persistent habituation to female odor by male sweat bees (*Lasioglossum zephyrum*). J Kansas Entomol Soc 55:525-531
Greenberg L (1982b) Year-round culturing and productivity of a sweat bee, *Lasioglossum zephyrum*. J Kansas Entomol Soc 55:13-22

Greenberg L, Buckle GR (1981) Inhibition of worker mating by queens in a sweat bee, *Lasioglossum zephyrum*. Insect Soc 28:347-352

Hölldobler B, Michener CD (1980) Mechanisms of identification and discrimination in social Hymenoptera. In: Markl H (ed) Evolution of social behavior: Hypotheses and empirical tests. Verlag Chemie, Weinheim, Deerfield Beach, Fla. and Basel: pp 35-58

Kamm DR (1974) Effects of temperature, day length, and number of adults on the sizes of cells and offspring in a primitively social bee. J Kansas Entomol Soc 47:8-18

Knerer G (1980a) Biologie und Sozialverhalten von Bienenarten der Gattung *Halictus* Latreille. Zool Jb Syst 107:511-536

Knerer G (1980b) Evolution of halictine castes. Naturwissenschaften 67:133-135

Knerer G (1981) Biologie von vier kleinen *Evylaeus*-Arten im Mittelmeergebiet. Zool Anz 206:1-10

Knerer G (1983) The biology and social behavior of *Evylaeus linearis* (Schenck). Zool Anz 211:177-186

Knerer G (1987) Photoperiod as cue for voltinism and caste regulation in halictine bees. In: Eder J, Rembold H (eds) Chemistry and biology of social insects. Peperny, München: p 305

Knerer G, Plateaux-Quénu C (1967) Sur la production de males chez les Halictinae. CR Acad Sci, Paris, (D)264:1096-1099

Knerer G, Plateaux-Quénu C (1970) The life cycle and social level of *Evylaeus nigripes* (Hymenoptera: Halictinae), a Mediterranean halictine bee. Canad Entomol 102:185-196

Kukuk PF (1980) Diurnal activity of a primitively eusocial bee, *Lasioglossum zephyrum*, within the nest. J Kansas Entomol Soc 53:149-156

Kukuk PF (1985) Evidence for an antiaphrodisiac in the sweat bee *Lasioglossum* (*Dialictus*) *zephyrum*. Science 227:656-657

Kukuk PF, Decelles PD (1986) Behavioral evidence for population structure in *Lasioglossum* (*Dialictus*) *zephyrum* female dispersion patterns. Behav Ecol Sociobiol 19:233-239

Kukuk PF, Eickwort GC (1987) Alternative social structures in halictine bees. In: Eder J, Rembold H (eds) Chemistry and biology of social insects. Peperny, München: pp 555-556

Kukuk PF, Eickwort GC, May B (1987) Multiple maternity and multiple paternity in first generation brood from single foundress colonies of the sweat bee, *Dialictus zephyrus*. Insect Soc 34:131-135

Kukuk PF, May B (1985) A re-examination of genetic variability in *Dialictus zephyrus*. Evolution 39:226-228

Kukuk PF, Schwarz M (1987) Intranest behavior of the communal sweat bee *Lasioglossum* (*Chilalictus*) *erythrurum*. J Kansas Entomol Soc 60:58-64

Kukuk PF, Schwarz M (1988) Macrocephalic male bees as functional reproductives and probable guards. Pan-Pac Entomol 64:131-137

Kumar S (1975) Relations among bee size, cell size, and caste in *Lasioglossum zephyrum*. J Kansas Entomol Soc 48:374-380

Litte M (1977) Aspects of the social biology of the bee *Halictus ligatus* in New York State. Insect Soc 24:9-36

Michener CD (1974) The social behavior of the bees. Harvard Univ Press, Cambridge, Massachusetts

Michener CD (1977) Aspects of the evolution of castes in primitively social insects. Proc Int Congr IUSSI, Wageningen, 8:2-6

Michener CD (1978) The classification of halictine bees: Tribes and Old World nonparasitic genera with strong venation. Univ Kansas Sci Bull 51:501-538

Michener CD (1982) Early stages in insect social evolution: Individual and family odor differences and their functions. Bull Entomol Soc Amer 28:7-11

Michener CD (1985) From solitary to eusocial: need there be a a series of intervening species? In: Hölldobler, Lindauer M (eds) Experimental behavioral ecology and sociobiology. Fortschr Zool 31:293-305

Michener CD, Benett FD (1977) Geographical variation in nesting biology and social organization of *Halictus ligatus*. Univ Kansas Sci Bull 51:233-260

Michener CD, Breed MD, Bell WJ (1979) Seasonal cycles, nests, and behavior of some Colombian halictine bees. Rev Biol Trop 27:13-34

Michener CD, Brothers DJ (1974) Were workers of eusocial Hymenoptera initially altruistic or oppressed? Proc Nat Acad Sci USA 71:671-674

Michener CD, Brothers DJ, Kamm DR (1971) Interactions in colonies of primitively social bees: Artificial colonies of *Lasioglossum zephyrum*. Proc Nat Acad Sci USA 68:1241-1245

Michener CD, Smith BH (1987) Kin recognition in primitively eusocial insects. In: Fletcher DJC, Michener CD (eds) Kin recognition in animals. Wiley, Chichester and New York: pp. 209-242

Michener CD, Wille A (1961) The bionomics of a primitively social bee, *Lasioglossum inconspicuum*. Univ Kansas Sci Bull 42:1123-1202

Nye WP (1980) Notes on the biology of *Halictus* (*Halictus*) *farinosus* Smith. US Dept Agric, Agric Research Results ARR-W-11, [iv] + 29 pp. [Bee Biol and Syst Lab, Utah State Univ, Logan]

Ordway E (1965) Caste differentiation in *Augochlorella*. Insect Soc 12:291-308

Oster GF, Wilson EO (1978) Caste and ecology in the social insects. Princeton Univ Press, [xv] + 352 pp

Packer L (1983) The nesting biology and social organization of *Lasioglossum* (*Evylaeus*) *laticeps* (Hymenoptera, Halictidae) in England. Insect Soc 30:367-375

Packer L (1985) Two social halictine bees from southern Mexico with a note on two bee hunting philanthine wasps. Pan-Pac Entomol 61:291-298

Packer L (1986a) The biology of a subtropical population of *Halictus ligatus* Say. II: Male behavior. Ethol 72:287-298

Packer L (1986b) The biology of a subtropical population of *Halictus ligatus*. IV: A cuckoo-like caste. J New York Entomol Soc 94:458-466

Packer L (1986c) The social organization of *Halictus ligatus* (Hymenoptera; Halictidae) in southern Ontario. Canad J Zool 64:2317-2324

Packer L (1986d) Multiple foundress associations in a temperate population of *Halictus ligatus*. Canad J Zool 64:2325-2332

Packer L (1986e) Aspects of the social biology of *Halictus ligatus*. Ph D thesis, Univ of Toronto, xx+219 pp

Packer L, Knerer G (1985) Social evolution and its correlates in bees of the subgenus *Evylaeus*. Behav Ecol Sociobiol 17:143-149

Packer L, Knerer G (1986a) The biology of a subtropical population of *Halictus ligatus* Say. I: Phenology and social organization. Behav Ecol Sociobiol 18:363-375

Packer L, Knerer G (1986b) An analysis of variation in the nest architecture of *Halictus ligatus* in Ontario. Insect Soc 33:190-205

Packer L, Knerer G (1987) Biology of a subtropical population of *Halictus ligatus* Say (Hymenoptera; Halictidae). III: The transition between annual and continuously brooded cycles. J Kansas Entomol Soc 60:510-516

Plateaux-Quénu C (1973) Construction et évolution annuelle du nid d'*Evylaeus calceatus* Scopoli avec quelques considérations sur la division du travail dans les sociétés monogynes et digynes. Insect Soc 20:288-320

Plateaux-Quénu C (1974) Comportement de sociétés orphelines d'*Evylaeus calceatus* (Scop.). Insect Soc 21:5-12

Plateaux-Quénu C (1978) Les sexués de remplacement chez *Evylaeus calceatus* (Scop.). Insect Soc 25:227-236

Plateaux-Quénu C (1979) Qui remplace la reine dans un groupe d'ouvrières orphelines d'*Evylaeus calceatus* (Scop.)? Ann Sci Nat Zool (13)1:213-218

Plateaux-Quénu C (1983) Le volume d'un pain d'abeille influence-t-il le sexe de l'oeuf pondu sur lui? Étude expérimentale, portant sur la première couvée d'*Evylaeus calceatus* (Scop.). Ann Sci Nat Zool (13)5:41-52

Plateaux-Quénu C (1984) Fondation tetragyne chez *Evylaeus calceatus* (Scopoli). Actes Colloques Insectes Sociaux, Vol 1:225-231

Plateaux-Quénu C (1985) Seconde couvée d'*Evylaeus calceatus* (Scop.) (Hym., Halictinae) - Les fondatrices sont-elles seules capables d'engendrer des fondatrices? Ann Sci Nat Zool (13)7:13-21

Plateaux-Quénu C, Plateaux L (1979) Action de la température sur la taille d'ouvrières d'*Evylaeus calceatus* (Scop.). Ann Sci Nat Zool (13)1:63-67

Plateaux-Quénu C, Plateaux L (1980a) Analyse des constituants mâles et femelles de la première couvée chez *Evylaeus calceatus* (Scop.). Ann Sci Nat Zool (13)2:209-214

Plateaux-Quénu C, Plateaux L (1980b) Action de la température sur la taille, le sexe et le cycle des individus de première couvée chez *Evylaeus calceatus* (Scop.) (Hym. Halictinae): première étude expérimentale. Ann Sci Nat Zool (14)2:27-33

Plateaux-Quénu C, Plateaux L (1985) La variatiion (sic) individuelle d'*Evylaeus villosulus* (K.), espèce solitaire (Hym., Halictidae). Comparaison des fondatrices de printemps et de leurs filles, appartenant à la première génération. Actes Colloques Insectes Sociaux, Vol 2:293-302

Plateaux-Quénu C, Plateaux L, Packer L (1987) Preadaptations to sociality in the solitary, bivoltine halictine bee *Evylaeus villosulus*. In: Eder J, Rembold H (eds) Chemistry and biology of social insects. Peperny, München: p 391

Sakagami SF (1974) Sozialstruktur und Polymorphismus bei Furchen- und Schmalbienen. In: Schmidt GH (ed) Sozialpolymorphismus bei Insekten. Stuttgart: Wissenschaftliche Verlagsgesellschaft mbH (reprinted in 1987): pp 257-293

Sakagami SF (1977) Seasonal change of nest survival and related aspects in an aggregation of *Lasioglossum duplex* Dalla Torre, a eusocial halictine bee. Res Popul Ecol 19:69-85

Sakagami SF, Hirashima Y, Maeta Y, Matsumura T (1982) Bionomic notes on the social halictine bee *Lasioglossum affine*. Esakia 19:161-176

Sakagami SF, Hoshikawa K, Fukuda H (1984) Overwintering ecology of two social halictine bees, *Lasioglossum duplex* and *L. problematicum*. Res Popul Ecol 26:363-378

Sakagami SF, Munakata M (1972) Distribution and bionomics of a transpalaearctic eusocial halictine bee, *Lasioglossum* (*Evylaeus*) *calceatum*, in northern Japan, with reference to its solitary life cycle at high altitude. J Fac Sci Hokkaido Univ (VI, Zool) 18:411-439

Sakagami SF, Okazawa T (1985) A populous nest of the halictine bee *Halictus* (*Seladonia*) *lutescens* from Guatemala. Kontyu 53:645-651

Smith BH (1983) Recognition of female kin by male bees through olfactory signals. Proc Nat Acad Sci USA 80:4551-4553

Smith BH (1987) Effects of genealogical relationship and colony age on the dominance hierarchy in the primitively eusocial bee *Lasioglossum zephyrum*. Anim Behav 35:211-217

Smith BH, Ayasse M (1987) Kin-based male mating preferences in two species of halictine bees. Behav Ecol Sociobiol 20:313-318

Smith BH, Carlson RG, Frazier J (1985) Identification and bioassay of macrocyclic lactone sex pheromone of the halictine bee *Lasioglossum zephyrum*. J Chem Ecol 11:1447-1456

Smith BH, Wenzel J (1988) Pheromonal covariation and kinship in the social sweat bee *Lasioglossum zephyrum*. J Chem Ecol 14:87-93

Snyder TP (1975) Lack of allozymic variability in three bee species. Evolution 28:687-689

Wcislo WT (1987) The role of learning in the mating biology of a sweat bee *Lasioglossum zephyrum*. Behav Ecol Sociobiol 20:179-185

Wille A, Michener CD (1971) Observations on the nests of Costa Rican *Halictus* with taxonomic notes on Neotropical species. Rev Biol Trop 18:17-31

Yanega D (1988) Social plasticity and early diapausing females in a primitively social bee. Proc Nat Acad Sci USA 85:4374-4377

Yanega D (1989) Caste determination and differential diapause within the first brood of *Halictus rubicundus*. Behav Ecol Sociobiol 24:97-107

Added in proof: The following important paper is relevant to several sections of this chapter:

Greenberg L (1988) Kin recognition in the sweat bee, *Lasioglossum zephyrum*. Behavior Genetics 18:425-438

Chapter 5

Castes in Xylocopine Bees*

CHARLES D. MICHENER[1]

CONTENTS

1 Introduction

XYLOCOPINAE are not usually thought of as having castes and two of the three major included tribes are not or scarcely mentioned in accounts of social behavior in bees (e.g., Michener 1974a). The methods and objectives of this review are the same as for the HALICTINAE treated in the preceding chapter. The only xylocopine tribe treated in detail in 1974 is now called the ALLODAPINI (Michener 1974a, Chapter 27), then called simply the allodapine group of CERATININI. We now know that some species in the other tribes are also, in varying degrees, social. Such species have, in some nests, two or more adult females with division of activities among them and sometimes behaviorally recognizable castes. This chapter is devoted largely to a descriptive account of the evidence for castes in xylocopine bees, with enough general natural history to show the relations of the castes to the life cycles of the species. So far as known no xylocopines are obligately eusocial.

---

[1] *Departments of Entomology and of Systematics & Ecology, and Snow Entomological Museum, University of Kansas, Lawrence, Kansas, USA*

* Contribution number 1905 from the Department of Entomology, University of Kansas, Lawrence

The XYLOCOPINAE are usually regarded as a subfamily of Anthophoridae. These are long-tongued bees, entirely unrelated to the HALICTIDAE, but comprise the sister group of the family APIDAE (Sakagami and Michener 1987) which contains the highly social bees.

The XYLOCOPINAE are divided into four tribes. The first, the MANUELIINI, contains small and rather slender bees. They nest either in stems or in somewhat decomposed wood, and constitute the relictual sister group to all other XYLOCOPINAE. The little that is known of their biology was summarized by Sakagami and Michener (1987) and Daly et al. (1987). The only genus, *Manuelia*, occurs in Chile and adjacent parts of Argentina. Since so little is known about the behavior of *Manuelia*, I ignore it in the rest of this chapter; for simplicity the word XYLOCOPINAE refers below to the three better-known tribes only.

These three tribes are as follows: (1) The XYLOCOPINI contain the usually large, robust carpenter bees, most of which nest in solid wood or thick stems in warm temperate and especially tropical parts of the world. The principal genus is *Xylocopa*. (2) The CERATININI consists of small, relatively slender bees (Fig. 1) that nest in hollow or pithy stems in temperate and especially tropical regions. Major included genera are *Ceratina* and *Pithitis*. (3) The ALLODAPINI, often included in (2) but abundantly distinct, are like the CERATININI in usually being small and slender (Fig. 2) and nesting in hollow or pithy stems. They are found in tropical and south temperate parts of the Eastern Hemisphere, but range north in Asia to Turkey, Punjab, central Sichuan (China) and Taiwan. Allodapine genera mentioned below are *Allodape*, *Allodapula*, *Braunsapis*, *Exoneura*, *Exoneurella*, and *Halterapis*

The ALLODAPINI have long been regarded as primitively social because in most genera the larvae are fed progressively. A single female with her eggs, larvae and pupae thus constitutes a subsocial unit. Reviews of allodapine behavior and references to much of the literature therefore appear in Michener (1974a, b) and other works on social insects. Indications of social behavior in the other tribes have only recently been widely appreciated.

The term colony is used in this chapter for a group of two or more nest-making or nest-tending adult females. Subsocial groups (e.g., mother and larvae) and prereproductive assemblages (see below) are not considered colonies.

XYLOCOPINAE are unusual among bees in several respects, as follows:

(1) Long life of the adults, some females living two or three years. [Not verified, although probable, for ALLODAPINI; the longest record is 16 months (Maeta, Sakagami, and Michener 1985).]

(2) Prereproductive assemblages. Adults of both sexes often live together before the reproductive stage. Such an assemblage probably consists of the bees that have matured in a nest plus sometimes the

mother or others; they clean out the nest and until they disperse, remain in it, in temperate climates usually through the winter. In some cases unrelated bees are known to join such assemblages.

(3) Mutual tolerance, such that bees in a prereproductive assemblage or in a colony readily pass one another and do not eject one another from their burrow.

(4) More than one female is often in a nest, forming a colony during the time when young are being produced and reared. There is commonly division of labor among these females with a principal egg layer (queen) and others of reduced productivity and often unmated (workers). Alternatively the colony may consist of one active, foraging female and one or more inactive individuals, but without permanent caste differences. Colonies occur in all three tribes but there are also species in each that are ordinarily solitary.

(5) Mouth to mouth nectar transfer occurs in both prereproductive assemblages and active nesting colonies (Michener 1972).

(6) Pollen and less commonly nectar are deposited in the nest (on burrow walls, in a cell, or on bodies of larvae) and subsequently eaten by the same or other individuals of the assemblage or colony.

The only other bees where (5) occurs are in the APIDAE (except for one report in HALICTINAE, see preceding chapter); (6) is also suggestive of APIDAE. In other bees, such as HALICTIDAE, pollen is only deposited in cells. Probably the potential for social polymorphism (including such features as 1, 3, 4, and 6 above) arose in a common ancestor of XYLOCOPINAE and APIDAE and was passed along to all xylocopine tribes, and also to the APIDAE (Sakagami and Michener 1987).

2 Tribe XYLOCOPINI

There are various records of more than one bee in a *Xylocopa* nest, as reviewed by Sakagami and Laroca (1971); see also Watmough (1974). Some such records relate to prereproductive assemblages, but others concern functioning colonies. Knowledge of caste-like division of activities, however, has been lacking or tenuous. Only since X-ray photography of nests in wood has been used, do we have the beginnings of adequate studies of xylocopine social biology, nearly all of them made under the impetus of Dan Gerling of Tel-Aviv University.

The nests consist of branching or simple burrows, transversely divided into cells by partitions made of bits of wood. The bees never destroy (and rebuild) the partitions to gain access to immature stages, as do some CERATININI.

Apparently in some essentially solitary species offspring of a female may emerge before their mother's death. The mother then forages and feeds nectar to the juvenile adults, who assiduously meet her in the nest when she returns. The mother soon dies, however, and may die before her younger progeny mature. Several African species of *Xylocopa* studied by Anzenberger (1977) and Brazilian species studied by Camillo and Garófalo (1982) appear to have such minimal interactions between adults of succeeding generations.

In *Xylocopa pubescens* (subgenus *Koptortosoma*) a single female occupies each nest in the spring and rears several offspring. Until maturation of the offspring, the mothers are solitary. After emergence of the young, the mother forages and feeds them; they block her return into her nest and her passage through it until they have been fed. Being a guard is advantageous because the guard is the first bee that the returning forager meets and feeds. When several young females are in the nest, they appear to compete with one another for this position. The mother now constructs and provisions new cells; her sons leave the nest after two or three weeks but some of the daughters remain and one (or more) functions as a guard. Guarding appears to be important, and the mother-daughter colony might be called temporarily eusocial, or at least the daughters could be called temporary workers. It seems more appropriate, however, to speak of the active bee (mother) and inactive bees (daughters). The latter differ from workers in ordinary eusocial colonies in that they are not long-term workers; they do not forage until they mate and establish nests of their own or re-use that of the mother. The young females also clean the nest and lick the cell walls, presumably placing the thin lining on them; such activities, however, appear to be preparation for their own reproduction rather than helping their mother. Some of the daughters' nests, instead of being independent, are branches from the mother's nest. Guarding is then shared among occupants of the nest and is at the common entrance instead of at the entrance to the branch. So far as known, no branch burrow is used by more than one bee (Ben Mordechai et al. 1978; Gerling, Hurd, and Hefetz 1981, 1983; Velthuis and Gerling 1983).

The arrangement described above would be communal according to the terminology of Michener (1974a) except that communality as well as other types of parasociality are supposed to involve only one generation. Gerling, Hurd, and Hefetz (1983) prefer to broaden the term parasocial to include the *Xylocopa* situation but alternatively propose the word metasocial for it. The term eosocial was proposed by Sakagami and Maeta (1977) for one type (like quasisocial) of "parasocial" behavior involving two generations. In XYLOCOPINAE these terms mean little since any two (or more) bees living in a nest seem able to work together, regardless of age differences. The word parasocial seems adequate although the other terms are convenient brief ways of indicating relationships.

In *Xylocopa sulcatipes* (subgenus *Ctenoxylocopa*), *X. sonorina* (subgenus *Neoxylocopa*), and *X. virginica* (subgenus *Xylocopoides*) social relationships similar to those described for *X. pubescens* can be found. Some nests of all these species are occupied by solitary females, like the spring nests of *X. pubescens*. However, even in spring, some nests contain two or three bees. Such colonies can contain bees that were in the nest during the prereproductive assemblage, probably young bees reared in the nest or the old mother and one or two of her daughters; alternatively for *X. sulcatipes* they can consist of a solitary nestmaker joined by another bee (Velthuis 1987).

In *X. sulcatipes* the break-up of the prereproductive assemblage has been studied (Gerling and Hefetz 1981; Velthuis 1987). A female becomes active, shares nectar with nestmates, and deposits pollen at the back of the nest where it is soon eaten. Aggressive interactions between the forager and other nestmates may lead to departure of the active individual, which occurs after a few days. This process continues as one female after another becomes active, then leaves. (Males also leave.) When only two females are left the pollen consumption becomes less than the amount collected, and a cell may be provisioned by an active female. Aggressiveness may subside, in which case a two-bee colony exists with an active bee and an inactive guard. Or the inactive bee may be ejected from the nest.

While the preceding paragraph presents the usual situation, Velthuis (1987) reports that in a minority of nests both females forage and to judge by the ovaries, probably lay eggs; new cells are built and provisioned in rapid succession. Gerling, Hurd, and Hefetz (1983) regard such colonies as probably quasisocial.

In *X. virginica* there is only one reproductive period per year. The young adults remain in prereproductive assemblages with surviving older adults from late June through the winter (Gerling and Hermann 1978). In the spring two females probably remain in the old nest; at any rate most nests contain two adult females. One, believed to have survived the two preceding winters, is inseminated and has enlarged ovaries. The other, unworn and having survived only one winter, is often uninseminated, presumably mates in the spring with an overwintered male, and has slender ovaries. The older female is the active one; she forages, makes and provisions cells and lays eggs. Such bees die after the nesting season. The younger or inactive bee rarely leaves the nest; she acts as a guard when the older bee is away but does little else. She is, however, a temporary worker in the same sense as comparable bees in summer nests of *X. pubescens* and most two-bee nests of *X. sulcatipes*. She remains in the nest with the juvenile adult progeny of her now dead older nestmate and lives in the prereproductive assemblage. Probably the next year such a bee inherits the nest and an associate from among the female progeny of her older nestmate. Some *X. virginica* females may reproduce after their first winter but there is no evidence that this is so. Given

Fig. 1 - *Ceratina timberlakei*, female (from Daly 1973; drawing by Barbara B. Daly, courtesy of Univ. of California Press).

its abundance and wide distribution, *X. virginica* should be studied at other sites to see if the account by Gerling and Hermann represents typical behavior for the species.

Similar small colonies also occur in *X. sonorina*. Typically one bee forages and not only provisions cells and lays eggs in them but also feeds any other adults in the nest; another (inactive) bee guards at the nest entrance. Worn, uninseminated females of *X. sonorina* are sometimes found (Gerling 1982); it is not clear whether such bees are permanent (or long term) workers.

In summary, several and presumably many *Xylocopa* species have small colonies (most commonly only two females) in at least some of the nests. The individuals of the colony are usually an active individual (sometimes the mother) and an inactive, largely nest-bound individual whose principal activity is guarding. The active bee not only makes and provisions cells and lays eggs but forages and feeds other adults. The inactive bees presumably do not average smaller than active ones, for inactive bees become active in the next reproductive season. These "castes" are ontogenetic stages in the lives of individuals rather than classes of individuals. Less commonly quasisocial colonies (two females, both active) exist. When juvenile adults emerge, they remain in the nest for some time, are fed by the mother, and sometimes with the mother or foreign bees, form a prereproductive assemblage.

3 Tribe CERATININI

The principal genus of the tribe is *Ceratina* (Fig. 1). Like other genera in the tribe, it is ordinarily regarded as consisting of solitary bees. As in the XYLOCOPINI, males are often present in the prereproductive assemblages with females. Mating and dispersal usually occur after such assemblages, in temperate regions usually in the spring. Then new nests are made, mostly in dead, dry, pithy stems. The nests consist of simple burrows; young are reared in cells made by transverse partitions of bits of pith (Fig. 3a). In some species the mothers die in the autumn but in others they often survive with their offspring in the prereproductive assemblages and use the nest again in a second, and rarely even a third, summer. Reviews of *Ceratina* biology are by Sakagami and Laroca (1971), Michener (1985), and Sakagami and Maeta (1987a).

In various species of *Ceratina* such as *C. calcarata* and *japonica* (Kislow 1976; Sakagami and Maeta 1977), the mothers regularly open the cells containing their progeny. A female can work her way down a nest burrow and then back again, inspecting her offspring, destroying partitions and rebuilding them behind her. She removes larval fecal pellets and other debris, even dead larvae, from the cells by incorporating them into the partitions. It is remarkable that all this can happen without damage to the delicate stages, especially eggs and young larvae. In *C. megastigmata* partitions are not removed until the youngest larva finishes feeding; at that time the mother removes partitions and larval feces, and subsequently, additional feces as they are produced (Maeta and Katayama 1978). In *C. iwatai*, larval feces are not removed until the beginning of the prereproductive phase (Sakagami and Maeta 1977). It is probable that in such species the mother does not open the cells although she remains in the upper part of the nest, above the cells. Thus *C. iwatai* (subgenus *Ceratina*) is an ordinary solitary bee while *C. japonica* (subgenus *Ceratinidia*) and *C. calcarata* (subgenus *Zadontomerus*), among others, can be called subsocial because of the continuing attention given by the mother to her offspring.

Seemingly quite independently from the evolution of long-term larval care, some species of *Ceratina* facultatively develop colonies of adults. Such behavior is probable in species of several subgenera (*Calloceratina*, *Ceratina* s. str., *Ceratinidia*) but is not found in all species of at least the last two. For example, in Japan, *C.* (*Ceratinidia*) *flavipes* almost never develops colonies whereas *C.* (*C.*)

japonica does. Most of our knowledge of such colonies is a result of the work of Sakagami and Maeta (1977, 1982, 1985, 1987a, b), summarized by Michener (1985).

Out of 230 new nests of *Ceratina japonica*, 1.3% contained more than one adult female, while of 203 reused nests, 31.0% contained more than one adult. Another set of data (188 new and 114 reused nests) gave 1.6 and 35.6% respectively (Sakagami and Maeta 1987a). Presumably such data mean that two or more bees sometimes remain behind after dispersal from a prereproductive assemblage, but only rarely do extra bees join a new nest. For all nests combined, 19.2% contained colonies. Most of them consisted of two adult females but a few contained three and one, four.

The interactions among the bees of a colony were diverse. To facilitate expression, colonies are described as though all contained only two adult females. Usually (81% of field collected colonies) both bees were in their first reproductive summer, i.e., they matured in the preceding autumn and were probable sisters. The others contained one such bee and one in its second reproductive summer, likely to be the mother of the first. About half the colonies (of each age type) were quasisocial (or eosocial if one wishes a term for such colonies with bees of two generations), i.e., both bees were presumably reproductive for both had enlarged ovaries. In the other half of the colonies, one bee had enlarged ovaries while the other had less enlarged or slender ovaries (Sakagami and Maeta 1985). This suggests division of labor and eusocial (if the bees are of two generations) or semisocial (if the bees are of the same generation) relationships. But even in colonies both members of which had enlarged ovaries, the two bees were not behaviorally equivalent; associations of two behaviorally equivalent bees were never established in artificially induced colonies (Sakagami and Maeta 1987a).

The most remarkable feature is that in the eusocial and semisocial colonies, the bee with slender ovaries also exhibited other characteristics of workers of social HYMENOPTERA. The worker was usually smaller than the queen, was usually uninseminated, did the foraging, and on return to the nest fed the queen. The queen rarely left the nest, frequently guarded the nest entrance, and seemingly laid most of the eggs. Presumably the queen can inhibit certain activities of her workers, and perhaps vice versa. Uninseminated bees seem to have a predilection for the worker role; 51 out of 112 bees in colonies contain no sperm whereas 137 solitary bees were all inseminated.

Thus in this species in which only about 10% of the nests have colonies made up of queens and workers, the division of labor in such colonies attains a level similar to that described in the preceding chapter for *Lasioglossum zephyrum*, an obligately eusocial halictid.

The situation differs from that in a colony of *Xylocopa* in which there is an inactive female because the latter does little but guard the nest and is not a terminal caste but is destined to become a reproductive next season. In *Ceratina* the workers work, and are commonly not a developmental stage leading toward future reproductive activity. Unfortunately there are no adequate data on behavior in natural colonies of *Ceratina* in which both bees are reproductive. There are probably intergradations between the two types of colonies, for reproductivity is relative.

Studies of artificially produced colonies of *Ceratina* shed further light on social behavior and its genesis. Large numbers of adults of *C. japonica* were placed in a cage with few nest sites (pithy stems). The result was a few colonies, mostly of only two bees (Sakagami and Maeta 1985, 1987b). The artificial colonies differed from the eusocial and semisocial natural colonies in that both bees had enlarged ovaries and were probably inseminated. Usually the smaller bee started the nest and was joined later by the larger bee. Unlike natural colonies that probably contain close relatives (sisters, daughters, mothers), relationships of bees in the artificial nests were completely unknown, for the adults placed in the cages were of miscellaneous origins. In six artificial colonies analyzed by Sakagami and Maeta (1987b), division of labor was reminiscent of that in natural colonies except that the smaller bee, after some hours and frequently late at night, often ate the egg laid by the larger bee and replaced it with one of her own. In these colonies 39 "worker"-laid eggs survived compared to 21 queen-laid eggs. These findings may give a clue to the behavior in natural quasisocial nests inhabited by females both of which are inseminated and with functioning ovaries.

The study of *Ceratina flavipes* also suggests something as to the origin or loss of social behavior. This is a close relative of *C. japonica*; both were observed at the same localities (Sakagami and Maeta 1987b). Unlike *C. japonica*, prereproductive assemblages of *C. flavipes* break up in the autumn and the bees mostly occupy individual burrows in pithy stems where they overwinter, often moving to another site in the spring. Thus each female is usually alone when nesting begins. At one site out of more than 2,000 nests, none contained colonies; at another site, of 307 brood-containing nests, three contained colonies. The rarity of natural colonies is obvious; *C. flavipes* is an almost fully solitary bee.

In artificial colonies of *C. flavipes*, which could only be established when bees of different sizes were introduced into a nest, the social organization was often unstable - the larger or smaller bee left. The larger bee did most of the guarding in nests inhabited by colonies, but sometimes refused to admit the small bee, which therefore left. The bees often showed difficulty in passing one another. In *C. japonica* and other primitively social bees living in burrows, passing is always venter to venter. In *C. flavipes* a bee sometimes forces its way past another, venter to dorsum. Sometimes, however, two bees face one another for hours, often moving back and forth in tandem, without mutual passing. Nevertheless, in five artificial colonies Sakagami and Maeta (1987b) found cell construction, provisioning, and egg laying.

In such colonies there was some recognizable division of labor. In four out of five the large bee was the principal guard. In three out of five the small bee fed the large one more often than the reverse. In two out of four the small bee was the principal forager for pollen (later reversed in one nest). After emergence of the juvenile adults, the small bee was the principal one to forage, feed them, and clean the nest burrow. While both large and small bees laid eggs, the small bee, hours later, commonly ate the egg of its larger nestmate. The number of surviving eggs from small bees in two carefully watched nests (one with three bees, two of them small) was 22 compared to 2 eggs from large bees. Thus task allocation was more variable than in eusocial colonies of *C. japonica*, but there was some division of labor, competition among nestmates, etc., as in colonies of other primitively social bees. In *C. japonica*, in any one eusocial or semisocial colony, task allocation is rather complete, in contrast to *C. flavipes*.

C. flavipes possesses a behavioral inventory similar to that of *C. japonica*, especially in the prereproductive assemblages that occur in the ordinary solitary cycle. This inventory is not sufficient to bring about colony formation. Difficulties that frequently led to breakdown of forced associations were intolerance of the guard for the other bee, who therefore left; difficulty in passing (perhaps another aspect of intolerance); and difficulty in tolerating (agreeing on the position of) the bottom plug, which must be finished before provisioning of the first cell. It is remarkable that a caste system, however weak and variable, can arise in this essentially solitary species that usually does not tolerate any other individual in its nests.

The ontogeny of caste induction in the artificial colonies of *C. flavipes* (Sakagami and Maeta 1987b) is as follows: Guarding (usually by the larger bee) can arise even in the incipient stages of cohabitation before other nesting tasks begin. The guard may later perform other tasks, while her nestmate merely remains in the nest. Only later may this relationship change to one with division of labor, i.e., with different tasks for each bee. But the division of labor is not immutable; if either one of the bees disappears, the other goes on as a solitary bee.

Sakagami and Maeta (1987a) recount other indications of behavioral interactions and flexibility, both in colonies and prereproductive assemblages. In assemblages of both *C. japonica* and *flavipes*, the

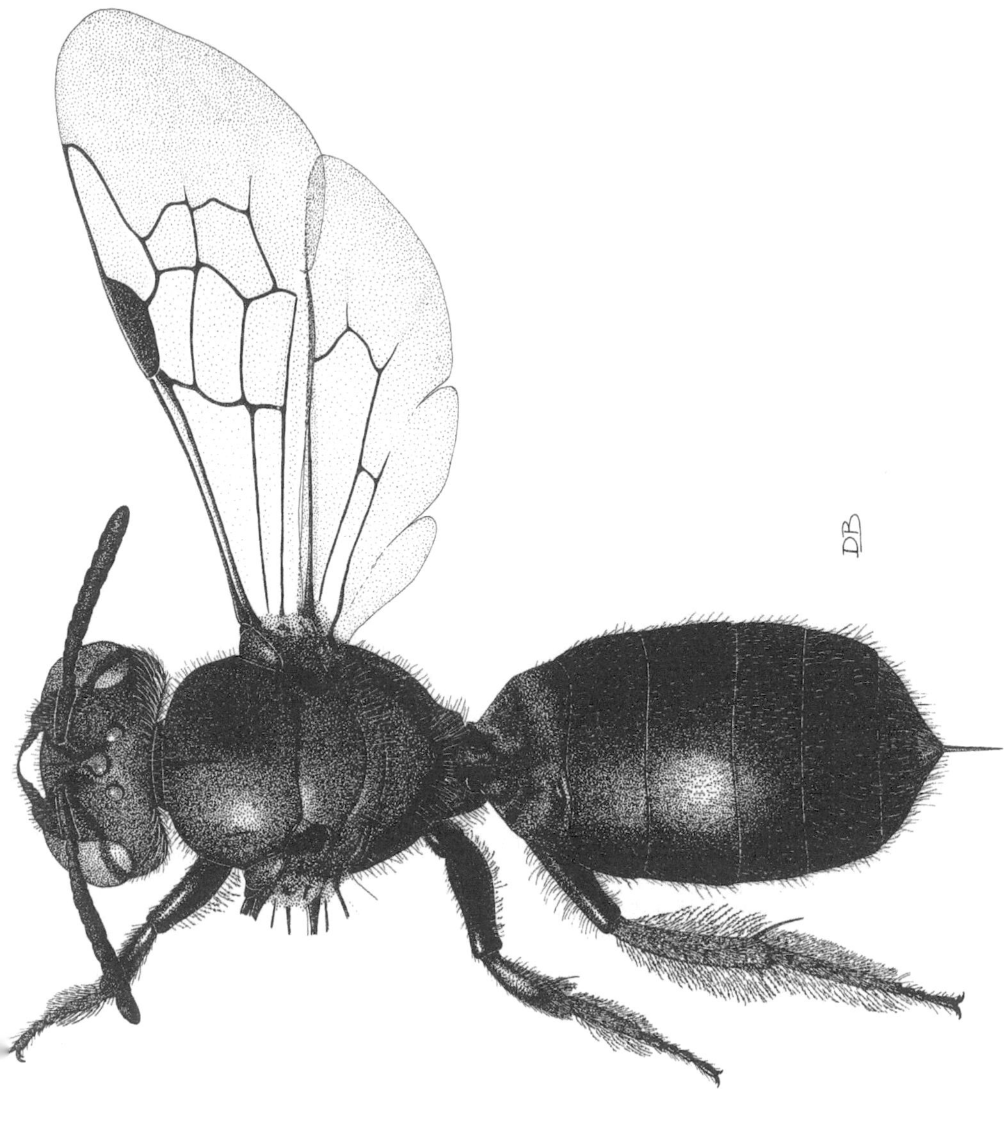

Fig. 2 - *Braunsapis facialis*, female (from Michener 1975; drawing by Denis J. Brothers).

mother ordinarily forages and feeds the juvenile adults. Sometimes, however, a daughter, usually a relatively old and small one, takes over this activity. When this happens, the mother abandons foraging and is fed by the foraging daughter. It is not clear whether the mother or foraging daughter initiates these changes, but it is clear that, as in a reproductive colony, activity of one individual can strongly influence behavior of another.

4 Tribe ALLODAPINI

Because of progressive feeding of larvae in all allodapine genera except *Halterapis*, adults (Fig. 2) have long been known to have more or less continuous contact with and to care for their immature stages. Earlier reviews are those of Sakagami (1960) and Michener (1971, 1974a, b). Like those of CERATININI, the nests are simple burrows in stems, but they are not divided into cells (Fig. 3b-d). In nests of most genera, larvae are in frequent contact with one another. Probably because they are not isolated in cells and therefore have contact with one another (Fig. 3) and perhaps compete with one another to obtain food from adults, the larvae have a variety of tubercles, projections, and hairs and often relatively elongate antennae for bee larvae. They are the morphologically most diversified of bee larvae; their characters are important in recognition of genera (Michener 1977).

Recent studies of allodapine biology include those of Maeta et al. (1984, 1985, and in prep.) concerning *Braunsapis sauteriella*, a species from Taiwan. This is the first species maintained for long periods in observation nests where behavior could be directly observed; various things only inferred previously from data on nests and dissections of nest contents were verified. Nests can be deepened even while eggs and larvae are in them. Young nests are mostly occupied by subsocial groups, a single foundress female and her immature progeny. Occasionally, however, one or two other bees join to form a semisocial colony (Fig. 4). Older nests in which young have been reared are deepened and commonly contain a colony of two to several adults. Some daughters stay with their mothers as workers and form eusocial colonies of two to four females. Perhaps partly because of weakening and destruction of stems by decomposition, weather, fire, etc., most nests never are occupied by colonies of adults.

Earlier studies indicated that in most and perhaps all allodapine species, a minority of the nests contained colonies of two to several females during the reproductive phase. The minority ranges from 3% in *Allodapula dichroa* to 34% and 38% in *Braunsapis facialis* and *leptozonia*, respectively (Michener 1971, p. 228). Probably more meaningful figures consider only nests with eggs and young, thus

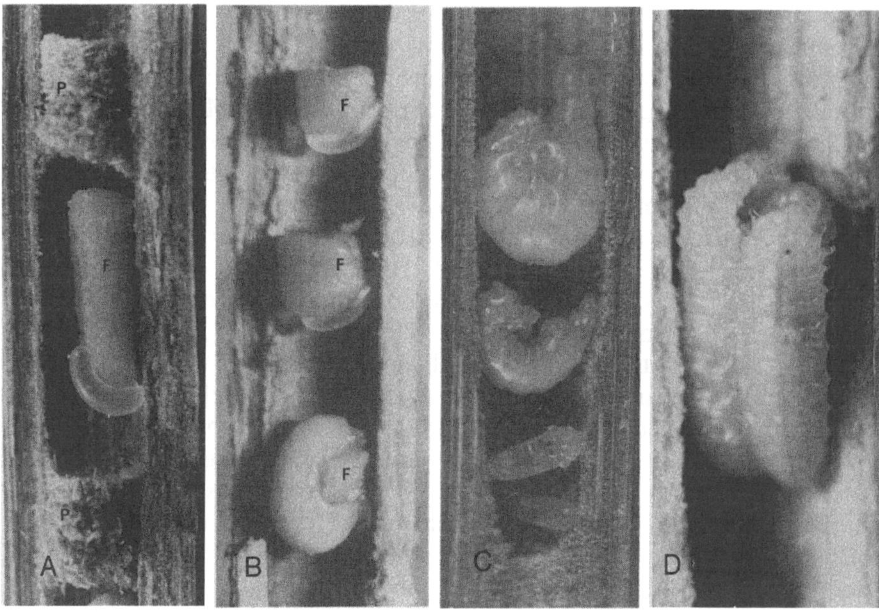

Fig. 3 - Longitudinal sections of portions of nests of CERATININI (A) and ALLODAPINI (B-D) in stems. P = partitions of pith fragments between *Ceratina* cells; F = food masses (pollen and nectar). **A.** *Ceratina* sp., an egg at the base of the food mass in a cell. **B.** *Halterapis nigrinervis*, a large larva feeding, above which are two eggs, each supporting a food mass. **C.** *Allodape ceratinoides*, egg and three larvae, which are fed progressively and therefore lack or only have small food masses. (Nests of *Braunsapis* are similar to those of *Allodape*.) **D.** *Allodapula acutigera*, large larvae which surround a common and progressively provided food mass which is not visible. (Photographs by C. D. Michener.)

excluding females just starting burrows. Of 123 such nests of *Allodapula dichroa*, 4% contained colonies of adults as did 43 and 44% of *Braunsapis facialis* and *leptozonia*, respectively. Juvenile adults, probably not yet working and that might soon disperse, were not included in these figures. Unfortunately it is not possible to determine, on the basis of a nest census, which females in a nest will disperse and which will remain as colony members; interpretations of census data are necessarily somewhat speculative on this point. However, melanization in these bees is usually slow, and the often numerous somewhat pallid or "callow" females in nests are not ordinarily foragers or guards, nor are they in reproductive condition. They are unmated, the ovaries are slender, and the wings unworn. Similarly nonreproductive unworn bees, although fully colored, might be young workers but near the end of the reproductive season are usually regarded as part of a prereproductive assemblage. Colonies, as indicated earlier, are recognized as two or more adult females that live together for an extended period during the reproductive phase.

The small size of allodapine colonies is noteworthy when one counts only mature, working females. Among African and Asiatic species the maximum numbers of such females range from 2 to 6, while mean numbers of mature females per nest range from 1.02 to 1.9 (Michener 1971, p. 231). If one includes young adults, however, much larger numbers are often found in nests, and prereproductive assemblages of a dozen or more bees in a stem are well known. Earlier studies, based mostly on censuses, were careful not to claim a reproductive social organization when the data could be explained on the basis of juveniles not yet dispersed, or prereproductive assemblages. Actual observations of behavior (Maeta, Sakagami, and Michener 1985) generally support the earlier conclusions.

Studies of African allodapines investigated by Michener (1971) are now being extended by Mason (1987). In observation nests of both *Allodape exoloma* and *Braunsapis foveata* only one of the colony members did most of the foraging. In the seven nests of *B. foveata* the principal forager was the oldest worker. In the nine nests of *A. exoloma*, however, the caste of the principal forager varied. Sometimes it was the oldest worker, but surprisingly it was sometimes the queen. In both species all nestmates participated in guarding but the queen was more involved in this activity in *A. exoloma* than in *B. foveata*. As to nectar exchange, 93% donated as well as received; while primary foragers (both species) were the principal donors, nonforagers were also frequent donors, showing that there is circulation of food among colony members. Nectar is stored by placing it on larvae (where it is not absorbed as it would be on the pithy walls of the nest). Principal foragers were less active in removing nectar from storage than were other bees. In summary, the distinction in activities between the principal forager and other colony members was quite sharp, that between queen and workers was less so, especially in *A. exoloma* in which the queen sometimes was also the principal forager.

Studies by Houston (1977) and Schwarz (1986) in Australia concerned species with colonies in a higher percentage of the nests than was found by Michener (1971) among African species. Houston gave an account of *Exoneurella*, a close relative if not a subgenus of *Exoneura*. Two species (*E. lawsoni*, *setosa*), and probably also a third, *E. eremophila*, are almost totally subsocial, each female rearing her offspring by progressive feeding. Probably sometimes a female may care for her younger siblings after death of her mother. In *E. eremophila* there were indications that two or more females sometimes oviposit together and rear a common brood, but there were no castes. In *E. tridentata*, however, there are regularly colonies with morphologically different individuals. Of 44 nests, 91% contained more than one adult female; some might soon disperse but obviously many nests contained working groups of females. Houston

(1977) showed that nests can be founded by solitary females and postulated that they later become semisocial with two or three sisters (daughters of the now dead foundress) living together, one of them the queen. In many colonies, however, one female is worn (queen), one or more others little worn and with less enlarged or slender ovaries (workers). Such colonies are probably matrifilial (eusocial).

The most remarkable feature of *E. tridentata* is the size variation among females. While head width is unimodal with a long tail to the right, the ratio of an abdominal measurement to head width (i.e., the metasomal index) is strongly bimodal, differentiating groups of females called minors and majors by Houston. About one fourth of the bees found in nests were majors. Because of an omission of some lines in Houston's (1977) account, an important paragraph (upper left, p. 107) is incomprehensible. It seems appropriate to give in full the correction supplied by Houston: "Majors, characterized by head widths of more than 1.67 mm and metasomal indices of 58 or more, all mate and lay eggs. Minors, characterized by head widths up to 1.67 mm and metasomal indices less than 58, function (at least in some cases) as infertile workers, although up to half of them may mate and/or produce eggs." In summary, minors can be queens or workers, while majors all become queens. Houston postulates that new nests are started by fertile minors, since majors not only have such large abdomens that they look incapable of flight, but in his experience did not fly when dropped. Old majors, moreover, have extremely worn wings like queens of MELIPONINAE and certainly could not fly. Females captured on flowers were all minors. Thus majors are probably a nondispersing fraction of the queen caste.

The most detailed field studies of an allodapine bee are those of Schwarz (1982, 1986, 1987, 1988a, c) and Schwarz, Scholz, and Jensen (1987), on *Exoneura bicolor* in southeastern Australia. Most nests at all times of year contain two or more females. In prereproductive assemblages, mated females forage, and feed younger associates so that in the autumn even callows that cannot fly have pollen and nectar in their guts. This is much as in other XYLOCOPINAE. In spring the one or two mated foundresses in the assemblage begin egg laying even before foraging. Other females soon mate so that there may be several egg layers. Later most unmated females disperse, mate, and establish new nests. Thus for a time the old nests contain both bees in the reproductive phase and overwintering nestmates. In the old nests there remain one to several unmated foundresses that presumably function as workers along with one to several mated egg laying foundresses. On the average, mated females are larger than unmated ones. The foundresses in new nests are nearly all mated; most of the nests contain several females.

Intranidal genetic relationship among the foundresses in new nests averages higher than for foundresses in overwintered nests, but is similar to that for brood in overwintered nests. Since swarm founding is unlikely, it is almost certain that new nests are started near the overwintered nest by lone foundress females that are soon joined by other dispersing foundress females. We know that halictids like *Lasioglossum zephyrum* have the ability to recognize nestmates (Michener and Smith 1987), and *Exoneura* can probably do the same thing. Possibly because of inclusive fitness benefits, potential joiners may seek out kin preferentially during colony initiation. Schwarz (1988b) has found that two other species of *Exoneura* also often have two or more foundresses in new nests.

By late November (late spring) the mean number of foundresses in new nests of *E. bicolor* was 2.7, range 1 to 8. In a sample of 118 new nests, 84.7% contained more than one female. In old (overwintered) nests the mean number of foundresses was 3.8, range 1 to 9 (from data in Schwarz 1988a). The colonies, in both old and new nests, produce a brood of offspring that matures in late summer and autumn (early February to late March). Most of the foundresses are dead by late March. There are no matrifilial colonies and there is no overlap of generations during the brood-rearing season. The species is univoltine.

As indicated above, in new nests nearly all foundresses (103 out of a sample of 106) are mated although in overwintered nests there are frequently unmated nonreproductives. In new nests there is no apparent reproductive differentiation among foundresses although there is variation in ovarian size; the eggs are placed in a single large clump at the bottom of the nest. The colony could be called quasisocial. In overwintered nests, since there are nonreproductive females, there is division of labor and presumably a worker caste although behavioral data are lacking. Such colonies could be called semisocial, sometimes with more than one queen.

Offspring of colonies in new nests on the average are genetically less closely related to one another than offspring of colonies in overwintered nests (Schwarz 1987). Presumably this is a result of the different types of social organization prevalent in the two types of colonies, those in new nests being quasisocial with the several females commonly contributing eggs, those in overwintered nests being semisocial, more often with only one egg layer.

Division of labor begins in the prereproductive assemblages. During the winter most nests contain one or two mated females with ovarian enlargement (although no eggs are laid until spring), in addition to unmated females with slender ovaries. In early spring (September 14) when assemblages still seemed to be in overwintering condition, Schwarz, Scholz, and Jensen (1987) isolated bees or transferred them to form various artificial colonies and showed that mated potential egg-layers inhibit ovarian development of unmated nestmates. Isolated unmated females, however, showed ovarian development within sixteen days. Thus even in such a primitively social form inhibitory effects are evident; also, insemination and ovarian development can be independent, as in replacement queens of some halictine bees.

The study area for *E. bicolor* has a temperate climate, and *Exoneura* is a primarily temperate genus. *E. variabilis*, the species most intensively studied prior to the Schwarz study, was investigated at a point 10° latitude (nearly 1100 km) farther north, and therefore with a longer warm season (Michener 1965). This may be the primary factor that permits development of eusocial colonies, second broods, a second generation in some nests, etc., in *E. variabilis*.

Some important findings relevant to ALLODAPINI since 1974 do not concern allodapines directly. They are aspects of the colonial behavior of various species of CERATININI, that greatly reduce the differences in nesting biology that had been believed to exist between the two tribes. Indeed, the differences between the social behavior of allodapines and that of *Ceratina* seem largely due to a mechanical matter - absence of cell partitions in ALLODAPINI. Below is a list of some allodapine characteristics modified from a list of *Ceratina* attributes (Michener 1985). For comparative purposes the status of each variable in *Ceratina* and *Xylocopa* is indicated in parentheses.

(1) Nests are started by foundresses that ordinarily rear their first brood alone. (Same in *Ceratina*, *Xylocopa*, and other bees except the highly social forms like *Apis*.) Exceptions include *Exoneura bicolor* and others where joiners usually result in colonies; joiners are less common in *Xylocopa*, *Ceratina*, *Braunsapis*, etc.

(2) Colonies of adults are therefore rare in recently founded nests. (Same for *Xylocopa* and *Ceratina japonica*; same exceptions as for 1.)

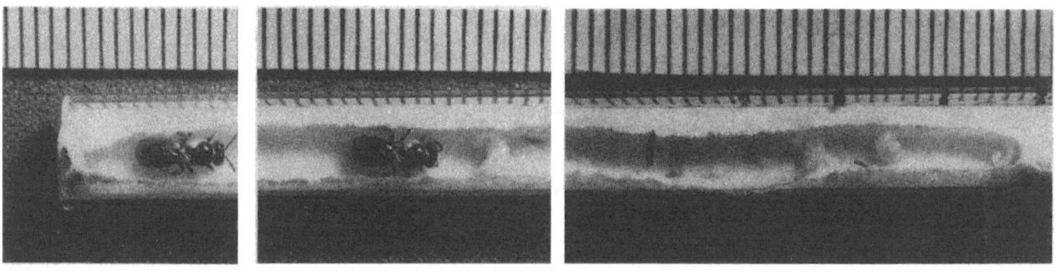

Fig. 4 - A nest of *Braunsapis sauteriella* with the entire colony of two adult bees. The one near the entrance (at left) is probably the queen. Immature stages are in the bottom of the nest (right). The scale is in millimeters. (Photograph by Y. Maeta.)

(3) After the reproductive phase, adults of both sexes inhabit the cleaned nest burrow and form a prereproductive assemblage. This sometimes includes the mother. While inhabitants of such assemblages are probably mostly siblings, bees from other nests are sometimes admitted. (I have seen *Ceratina* assemblages in North America and *Exoneura* assemblages in Australia with one or more bees of another species living with the predominant species; see also Schwarz 1982.) In temperate climates adults of both sexes usually overwinter in prereproductive assemblages (exception: *Ceratina flavipes*). In tropical climates prereproductive assemblages perhaps last for rather brief periods and adults may disperse soon after attaining mature coloration, so that the assemblages are small and perhaps often not developed at all. (Same in *Ceratina* and *Xylocopa*; in allodapines some colonies containing larvae pass the winter in warm temperate climates, or in the case of *Exoneurella tridentata*, even in the rather cool winter weather of southernmost Australia.)

(4) At least a few nests, perhaps in all species, contain colonies of two (Fig. 4) to several females during the reproductive phase. In *Exoneura bicolor* and probably *Exoneurella tridentata* a majority of nests contains colonies. (Some *Ceratina* species do not form colonies while others do so in a minority of nests; most *Xylocopa* species do so at least at certain seasons and some species may have colonies in most nests.)

(5) A colony probably usually arises when (a) two or more bees [commonly sisters or a mother and daughter(s)] in a prereproductive assemblage remain together in the old nest after the other bees disperse, or (b) daughters remain with their mother without a prereproductive break in activity. Joining a lone nest initiator (c), however, also produces occasional colonies (regularly in *Exoneura*). [Both (a) and (b) are known in *Xylocopa*, (a) is known in some *Ceratina* and (b) may occur in *Ceratina* species with more than one reproductive phase per year; (c) may occur in all groups and is known in *Xylocopa* and as a rare occurrence in *Ceratina japonica*, etc.]

(6) Adult females in some colonies may be similar in reproductive condition, e.g., inseminated with enlarged ovaries. Such females are probably similar in size and behavior, i.e., there are no castes. Such quasisocial (or eosocial) colonies are not common except in new nests of *Exoneura bicolor*, perhaps because they are temporary. (Same for some species of *Ceratina* and for *Xylocopa sulcatipes*.)

(7) In other colonies of the same species females differ in reproductive condition. One (rarely more) has enlarged ovaries and is inseminated. One or more others have slender ovaries and are commonly not inseminated. The former, a queen, is usually larger (not so in some species, see Michener 1971, p. 229), is the principal egg-layer, sometimes the principal guard, and the minimal forager (in *Braunsapis sauteriella* and *foveata*) or sometimes principal forager in *Allodape*

exoloma. The latter, workers, are usually smaller, often lay few eggs, and some of them are usually active foragers. The queen may be the same age as and probably the sister of the worker(s) - a semisocial colony - or older than and probably the mother of the workers - a eusocial colony. (Same for some *Ceratina*; probably not applicable to *Xylocopa* whose "castes" are temporal.)

(8) If the queen in a semisocial or eusocial colony dies, one of her daughters presumably becomes a replacement queen with the others being workers (semisocial) or more than one become layers (quasisocial). As the workers die and are replaced by progeny of the new queen (if any), the colony is again eusocial. (Probably not applicable to *Exoneura bicolor* or to *Ceratina* or *Xylocopa*.)

(9) In ovarian development, behavior and size, there are all intermediates between queens and workers. (Same for *Ceratina*, not applicable to *Xylocopa*.)

(10) A queen can inhibit ovarian development of workers, at least in *Exoneura bicolor*. (Probably applicable also in *Ceratina* colonies.)

(11) Workers may be reproductively disadvantaged and hence prone to join other bees, but at least some of them are capable of provisioning and laying if orphaned. (Same in *Ceratina japonica*; not applicable to *Xylocopa*.)

(12) Loss of workers leads to foraging by the former queen like any other lone individual. (Same in *Ceratina japonica*; not applicable to *Xylocopa*.)

(13) If adults in a colony with division of labor differ in age, the older one is ordinarily the queen. (Same in *Ceratina*; usually the older one is the active individual in *Xylocopa*.)

(14) There is little obvious agonism between colony members but queens (in *Braunsapis sauteriella*, sometimes eat eggs of workers and vice versa. (Little agonism in *Ceratina* and *Xylocopa*; egg eating occurs also in *Ceratina* colonies.)

(15) Food transfer by regurgitation between colony members is common. (Same for *Ceratina* and *Xylocopa*.)

(16) Pollen brought into the nest is eaten by other colony members. (Same for *Ceratina* and *Xylocopa*.)

(17) Larvae are fed progressively or in *Halterapis*, mass provisioned. (Mass provisioned in *Ceratina* and *Xylocopa*; continuing care provided by the mother in some *Ceratina* species.) Thus lone foundresses become subsocial in allodapines and some *Ceratina*.

(18) Larvae are not separated by partitions. (Larvae individually isolated in *Ceratina* and *Xylocopa*.)

(19) Statements 14 (in regard to agonism), 15, and 16 are also applicable to prereproductive assemblages. (Same for *Ceratina* and *Xylocopa*.)

Many of the statements above apply only when colonies exist. For example, the statements about allodapines under (6) and (7) above do not apply to all species; they are based especially on *Braunsapis*. Some allodapines are essentially subsocial and form colonies of two adults only rarely. Examples are *Allodapula dichroa* and *Allodape mucronata*. In the former, even in the rare colonies (3% of 150 nests) with two adults, both were inseminated, but in the latter, in each such colony (6% of 106 nests) there was one inseminated and one uninseminated bee. Other species were intermediate, i.e., variable in this respect (Michener 1971, pp. 228-231).

5 Discussion

Little is known about caste induction in XYLOCOPINAE but it may be similar to replacement queen vs. worker differentiation in HALICTINAE. In *Exoneura bicolor* and likely in many other forms, a mated female with enlarged ovaries can inhibit ovarian development in nestmates; such development occurs promptly in artificially isolated females. However, in this and most other social ALLODAPINI and CERATININI, the queens average larger in size than workers. This suggests that some sort of caste bias arises during larval growth, probably as a result of amount of food received. Perhaps it is only that larger individuals are likely to be dominant over small individuals. Age, however, is even more important than size at least in *Ceratina* colonies and probably also in allodapines. The acquisition of ability of older and/or larger bees to dominate younger and/or smaller ones must be crucial; selection must strongly favor it as shown by its evolution along with other queenly features in *Ceratina* and ALLODAPINI where for most species only a minority of the nests contain colonies.

Sex ratios have been recorded for various species (e.g., Michener 1971; Houston 1977; Schwarz 1988) and are highly variable, often female biased, whereas many halictine reproductive sex ratios are male biased. There is no noticeable correlation between sex ratio and social level attained, but in *E. bicolor* the degree of female bias is inversely related to brood size (Schwarz 1988).

Parthenogenetic reproduction is suggested by species of ALLODAPINI whose males are uncommon. It is known for wholly female populations of *Ceratina acantha* in southern California (Daly 1973) and for *Ceratina dallatorreana* both in the Palearctic Region (Daly 1966, 1983) and in California where it is adventive. Only one male is known for the latter species.

Since prereproductive assemblages are widespread in both solitary and social XYLOCOPINAE, they probably evolved before colonial behavior. Mutual tolerance, defense, food transfer, and foraging by nestmates are characteristic of assemblages and may be preadaptations for reproductive colony formation, like the unusual longevity characteristic of adult XYLOCOPINAE.

Reproductive division of labor is common when as few as two individuals coexist in a colony. In some species of all tribes, it is as though whatever two or more females join or remain together, whether sisters, mother and daughters, or perhaps unrelated bees, cooperative interactions result. The colonies can be quasisocial, eosocial, semisocial, or eusocial, all at low frequencies, in the same species, while the majority of nests are occupied by lone females. As nest occupants die, leave, or mature, the social organization switches among these categories, which in fact are not very useful for XYLOCOPINAE.

In *Xylocopa* (e.g., *X. virginica*) it is the active bee who lays eggs and does virtually all the work including the most dangerous tasks (foraging). The inactive bee helps only or principally by guarding. This finding, combined with the eventual reproductivity of the inactive bee, suggests that its behavior is more selfish and less likely to result from parental manipulation than in other bee family groups. Since nest burrows are hard to make, inheritance of the nest may be important, just as the inheritance of a territory is important for helpers in certain bird species.

In semisocial and eusocial colonies where one bee (rarely more) is the queen, especially in CERATININI and ALLODAPINI, the attributes of the castes are remarkably similar to those of HALICTINAE and other social bees. It is noteworthy that there is an association of queen-like features (reproductivity, average large size, insemination, reduced foraging, receipt of food from the worker(s), etc.) in some individuals and of worker-like features in others. Even though such colonies constitute a small minority of the population in nearly all species that have colonies, selection somehow must have associated the caste characteristics, as noted by Michener (1985). Behavioral polymorphism (lone females vs. colonies of various sorts) persists in many species of XYLOCOPINAE but has not gone to fixation so as to produce colonial populations or species except for *Exoneura bicolor*. Many species, however, remain in or have returned to a nearly fixed solitary or subsocial (lone female with larvae, etc.) condition. Selection for colonial life must therefore have been weak in XYLOCOPINAE, in contrast to HALICTINAE in which many species have life cycles that involve obligate colonial activity.

Michener (1985) has suggested that if an important factor in promoting colonial behavior in primitively social bees is defense of nests, then defense may be less important for XYLOCOPINAE than for HALICTINAE. Predators and parasites should have a relatively difficult time finding nests of XYLOCOPINAE scattered in three dimensional space and often in a maze of diverging stems. Halictine nests, with entrances on the two dimensional soil surface, should be easier to find. Therefore defense should be more critical for them, and once colonial behavior arises, it might quickly go to fixation. In contrast, in XYLOCOPINAE the advantages of colonial life should be less because of lower parasite-predator pressure, and might well be counterbalanced by disadvantages, so that obligate social life has not arisen.

The XYLOCOPINAE show that castes do not necessarily arise through a series of populations or species intermediate between solitary and eusocial. Instead caste behavior can arise as a minority variant within a population. In more general terms, socially important attributes can arise and become associated with one another and improved through selection to develop a significant new behavior while still at low frequency within a population consisting largely of individuals having the ancestral attributes. For both *Ceratina* and allodapines, although most nests contain only a single bee, selection must have operated in the minority of nests containing colonies to produce division of labor, involving castes with activities similar to those of castes in HALICTINAE and familiar highly social insects. A shift from solitary to eusocial and back again can occur in a single population, even in a single nest, with no intermediate species (Michener 1985). Such shifts happen repeatedly and, especially in temperate climate species, may be part of the normal seasonal cycle. If the new behavior becomes fixed in some populations, the result may ultimately be species characterized by it, e.g., eusociality in many HALICTINAE.

West-Eberhard (1986) has gone further and postulated that frequently a character, either behavioral or morphological, arises in this way. Thus an uncommon morph determined by conditions or genetically, may be improved by selection while protected from extinction within a population most members of which do not exhibit the morph. Once sufficiently improved, the feature may come to characterize the population or species. This process could account for some apparently rapid evolutionary events; i.e., the probably slow process of assembling favorable modifications for a character or favorable combinations of characters could go on inconspicuously before the apparently rapid evolutionary event.

6 References

Anzenberger G (1977) Ethological study of African carpenter bees of the genus *Xylocopa*. Z Tierpsychol 44:337-374

Ben Mordechai Y, Cohen R, Gerling D, Moscovitz E (1978) The biology of *Xylocopa pubescens* Spinola (Hymenoptera, Anthophoridae) in Israel. Israel J Entom 12:107-121

Camillo E, Garófalo CA (1982) On the bionomics of *Xylocopa frontalis* (Oliver) (sic) and *Xylocopa grisescens* (Lepeletier) in southern Brazil. I: Nest construction and biological cycle. Rev Brasil Biol 42:571-582

Daly HV (1966) Biological studies on *Ceratina dallatorreana*, an alien bee in California which reproduces by parthenogenesis. Ann Entom Soc Amer 59: 1138-1154
Daly HV (1973) Bees of the genus *Ceratina* in America north of Mexico. Univ Calif Publ Entom 74:1-113, pls. 1-2
Daly HV (1983) Taxonomy and ecology of Ceratinini of North Africa and the Iberian Peninsula. Syst Entom 8:29-62
Daly HV, Michener CD, Moure JS, Sakagami SF (1987) The relictual bee genus *Manuelia* and its relation to other Xylocopinae. Pan-Pac Entom 63:102-124
Gerling D (1982) Nesting biology and flower relationships of *Xylocopa sonorina* Smith in Hawaii. Pan-Pac Entom 58:336-351
Gerling D, Hefetz A (1981) The ecology of the carpenter bee, *Xylocopa sulcatipes* Maa in Israel. In: Shuval H (ed) Developments in arid zone ecology and environmental quality. Balaban ISS, Philadelphia, Pa, pp 71-74
Gerling D, Hermann HR (1978) Biology and mating behavior of *Xylocopa virginica* L. Behav Ecol Sociobiol 3:99-111
Gerling D, Hurd PD, Hefetz A (1981) In-nest behavior of the carpenter bee, *Xylocopa pubescens* Spinola. J Kansas Entom Soc 54:209-218
Gerling D, Hurd PD, Hefetz A (1983) Comparative behavioral biology of two middle east species of carpenter bees (*Xylocopa* Latreille). Smithsonian Contrib Zool 369:i-iv + 1-33
Houston TF (1977) Nesting biology of three allodapine bees in the subgenus *Exoneurella* Michener. Trans Royal Soc S Australia 101:99-113
Kislow CJ (1976) The comparative biology of two species of small carpenter bees, *Ceratina strenua* F. Smith and *C. calcarata* Robertson. iii + 221 pp., PhD Thesis, Univ of Georgia, Athens
Mason CA (1987) Division of labor and nestmate interactions in two species of allodapine bees. In: Eder J, Rembold H (eds) Chemistry and biology of social insects. Peperny, München, p 276
Maeta Y, Katayama E (1978) Life history and nesting habits of a small carpenter bee, *Ceratina megastigmata*. Bull Tohoku Natl Agric Expt Sta 58:231-260
Maeta Y, Sakagami SF, Michener CD (1985) Laboratory studies on the life cycle and nesting biology of *Braunsapis sauteriella*, a social xylocopine bee. Sociobiol 10:17-41
Maeta Y, Shiokawa M, Sakagami SF, Michener CD (1984) Field studies in Taiwan on nesting behavior of a social xylocopine bee, *Braunsapis sauteriella*. Kontyu 52:266-277
Michener CD (1965) The life cycle and social organization of bees of the genus *Exoneura* and their parasite *Inquilina*. Univ Kansas Sci Bull 46:317-358
Michener CD (1971) Biologies of African allodapine bees. Bull Amer Mus Nat Hist 145:219-302
Michener CD (1972) Direct food transferring behavior in bees. J Kansas Entom Soc 45:373-376
Michener CD (1974a) The social behavior of the bees. Harvard Univ Press, Cambridge, Massachusetts
Michener CD (1974b) Polymorphismus bei allodapinen Bienen. In: Schmidt GH (ed) Sozialpolymorphismus bei Insekten. Wissenschaftliche Verlagsgesellschaft mbH, Stuttgart, pp 246-256 (reprinted 1987)
Michener CD (1977) Discordant evolution and the classification of allodapine bees. Syst Zool 26:32-56
Michener CD (1985) From solitary to eusocial: need there be a series of intervening species? In: Hölldobler B, Lindauer M (eds) Experimental behavioral ecology and sociobiology. Fortschr Zool 31:293-305
Michener CD, Smith BH (1987) Kin recognition in primitively eusocial insects. In: Fletcher DJC, Michener CD (eds) Kin recognition in animals. John Wiley & Sons, Chichester and New York, pp 209-242
Sakagami SF (1960) Ethological peculiarities of the primitive social bees, *Allodape* Lepeltier (sic) and allied genera. Insect Soc 7:231-249

Sakagami SF, Laroca S (1971) Observations on the bionomics of some neotropical xylocopine bees, with comparative and biofaunistic notes. J Fac Sci, Hokkaido Univ 6, (Zool) 18:57-127

Sakagami SF, Maeta Y (1977) Some presumably presocial habits of Japanese *Ceratina* bees, with notes on various social types in Hymenoptera. Insect Soc 24319-343

Sakagami SF, Maeta Y (1982) Further experiments on the artificial induction of multifemale associations in the principally solitary bee genus *Ceratina*. In: Breed MD, Michener CD, Evans HE (eds) The biology of social insects. Proc Congr IUSSI, Westview Press, Boulder, Colorado, 9:171-174

Sakagami SF, Maeta Y (1985) Multifemale nests and rudimentary castes in the normally solitary bee *Ceratina japonica*. J Kansas Entom Soc 57:639-656

Sakagami SF, Maeta Y (1987a) Sociality, induced and/or natural, in the small basically solitary carpenter bees (*Ceratina*). In: Ito Y, Brown JL, Kikkawa J (eds) Animal societies: theory and facts. Japan Sci Soc Press, Tokyo, pp 1-16

Sakagami SF, Maeta Y (1987b) Multifemale nests and rudimentary castes of an "almost" solitary bee *Ceratina flavipes*, with additional observations on multifemale nests of *C. japonica*. Kontyu 55:391-409

Sakagami SF, Michener CD (1987) Tribes of Xylocopinae and origin of the Apidae. Ann Entom Soc Amer 80:439-450

Schwarz MP (1982) Life cycle and social organization in the allodapine bee *Exoneura* (*Exoneura*) *bicolor* Smith. i-viii + 1-95 pp., BS Thesis, Monash Univ, Clayton, Australia

Schwarz MP (1986) Persistent multi-female nests in an Australian allodapine bee, *Exoneura bicolor*. Insect Soc 33:258-277

Schwarz MP (1987) Intra-colony relatedness and sociality in the allodapine bee *Exoneura bicolor*. Behav Ecol Sociobiol 21:387-392

Schwarz MP (1988a) Local resource enhancement and sex ratios in a primitively social bee. Nature 331:346-348

Schwarz MP (1988b) Notes on cofounded nests in three species of social bees in the genus *Exoneura*. Victoria Naturalist 105:212-215

Schwarz MP (1988c) Intraspecific mutualism and kin association of cofoundresses in allodapine bees. Monitore Zool Italiano 22:245-254

Schwarz MP, Scholz O, Jensen G (1987) Ovarian inhibition among nestmates of *Exoneura bicolor* Smith. J Australian Entom Soc 26:355-359

Velthuis HHW (1987) The evolution of sociality: ultimate and proximate factors leading to primitive social behavior in carpenter bees. Experientia, Suppl (Behavior in Social Insects) 54:405-430

Velthuis HHW, Gerling HD (1983) At the brink of sociality: interactions between adults of the carpenter bee *Xylocopa pubescens* Spinola. Behav Ecol Sociobiol 12:209-214

Watmough RH (1974) Biology and behaviour of carpenter bees in southern Africa. J Entom Soc S Africa 37:261-281

West-Eberhard MJ (1986) Alternative adaptations, speciation, and phylogeny (a review). Proc Nat Acad Sci USA 83:1388-1392

Chapter 6

Castes and Reproduction in Bumblebees

PETER-FRANK RÖSELER[1], COR G.J. VAN HONK[2]

CONTENTS

1 Introduction

In bumblebees, the females are differentiated into two castes: queens and workers. In the rule, queens are larger than workers. Apart from this size dimorphism a main difference between the castes is the lifespan: queens are characterized by their longevity. Only young queens are able to hibernate and to start new nests, a characteristic workers lack. This caste characteristic is imposed on female larvae by their differentiation into queens and workers during their

[1] *Zoologisches Institut der Universität Würzburg, BRD*
[2] *Laboratorium voor Vergelijkende Fysiologie, Rijksuniversiteit Utrecht, Nederland*

preadult development. Another difference between the castes is the extent to which both of them contribute to the reproduction of a colony as a whole. As only queens mate, they are the only to produce female offspring. But to the male offspring workers are also able to make a contribution. The queen, however, tries to monopolize her reproduction by preventing oviposition of others, and succeeds in doing so as long as she is able to maintain her dominant position. But when the queen loses her dominance at the end of colony development, workers start egg-laying as well. They compete with their queen and, finally, they often push her off the nest and sometimes even kill her.

The worker caste, however, is not uniform. Among workers a division of labour exists, some workers are specialized as foragers, others as nest bees or guard bees. But it is a weak polyethism within the worker caste, since the relatively small colonies require a great flexibility of workers. The different tasks can be performed by nearly all workers according to demands of the colony. Moreover,

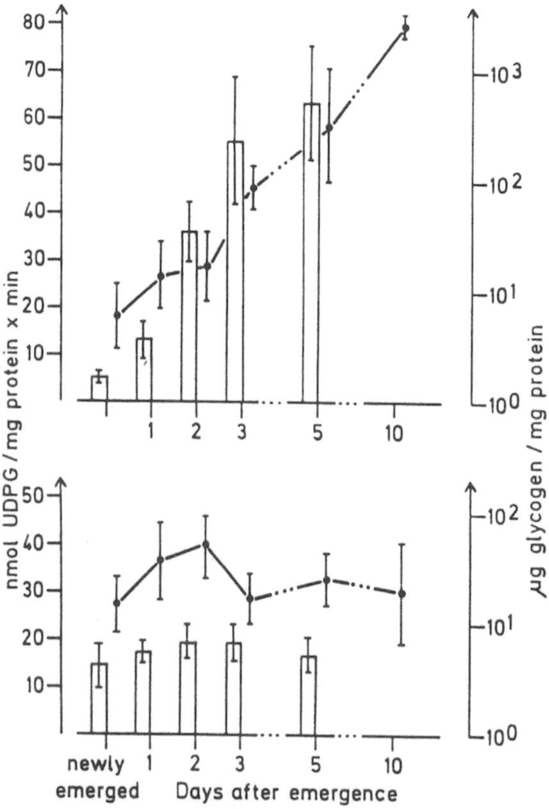

Fig. 1 - Glycogen content (lines) and activity of glucosyltransferase (columns) of fat body in queens (above) and workers (below) of *Bombus terrestris* during the first days after eclosion. Vertical bars indicate SD. (Data taken from Röseler and Röseler 1986).

among workers a dominance hierarchy is established which is partly connected to the tasks performed. Only the most dominant bees become egg-layers. These workers also display agonistic behaviour towards oneanother and towards the queen, and they compete for opportunities to lay eggs.

The sterility of workers, therefore, is only a transitory characteristic of their caste and is controlled by the presence or absence of a dominant queen as well as by dominant workers. In this review we show the contribution of both castes to reproduction, the factors influencing the reproductive success and the different strategies which queen and workers can follow to pass on their genes to the next generation.

2 Caste Differences

Differences in metabolism and behaviour between both castes already exist on the very first day after eclosion (Röseler 1975). Workers start to feed larvae and to build on the comb on the second day and sometimes they start to collect food even at that age. Young queens, however, do not participate in these tasks, and consequently they do not help to raise their sisters, except they might contribute to warming up the brood. They do not even defend the nest when it is disturbed, but try to escape under the comb. About the third-fifth day young queens leave the nest for the first time. During the following days they are attractive to drones and they are willing to mate. Shortly after copulaton they set out to seek a place suitable for hibernation.

During the first days after eclosion the metabolism of young queens is adjusted to accumulate reserves for hibernation, and the fat body becomes very voluminous (Alford 1969; Pouvreau 1976; Röseler and Röseler 1986). The differences between the metabolism of both castes can be seen in the lipid and glycogen content of the fat body as well as in the lipid content of the haemolymph. On the second day after emergence the content of the reserves in queens starts increasing rapidly (Röseler 1989). On the day ten the lipid content of fat body is tenfold higher, the glycogen content 100fold higher than in workers. The accumulation of glycogen in queens is brought about by a higher activity of glucosyl-transferase, the key enzyme of glycogen formation (Fig. 1).

3 Juvenile Hormone Control of Caste Characteristics

The caste characteristics listed above are not influenced by environmental factors such as daylength or temperature, but they are controlled by an endogenous program switched on during caste differentiation in the larval development. The synthesis of reserves in queens is correlated with a low juvenile hormone titre in haemolymph. When the titre of juvenile hormone in young queens in increased by topical application or by injection of synthetic juvenile hormone, they do not produce reserves, but they form eggs and may start egg-laying after day five. The fat body content of lipid and glycogen does not increase after day two, but they both remain on a level as in workers at the same age. Juvenile hormone treatment also changes the behaviour of young queens. They start to participate in the different tasks in the nest like a worker, they feed larvae and they collect food.

Studies on ovariolectomized queens have shown that juvenile hormone primarily induces vitellogenin synthesis in fat body (Röseler and Röseler 1988). The increased demand for energy seems to inhibit the formation of reserves. The lowered glycogenesis and lipogenesis as well as the changes in behaviour are, therefore, not directly influenced by juvenile hormone, but they are most likely consequences of the induced vitellogenesis and oogenesis, though a direct influence of juvenile hormone on lipogenesis can not be excluded.

A characteristic to young queens is that they become ready to copulate from day three on. After overwintering, when queens form eggs and start new nests, they do not accept drones. Interestingly, prior to hibernation the juvenile hormone treated queens remain ready to mate despite the enhanced oogenesis so that egg-formation seems not to influence mating. Van Honk et al. (1978) showed that queens in their mandibular glands produce sexual pheromones which are species specific, and by means of which the queens communicate their readiness to mate. Workers do not attract drones by means of pheromones and defend themselves against too highly motivated drones which occasionally mount them in captivity. Since juvenile hormone treated queens remain attractive and ready to mate, the production of the sexual pheromones and the mating behaviour cannot exclusively be controlled by the injected hormone nor by oogenesis and metabolism. The basis underlaying the ability to mate is not yet understood.

4 The Regulation of Fertility

Studies on the regulation of reproduction in bumblebees colonies were carried out in *Bombus terrestris*. The queen contributes the main part to the reproductive output of a colony by inhibiting egg-formation in workers as long as she is dominant. In the presence of a queen, oocytes in a worker grow up to a length of 0.15 mm until the fifth day after eclosion (Fig. 2). When newly emerged workers are kept together in small queenless groups, rapid oogenesis takes place and

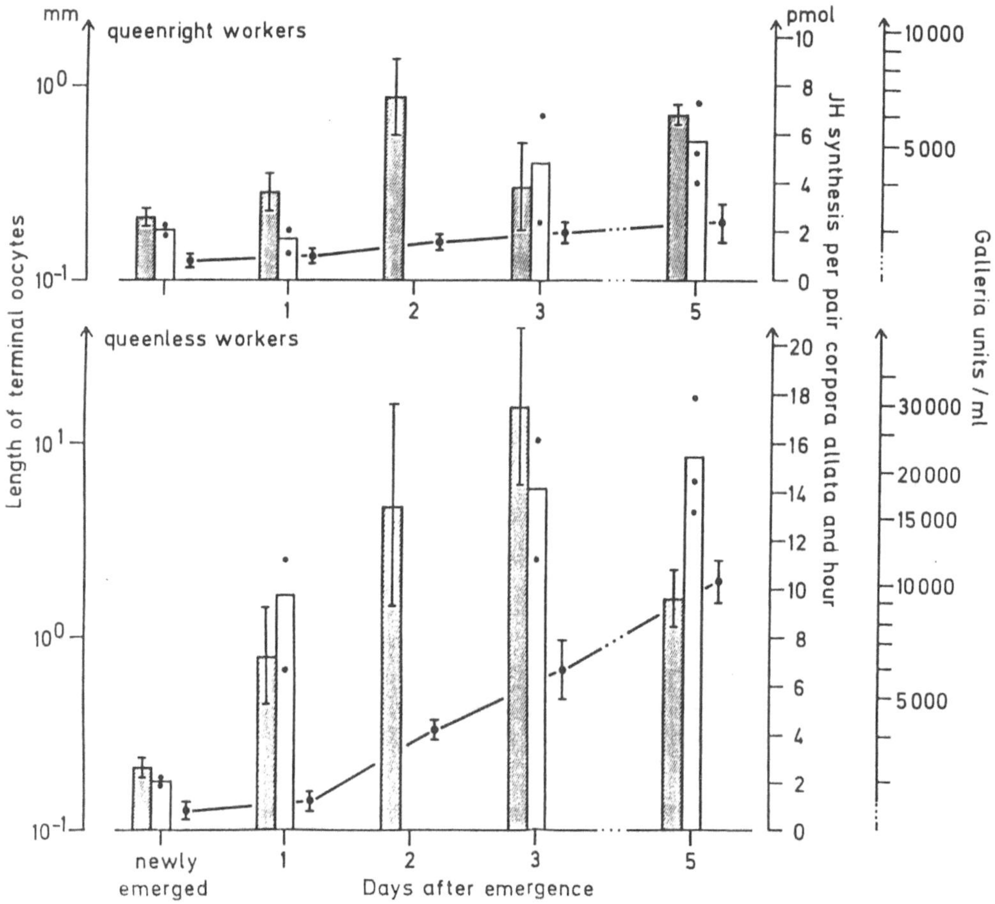

Fig. 2 - Synthesis of juvenile hormone by *corpora allata* in vitro (striped columns), juvenile hormone titer in haemolymph (white columns) and growth of terminal oocytes (lines) in queenright and queenless workers of *Bombus terrestris* during the first days after eclosion. Vertical bars indicate SD. (Data on juvenile hormone titer taken from Röseler 1977, data on juvenile hormone synthesis from Röseler and Röseler 1978).

worker ovaries can contain mature eggs (about 2.7 mm) on the fifth day, and by then the workers are ready for oviposition. Other factors inherent in the colony life, such as the presence or absence of larvae and pupae were not found to have any influence on oogenesis in workers. But even in the presence of a queen egg-formation in workers is not completely oppressed, and workers are able to lay eggs. Worker oviposition starts to occur at the climax of colony development, when the queen loses her dominance. As a rule, the oldest workers are the first to lay eggs; they are about 40 days old by then. The queen merely delays worker oviposition, she does not completely prevent it.

Several studies were performed to elucidate the way in which a queen inhibits worker reproduction. From the results we can offer the following model. The queen produces a pheromone which inhibits the activity of corpora allata in workers. Thus, juvenile hormone production is kept at a low rate and the titre in haemolymph remains relatively low so that no mature eggs are formed. Experimentally it was shown that completion of the process of oogenesis depends on a high juvenile hormone titre in haemolymph (Röseler 1977). When the titre is artificially elevated by exogenous juvenile hormone, workers form eggs even in the presence of a dominant queen. Oogenesis takes place as rapidly as in queenless workers and eggs are ready for oviposition on the fifth day. The speed of oogenesis depends on the amount of juvenile hormone injected.

The juvenile hormone titre in the haemolymph of queenright workers measured by means of the *Galleria* bioassay increases up to 5.000 GU/ml haemolymph until the fifth day (Fig. 2). In queenless workers the hormon titre is higher than in queenright workers, even on the very first day after eclosion, and it reaches a level up to 20.000 GU/ml haemolymph on the fifth day. The characteristic hormone titre in both worker groups is regulated by a difference in the synthetic activity of their *corpora allata*. This was determined by the juvenile hormone synthesis *in vitro* assay (Röseler and Röseler 1978). On the first day after eclosion the glands of queenless workers produce more juvenile hormone *in vitro* than the glands of queenright workers do (Fig. 2). The difference in the rate of juvenile hormone synthesis between the two groups until the fifth day corresponds with their different juvenile hormone titre in the haemolymph. The increased synthetic activity in queenless workers is also fairly correlated with an enlargement of their *corpora allata* volume.

The postulation that workers are influenced by the presence of the queen immediately after their eclosion seems justified as the *corpora allata* of queenless workers produce more juvenile hormone than those of queenright workers on the first day after eclosion. Newly born workers do not participate in any of the tasks in the nest on the first day, behave rather passively and do not interfere with the queen. Therefore, we postulated that the queen's signal should be a pheromone. In some investigations we have shown that this pheromone or at least an active component is produced in the queen's mandibular glands (Röseler et al. 1981). Extirpated and homogenized glands of queens inhibited the juvenile hormone synthesis in *corpora allata* of one day old workers in the same extent as a living queen does. Extracts obtained by washing freshly killed queens showed the same inhibitory effect. The pheromone seems to be spread over the body surface, poosibly during grooming. It seems not to be very persistent because its effect in *corpora allata* activity cannot be demonstrated with extracts made one day after the queen was killed. The results on the role of mandibular glands in preventing worker ovipositions were confirmed by observations in colonies headed by queens which were deprived of their mandibular glands (van Honk et al. 1980). In such colonies the workers were about 20 days old at their first ovipositions and with it clearly younger than workers in colonies headed by unmutilated queens. Pomeroy (1981), in contrast, could not detect an influence of mandibular glands on oogenesis in workers of *Bombus impatiens*; but a behavioural control by the operated queens cannot be excluded in the small groups of five workers used in the experiments.

5 Egg laying Workers

Regarding the frequent occurrence of ovipositions of old workers in *Bombus terrestris* colonies the question arises how these old workers become able to lay eggs? By autoradiographic studies it was shown that oogenesis in workers is not completely suppressed, even in the presence of a dominant queen, but that the processes related to egg-formation are going on at a low level (Röseler 1974b). The retarded oogenesis could cause only old workers to have mature eggs. On the other hand, old workers could become less sensitive to the queen's pheromone and, furthermore, in a large colony it becomes more and more difficult for a queen to distribute her pheromone adaequately. Experiments with screens showed that the influence of queens on their workers is lost at a distance of 3-5 cm (Röseler 1970).

In bumblebees, the behaviour of the dominant bee also plays an important role in safeguarding her dominant position (Free 1955; Katayama 1971; Röseler and Röseler 1977; van Honk et al. 1980; van Doorn and Hogeweg 1985; van Doorn and Heringa 1986). Queens severely attack those workers which have developed ovaries or try to lay eggs. In addition, queens eat the eggs laid by workers and destroy their egg cells; they defend their own egg batches against workers. Egg-laying workers defend their oviposition sites and their eggs in a way similar to the queen, and behave agonistically towards the queen as well as towards other egg-laying workers and try to eat the eggs laid by them.

5.1 Dominance and Egg-laying in Queenless Groups

A dominance hierarchy is also established among workers of queenless groups by means of agonistic behaviour. It results in the inhibition of egg-formation in young workers to the same extent as is found in queenright colonies (Röseler et al. 1981).

In groups of four newly emerged workers, the hierarchy among them is established on day two after confinement (Röseler and Röseler 1977; van Doorn 1988). Aggressive interactions (rapid jerks, head butting, buzzing) start between 20 hrs and 40 hrs. The level and the duration of aggression were found to be different in different groups; sometimes the level can increase to 60 interactions/hr and the duration can last up to three days. After the hierarchy is established, the level of aggression decreases. The dominant worker starts constructing the first egg-cell and ovipositing after day five. During that initial period the dominant bee is the only egg-layer. It inhibits the reproduction in subordinated workers mainly by agonistic behaviour. Whether workers are also able to produce queenlike pheromones remains speculative until now.

Several factors may be responsible for achieving the dominant position (Röseler and Röseler 1977; van Doorn 1988). The size of a worker was found to be most important. In 75%-80% of the groups one of the two largest workers became the dominant bee. However, the activity of *corpora allata* and ovaries also contribute to dominance as well as differences in age greater than five hours. Thus several factors may act synergistically. Sometimes a change in dominance occurs prior to the first oviposition, e.g. when a younger worker is clearly larger than the older, up to date dominant worker.

The relation between activity of *corpora allata* and dominance was experimentally confirmed (van Doorn 1988). Juvenile hormone treatment increased the probability for achieving the dominant position. By contrast, dominance behaviour is not necessarily connected to the presence of ovaries. In groups of four workers ovariectomized after emergence the dominance order was established at the same time as in control groups, and the workers showed the same aggressive behaviour as intact workers did. Ovariectomized workers are also able to become dominant over intact workers. Dominance in ovariectomized workers is much more significantly connected to the activity of *corpora allata* than in intact controls indicating that the deficiency of one system is compensated by a higher activity of another one.

5.2 Egg-laying Workers in Queenright Colonies

It is rather difficult to analyze the factors favouring a worker to become an egg-layer in queenright colonies which are composed inhomogenously. The size of workers is not as important for becoming an egg-layer as in queenless groups, there is no significant difference in size between egg-laying and not egg-laying workers (van Honk et al. 1981; van Doorn and Hogeweg 1985; van Doorn and Heringa 1986). The sequence in which the workers have emerged during the colony development was found to be more important. The first workers which start ovipositing emerged from the first cluster, they belong to the oldest workers of the colony. But once the oviposition of workers has started in a colony, the following workers are younger at

Fig. 3 - "Retinue behaviour" in *Bombus terrestris*. The queen is surrounded by several workers of the "elite group".

the time of their first oviposition than the first workers were. The minimum age was found to depend on the growth rate of the colony. In slowly growing colonies the minimum age was about 25 days, in fast growing colonies about 7-8 days. This rapid development to an egg-layer is similar to that in queenless groups (Röseler 1974b), but in queenright colonies it was observed only in the initial phase of worker egg-laying; later on during the period of worker egg-laying young workers are very unlikely to become egg-layers. It seems that the increasing number of dominant workers more and more inhibits egg-formation in younger workers. Old workers, moreover, have occupied the oviposition sites in the nest and defend their territories vigorously. Consequently, young workers must conquer such a territorium, and have to cope with old workers which apparently are more successful, probably because of their greater experience in agonistic interactions. But all these factors need further clarification.

Workers who are to become egg-layers later on in a colony development show behavioural differences compared to their non reproductive sisters. They seek the proximity of the queen and follow her over the comb (Fig. 3). When the queen is incubating, these workers take up a position mostly at the head and the thorax of the queen and exhibit short wing vibrations and buzzes (van Honk et al. 1971; van der Bloom 1986; van Doorn and Heringa 1986; van Doorn and

Chrambach 1988). The arrangement of several workers around the queen resembles to retinue behaviour of honey bees. But bumblebee workers do not lick the queen, they only antennate her. The youngest workers which showed that behaviour were two days old.

During the first five days there is no difference between future egg-layers and foragers in exhibiting this behaviour. Thereafter, future egg-layers more often show the retinue behaviour and buzzing than foragers. The meaning of that behaviour is not yet clear, but its nature might be agonistic. Future egg-layers are the most dominant workers and all the older buzzers have developed ovaries. This indicates that the queen's pheromone no longer affects these workers, though they are mostly exposed to it.

The eliciting of retinue behaviour is not restricted to queens, but it is connected to ovary development. Sometimes, egg-laying workers are also attended by buzzing workers.

Other behaviour patterns characteristic for egg-laying workers are constructing egg cells, opening and inspecting egg cells, attempts to destroy egg cells, and egg eating. Finally, egg-layers are aggressive towards other egg-layers as well as they are attacked by them.

Daily observations on several bumblebee colonies revealed that foragers which started continuous foraging will never become egg-layers. All dissected foragers had undeveloped ovaries independent on the period of colony development. They did, of course, not show the behaviour patterns related to egg-laying. In free flying colonies about 50-70% of the workers become foragers and, consequently, they have no chance for direct reproduction. That high proportion of foragers seems due to the short lifespan of 20 days in the mean, for in colonies foraging in flight cages foragers lived more than 40 days and only about 20% of the workers were observed as foragers.

By injection of juvenile hormone immediately after eclosion a complete oogenesis can be induced in queenright workers (Röseler 1977a; van Doorn 1987). They have ripe eggs in their ovaries between day five and seven. However, these workers do not oviposit and the oocytes degenerate. Observations on three colonies have shown that juvenile hormone treatment did not enhance the probability to become an egg-layer notwithstanding the induced egg-formation (van Doorn 1987). The workers participated in all colony tasks, however they started foraging up to two days later than untreated workers. A difference in behaviour was only observed in later stages of colony development. During the period of worker egg-laying the juvenile hormone treated workers showed more interactions with the queen and other dominant workers than untreated controls did. As long as the queen is dominant and able to inhibit worker reproduction, the effects of experimentally induced oogenesis seem to be overridden by the queen and the colony. This was also seen in queenless dominant workers introduced into queenright colonies. Only at the end of colony development these workers were able to supersede the queen and to become dominant egg-layers (van Doorn 1987).

Investigations on colonies in which all the workers or the half of them were ovariectomized immediately after eclosion revealed that dominance is not connected to the presence of ovaries (van Doorn 1987). This corresponds to the findings in queenless groups. Ovariectomized workers performed the behaviours characteristic to egg-laying workers, unless constructing and testing egg-cells. The latter behaviours seem to be only induced by mature eggs in the ovaries.

6 Worker Sub-castes Within a Dominance Hierarchy

All these observations indicate that the worker caste in bumblebees is not homogenous. Since only some workers become egg-layers, these workers are less affected by the dominance of the queen either by escaping from the inhibitory signals of the queen or by becoming less

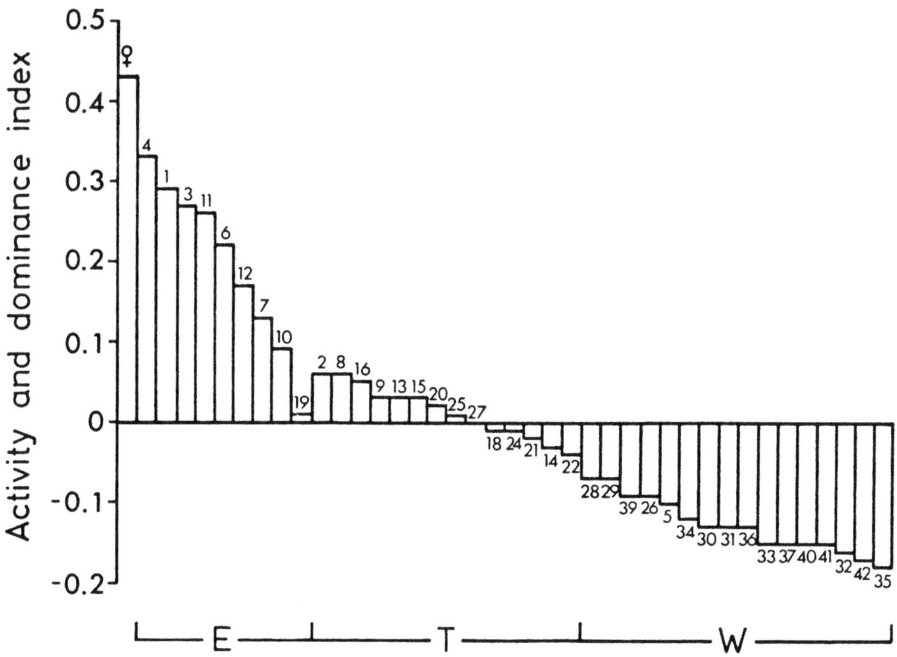

Fig. 4 - Social structure of a colony of *Bombus terrestris* between day 12 and 25 after eclosion of the first workers. The workers are ordered on the x-axis in groups (E, T, W) obtained by cluster analysis. The position of each worker is shown on the first principal coordinate (y-axis). The number of workers indicate the relative age. E = Workers of the "elite group"; T = workers of the "transition group"; W = workers of the "common worker group".

sensitive to them. In order to get further insight into the social structure of the worker caste, extensive studies were carried out (van Honk and Hogeweg 1981; van Doorn and Hogeweg 1985; van Doorn and Heringa 1986). During the total development of colonies 15.000-20.000 interactions of workers with their queen and fellow workers were gathered per colony by means of daily videorecording. The parameter for dominance used in the analyses was the "contact avoiding behaviour". When two workers encounter, the dominant will proceed after short antennation and the subordinate will retreat. The number of interactions is expressed as activity index, the number of proceedings per encounters as dominance index.

Among workers several groups with a high degree of similarity can be distinguished: The "elite group", the "transition group", and the "common worker group" (Fig. 4). At the top of the hierarchy is always the queen, she has the highest dominance index. Mostly she interacts with workers of the "elite group" and rarely with members of the "common worker group". The "elite group" contains workers with a high dominance and activity index ranking near the queen. The "elite workers" frequently interact with other "elite workers" and with the queen, but seldom with members of the "common worker group". Only "elite workers" lay eggs and can finally replace the queen. Workers of the "common worker group" have the lowest activity and dominance index. Between both groups are workers of the "transition group". The social structure of the worker caste is thought to be pyramidal (van Doorn and Heringa 1986), but this result could be due to the methods used comparing the behaviour of workers to all other members of the colony and not to another individual worker.

The hierarchical system is very dynamic. The proportion of workers of the different groups was found to be different during colony development. For instance, the proportion of "elite workers" is always less than 50% and it decreases with the increasing total number of workers. Then there is a greater competition between young workers and the "elite workers" which prevent the ascent of young workers. The "elite workers" defend their high position and inhibit oogenesis by dominant behaviour and/or by pheromones.

Workers started their lives at a position on the hierarchical ladder far from the position of the queen. They can enter the "elite group" either by an enhanced interaction frequency with the queen or by directing their behaviour towards members of the "elite group". These workers attempting to rise in the hierarchy are also characterized by a frequent retinue and buzzing behaviour.

Injection of juvenile hormone into newly emerged workers did not influence the rise in the hierarchy (van Doorn 1987). Only at the end of colony development the juvenile hormone treated workers showed a higher level of interactions wtih the queen and "elite workers" than untreated workers of the same age. Ovariectomized workers could also enter the "elite group". In the "elite$^+$ subgroup", however, they had a lower activity and dominance index compared to other workers of the same subgroup. Moreover, ovariectomized workers were more often found in the "common worker group" and less frequent in the "elite group". Whether this distribution is due to the absence of ovaries or a consequence of injuries by the surgery, is not yet clear.

An "elite worker", can again drop in the hierarchy when it becomes old or when it starts foraging. Foragers can be recruited from all groups. Most of them begin to collect food between day two and six after eclosion, the earlier the greater the need of the colony. Thus, an "elite worker" older than ten days is very unlikely to become a forager and to leave the group by switching to the task. Foraging workers have a very low dominance index and they never reach the "elite group" unless all nest bees were removed. Then they can switch the task, they can become nest bees and, finally, reach the "elite group" (van Doorn 1987).

7 Worker Polyethism

The extensive studies have clearly shown that a future egg-layer is characterized by several features early in its adult life. It is, therefore, tempting to speculate about the factors responsible for the differentiation of adult workers. Why does a certain worker start foraging, whereas its sister of the same age remains in the nest, seeks the proximity of the queen and will become an egg-layer? Is that differentiation only the result of the social environment, due to different demands of the colony and due to the competition between nestmates? Or are some workers predisposed to become egg-layers being more queenlike than others and less sensitive to the demands of the colony?

Some experiments have shown that workers maintain their flexibility to some degree in performing different tasks (van Doorn 1987). However, the higher the position of a worker in the hierarchy is and the older it is, the more the flexibility is reduced. When all the experienced foragers were removed, at first the remaining foragers reacted by intensifying their activities (Fig. 5). Then some housebees changed to foraging. Members of the "elite group", however, did not participate in this task. Even the total deprivation of honey and pollen for one day could cause only some few housebees to leave the nest and to seek food. As a rule, foragers have undeveloped ovaries and a low juvenile hormone titre in haemolymph. It was thought, therefore, that a low hormone titre could induce foraging. Injection of juvenile hormone into newly emerged workers had only an influence on the time of the first foraging, the bees started foraging up to two days later than untreated workers. This delay

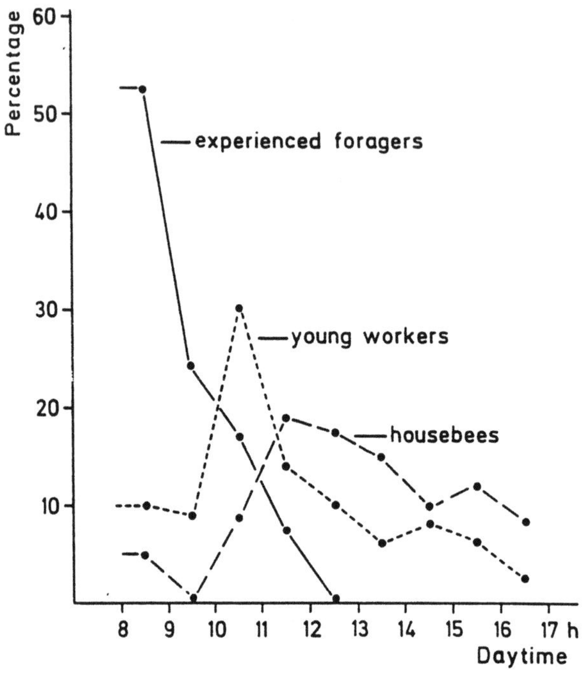

Fig. 5 - Percentage of experienced foragers, young workers and housebees of *Bombus terrestris* observed searching for food in the flight cage during the day. In the morning all the stored honey and pollen was removed from the nest. During observation time no foood was offered. (Data taken from van Doorn 1987).

could be caused by exogenous juvenile hormone which is rapidly excreted, so that the hormone titre might be again low at the time of first foraging. When, on the other hand, housebees are removed and with it all the members of the "elite group", then some foragers turned to nest tasks. They rapidly increased in dominance and could become egg-layers later on.

Summarizing the few results one can conclude that division of labour is influenced by the needs of the colony as well as by the structure of the society and by the physiological state (age, position in hierarchy, task) of the workers. Since some workers respond more sensitive to the demands of the colony than others, individual differences might also be involved in polyethism.

8 Queen Reproduction depending on Workers

In bumblebees, the castes are characterized by their difference in contribution to the reproductive output of a colony. Some attributes, e.g. the ability of queens to mate and hibernate, are adjusted during the preadult development by the differentiation into queens or

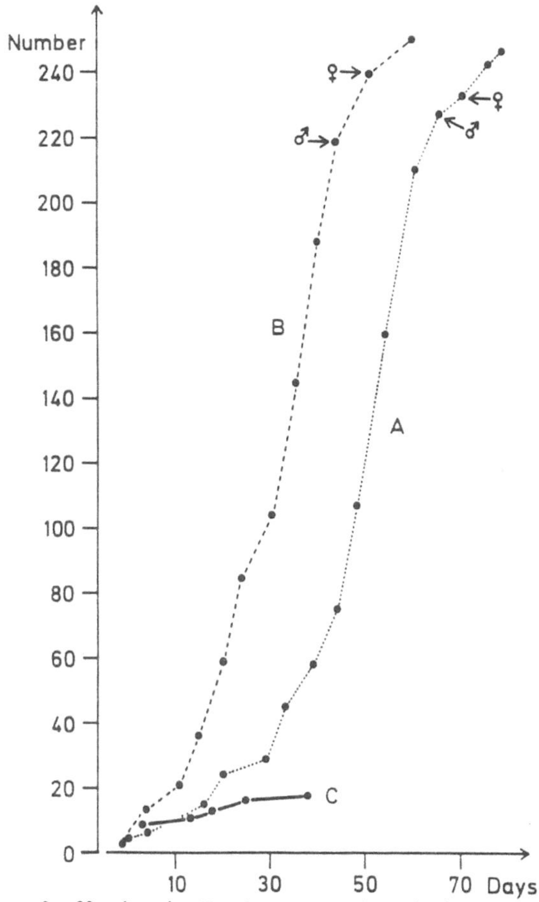

Fig. 6 - Increase in the number of offspring in *Bombus terrestris* colonies. A: The nest was started by a lone queen. B: The queen has started the nest together with three workers. C: The queen has started the nest alone, but all the workers were removed immediately after eclosion. ♂, ♀ : Emergence of the first drones resp. queens. (Data on A and C taken from Röseler 1967).

workers. This characteristic consists of a fixed physiological program which cannot be influenced by the worker caste. Other characteristics regarding reproduction are the results of the competition between the two castes. The transitory sterility of workers, for instance, is caused by the presence of a queen dominant to the workers.

The ability of the queen to lower the juvenile hormone titre in workers prevents not only worker oviposition, but presumably also induces the foraging behaviour, for in bumblebees foraging workers collected in the field were found to have a low juvenile hormone titre. Egg-laying workers very seldom forage. But whether only a low juvenile hormone titre releases or facilitates the foraging behaviour or whether it is the consequence of that energetically expensive task and to which degree the demand of the colony induces foraging is not clear. Nevertheless, by this behaviour workers support the reproduction of their queen. Whereas a solitary living queen lays the second cluster of eggs only after the first larvae have pupated, later on oviposition occurs daily, when workers are present to collect food and to feed larvae. Then the reproduction of the queens increases to an upper limit which is given by their reproductive capacity of about 15-20 eggs each day. Experimentally it was shown (Röseler 1967) that colonies grow more rapidly when the queen is helped by workers from the beginning of nest foundation (Fig. 6). Correspondingly, the reproduction is low and a queen perishes after a short time, if newly emerged workers are removed from the nest and the queen is kept in solitarity. The reproduction success of a queen, therefore, essentially depends on the presence of workers and on the cooperation of both castes.

9 Workers' Offspring

The relation between the number of workers and the number of offspring also exists in queenless groups (Röseler 1974a). But if we compare the number of eggs to the number of workers, the reproduction per worker decreases the greater the group becomes. This result might be due to the effect that the number of non-reproductives becomes larger and also due to the competition between egg-laying workers.

In general, the worker caste also contributes to the reproduction of a colony so that some drones might be workers' offspring. By this, the workers increase the variability of the gene pool of a population which could be very much reduced in case of a completely sterile worker caste. But there are conflicting data on the proportion of males produced by workers. Van Honk et al. (1981) estimated for a colony of *Bombus terrestris* that workers had produced 80% of the males. In contrast, crossing experiments between *Bombus terrestris terrestris* and *Bombus terrestris sassaricus* (Röseler, unpublished) and between *Bombus terrestris terrestris* and *Bombus terrestris xanthopus* (De Jonghe 1986) indicated by the colour differences that presumably all the males were queen's offspring as long as the queens were alive. Zucchi (1966; 1973 cited in Sakagami 1976) described that in *Bombus atratus* the workers have the parentage of 80-100% of the drones, and Owen et al. (1980) and Owen and Plowright (1982), on the other hand, observed in colonies of *Bombus melanopygus* that the queens produced 61-63% of the males. As long as the queens were present, only 19% of the males originated from workers. Katayama (1971, 1974) reported that in a colony of *Bombus ignitus* the workers laid 463 eggs in 26 days, and in a colony of *Bombus hypocrita* they laid 833 eggs in 21 days; but the percentage of survival is not known. Some of these discrepancies may be due the differences between the species studied and also between the colonies of the same species, other discrepancies may be due to the sampling error inherent in collecting exact data in the parentage of males.

The queen of *Bombus terrestris* switches to lay unfertilized eggs approximately two weeks before workers start ovipositing so that all the first males (up to 300) are exclusively produced by the queen (van Doorn and Heringa 1986). After the workers have started egg-laying only few eggs survive, since the queens and the workers mutually destroy egg cells and eat the eggs laid by others. Though the reproductive success of workers seems to be fairly low, workers, in general, contribute to the reproductive output of a colony.

10 Worker Reproduction and Male Origin

By the ability to lay eggs and to pass on their genes to the next generation the workers enter into competition with their queen and their sisters. Van Honk (1981) has analyzed the different strategies which both castes can follow. The best strategy of workers is to lay eggs and to inhibit the rearing of drones of their sisters and of the queen. This strategy is in accordance with the observations that workers compete for egg-laying and try to destroy egg-batches of the others. On the other hand, workers gain more profit by rearing new queens than the resident queen does, because of their greater relatedness to them compared to their mother. Workers, therefore, tend to rear female larvae into queens, which development in *Bombus terrestris* is inhibited by the old queen as long as she is dominant. But workers separated from their queen by a screen or kept under queenless conditions, immediately start to rear young queens (Rösler 1970). This behaviour, for instance, is used in bumbleebee rearing to obtain young queens even in the early stages of colony development. The best strategy for a queen is to produce drones of her own and to inhibit worker reproduction and queen rearing. In some captive colonies of *Bombus terrestris sassaricus* queens were able to monopolize male production completely, no workers got an opportunity to lay eggs.

Thus, a conflict exists between the interest of the queen and the interest of the worker caste (Trivers and Hare 1976). A successful control of the sex ratio by workers should result in a female biased offspring, a succesful control by the queen in a 1 : 1 ratio. But in most colonies a male biased offspring was observed, for instance 1 queen : 6 males in *Bombus melanopygus* and *Bombus terricola* (Owen et al. 1980; Owen and Plowright 1982). In *Bombus terrestris* colonies in captivity, the production of many drones occurs in all colonies, whereas the production of queens was found to be different, some colonies produced many queens, some only few queens, and others produced only drones. But also the offspring of a single queen was observed to be male biased, which ratio, in general, one should expect to ensure the mating success of young queens. Thus, bumblebees in their simplicity seem to ignore the rules stated by scientists and they are nevertheless very successful.

11 Colony Homeostasis and Output of Reproductives

Is the reproductive success of a colony guaranteed even after the queen is lost? Young colonies of *Bombus terrestris* from which we removed the queen sometimes perished rapidly. The workers noticed the absence of the queen after 20 minutes at the latest. They stopped working, did not forage or feed the larvae, and the behaved sluggishly as if paralized. They stood apart from the nest or in flight chamber. Only some, mostly younger workers, continued working after a certain time. This behaviour was also described for *Bombus pascuorum* (Meidell 1934). In larger colonies, however, the social life is rapidly stabilized by establishing a dominance order within the group of workers. In *Bombus terrestris*, a limit for the different development was found to exist at approximately 50 days after the first oviposition of the queen. After the 50th day all the colonies rapidly became stabilized. At the same time colonies contained about 40 workers, the oldest of them beeing about 25 days old. The development of a colony after removing of the queen probably depends of workers with developed ovaries who can take over the colony. This assumption is in accordance with the observation reported above that the oldest workers who started to oviposit are about 40 days old. In such queenless colonies female larvae are raised to become queens and many drones are reared. It seems that the reproductive success of colonies which have lost their queen is guaranteed only if a large investment in the reproduction was already made.

Acknowledgement - We thank Dr. A. van Doorn for permission to use a photograph for Fig. 3 and unpublished data for Fig. 4.

12 References

Alford DV (1969) Studies on the fat-body of adult bumblebees. J Apicult Res 8:37-48

Blom J van der (1986) Reproductive dominance within the colony of *Bombus terrestris* (L.). Behaviour 97:37-49

Doorn A van (1987) Investigations into the regulation of dominance behaviour and the division of labour in bumblebee colonies (*Bombus terrestris*). Netherlands J Zool 37:255-276

Doorn A van (1988) Factors influencing dominance behaviour in queenless bumblebee workers (*Bombus terrestris*). Physiol Ent: in press

Doorn A van, Chrambach A (1988) Retinue behaviour in bumblebee (*Bombus terrestris*) workers. J Apicult Res: in press

Doorn A van, Heringa J (1986) The ontogeny of a dominance hierarchy in colonies of the bumblebee *Bombus terrestris* (Hymenoptera, Apidae). Insect Soc 33:3-25

Doorn A van, Hogeweg P (1985) Die Entwicklung des agonistischen Verhaltens innerhalb der Arbeiterinnenkaste und zwischen Arbeiterinnen und der Königin während der Volksentwicklung bei der Erdhummel *Bombus terrestris*. Mitt dtsch Ges allg angew Ent 4:328-331

Free JB (1955) The behaviour of egg-laying workers of bumblebee colonies. Brit J Anim Behav 3:147-153

Honk CGJ van (1981) Queen pheromones and agonistic behaviour in *Bombus terrestris*. A view on the sociobiology of social Apidae based on intraspecific competition. Apidologie 12:96-100

Honk CGJ van, Hogeweg P (1981) The ontogeny of the social structure in a captive *Bombus terrestris* colony. Behav Ecol Sociobiol 9:111-119

Honk CGJ van, Velthuis HHW, Röseler P-F (1978) A sex pheromone from the mandibular glands in bumblebee queens. Experientia 34:838-839

Honk CGJ van, Velthuis HHW, Röseler P-F, Malotaux ME (1980) The mandibular glands of *Bombus terrestris* queens as a source of queen pheromones. Entomol Exp Appl 28:191-198

Honk CGJ van, Velthuis HHW, Hoogeveen JC (1981) Factors influencing the egg laying of worker in a captive *Bombus terrestris* colony. Behav Ecol Sociobiol 9:9-14

Jonghe R de (1986) Male parentage in a bumblebee *Bombus terrestrsi* (Linnaeus 1758) (Hymenoptera, Apoidea). Phegea 14:49-54

Katayama E (1971) Observations on the brood development in *Bombus ignitus* (Hymenoptera, Apidae). I. Egg-laying habits of queens and workers. Kontyû 39:189-203

Katayama E (1974) Egg-laying habits and brood development in *Bombus terrestris* (Hymenoptera, Apidae). I. Egg-laying habits of queens and workers. Kontyû 42:416-438

Meidell O (1934) Dagliglivet i et hoblebol. Naturen:85-95, 108-116

Owen RE, Plowright RC (1982) Worker-queen conflict and male parentage in bumble bees. Behav Ecol Sociobiol 11:91-99

Owen RE, Rodd FH, Plowright RC (1980) Sex ratios in bumble bee colonies: Complications due to orphaning? Behav Ecol Sociobiol 7:287-291

Pomeroy N (1981) Reproductive dominance interactions and colony development in bumble bees (*Bombus* Latreille; Hymenoptera, Apidae). Thesis, Univ Toronto

Pouvreau A (1976) Contribution à la biologie des bourdons. Etude des quelques paramètres écologiques et physiologiques en relation avec l'hibernation des reines. Thèse, Univ Paris

Röseler P-F (1967) Untersuchungen über das Auftreten der 3 Formen im Hummelstaat. Zool Jb Physiol 74:178-197

Röseler P-F (1970) Unterschiede in der Kastendetermination zwischen den Hummelarten *Bombus hypnorum* und *Bombus terrestris*. Z Naturforsch 25b:543-548

Röseler P-F (1974a) Größenpolymorphismus, Geschlechtsregulation und Stabilisierung der Kasten im Hummelvolk. In: Schmidt G (ed) Sozialpolymorphismus bei Insekten. Wiss Verlagsges Stuttgart, pp 298-335

Röseler P-F (1974b) Vergleichende Untersuchungen zur Oogenese bei weiselrichtigen und weisellosen Arbeiterinnen der Hummelart *Bombus terrestris* (L.). Insect Soc 21:249-274

Röseler P-F (1975) Die Kasten der sozialen Bienen. Inf Org 3:95 pp

Röseler P-F (1977) Juvenile hormone control of oogenesis in bumblebee workers, *Bombus terrestris*. J Insect Physiol 23:985-992

Röseler P-F (1989) Roles of morphogenetic hormones in caste polymorphism in bumblebees. In: Gupta AP (ed) Recent advances in comparative morphology, physiology, and development of arthropods. Vol. 3, pp (in press)

Röseler P-F, Röseler I (1977) Dominance in bumblebees. Proc Int IUSSI Congr, Wageningen, 8:232-235

Röseler P-F, Röseler I (1978) Studies on the regulation of the juvenile hormone titre in bumblebee workers, *Bombus terrestris*. J Insect Physiol 24:707-713

Röseler P-F, Röseler I (1986) Caste specific differences in fat body glycogen metabolism of the bumblebee, *Bombus terrestris*. Insect Biochem 16:501-508

Röseler PF, Röseler I (1988) Influence of juvenile hormone on fat body metabolism in ovariolectomized queens of the bumblebee, *Bombus terrestris*. Insect Biochem 18:557-563

Röseler P-F, Röseler I, Honk CGJ van (1981) Evidence for inhibition of *corpora allata* activity in workers of *Bombus terrestris* by a pheromone from the queen's mandibular glands. Experientia 37:348-351

Sakagami SF (1976) Specific differences in the bionomic characters of bumblebees. A comparative review. J Fac Sci Hokkaido Univ Ser VI, Zool 20:390-447

Trivers RL, Hare H (1976) Haplodiploidy and the evolution of the social insects. Science 191:249-263

Zucchi R (1966) Biological observations on *Bombus (Fervidobombus) atratus* Franklin. Pers comm (cf. Sakagami 1976)

Chapter 7

Caste Development, Reproductive Strategies, and Control of Fertility in Honey Bees and Stingless Bees

WOLF ENGELS[1], VERA L. IMPERATRIZ-FONSECA[2]

CONTENTS

[1] *Zoologisches Institut der Universität Tübingen, BRD*
[2] *Instituto de Biociências, Universidade de Sao Paulo, Brasil*

1 Introduction

In the highly eusocial honey bees (APINAE) and stingless bees (MELIPONINAE) marked caste syndromes have evolved. The typical queen and worker characteristics are different from one another not just with regard to the reproductive organs but include many other morphological, physiological, and behavioral differences as well. These striking caste distinctions are the result of postembryonic divergences in development which depend on ecological conditions, mainly on the external factor of larval nutrition. Caste development in social insects in general belongs to the widespread phenomena of insect polymorphism and, in accordance with environmental control, castes in bees were called ecomorphs (de Wilde and Wirtz 1974; de Wilde 1975). The trophogenic basis for caste induction is provided by the nursing workers, although in a colony brood rearing including gyne production is strongly influenced by the queen. The queen also rigidly controls all reproduction of the workers. On the other hand, her own reproductive output depends completely on the helper functions of the workers, her daughters.

Of the eusocial HYMENOPTERA, only in honey bees and stingless bees is the queen incapable of founding a colony alone and of rearing the brood on her own. Numerous individuals of the worker caste participate obligatoryly in both swarming and in the building of brood combs as well as in the subsequent nursing and attending activities. Theoretically, such a pronounced division of labor between reproductive and productive morphs could have resulted in a genetically determined infertility of the worker and an enhanced fecundity of the queen in relation to species-specific demands of the colony. But it is long known, at least for honey bees, that under queenless conditions a worker can also become fertile, although merely to a limited extent. Surprisingly, in stingless bees queenright workers were found to lay eggs (Sakagami et al. 1963; Sakagami and Zucchi 1963). During the 25 years following this important observation, many aspects of worker and queen fertility have been studied in both honey bees and stingless bees (Engels 1985, 1987a, 1988).

The modes of queen production, the manifold intracolonial relations between the two female castes, mating biology and the control of queen and worker fertility are the topics to be treated in this chapter. The mechanisms, in so far as they are known and understood, will be discussed under the aspect of reproductive strategies. The stingless bees will be compared with the honey bees in attempting an interpretation of the socioevolution of these two groups of strictly colonial bees. Since a greater diversity has evolved in the numerous species of stingless bees than in the four or five recent species of the genus *Apis*, such a comparison must concentrate on some of the better known meliponids.

2 Gyne Production: Colony Investment for Survival

In the highly eusocial bees, worker brood is produced more or less constantly. The only exceptions are during hibernation and during swarming in honey bees and during seasonal breaks caused by drought, heavy rainfalls or low temperatures in stingless bees. The life expectancy of a worker bee is related to such cyclic or occasional changes in colonial activities. For instance, overwintering honey bee workers may live 9 months whereas a summer bee after intensive foraging may die after only one month, or even less, of adult life. Under favorable conditions, brood is reared throughout the year and old workers are replaced daily by emerging bees.

In contrast to this adaptive formation of numerous workers, gyne production is a limited investment of the bee colony. In honey bees, queen rearing is seasonal and closely linked to swarming. Normally, a number of large queen cells is built at the same time. The tendency to construct swarm cells is a genetically fixed trait. Colonies of the Sicilian bee, *A. m. sicula*, and African races such as *intermissa* have been recorded as having often quite more than 100 queen cells at the same time (Ruttner 1988). Colonies of selected European strains normally build only a few royal cells. However, the heather bee of Northern Germany, *Apis mellifera mellifica*, which is kept in skeps in traditional apiaries where swarm beekeeping is still practized, sometimes has 30-50 queen cups when the prime swarm leaves. Besides this, emergency queens can be reared in case of accidental loss or of replacement of an old queen. This is done by changing the nutritional program in the progressive feeding of some young female larvae and by enlarging their comb cells to royal size. Since such a dietary switch is entirely effective only up untill the end of the third larval instar (von Rhein 1933; Weiss 1978), emergency queen rearing is impossible for a colony if eggs or young worker brood are lacking. Even if this is not the case, in *Apis cerana* emergency queens are rarely produced. Subsequently, laying workers are developed in the queenless colony (Velthuis, Ruttner, Crewe, chapter 8) which then can rear merely drone brood. The only exception known is *Apis mellifera capensis*. Laying workers of this South African race are capable of producing worker and queen offspring by automictic diploidization of unfertilized eggs (Verma and Ruttner 1983).

In stingless bees, queens are produced throughout the year. The number depends on the species and on the condition of the colony. Strong colonies with plenty of pollen and honey storage breed many queens at the same time, whereas weak colonies do so only on a small scale. A sudden change in weather with subsequent honey flow often induces the construction of royal cells. The rate of gyne production is influenced by the physiological state of the old physogastric queen and the reproductive cycle of the colony. In some species for instance an increased number of royal cells is found after supersedure or swarming. The mere building of enlarged queen cells does not necessarily lead to gyne production. In *Paratrigona subnuda* and in *Plebeia droryana*, the old queen sometimes failed to oviposit in already provisioned royal cells. Worker eggs were laid in them instead, probably resulting in giant males (see 3.2). Such an inconsistancy in the egg laying activity of the physogastric queen can likewise include the normal brood cells into which then workers oviposit, as was recorded for *Plebeia droryana*. This leads to an enhanced drone production of the colony (see 5).

In contrast to honey bees, the gynes of stingless bees are always reared within the normal brood cycles. Because of the mass provisioning system in larval nutrition, there is no possibility of influencing caste differentiation in a female larva once a brood cell has been sealed after oviposition. Therefore, in meliponids, queen replacement after loss of a queen depends either on the presence of sealed royal cells in the brood still under development or on already emerged young virgins. Normally quite a number of gynes will be available.

As the perennial colonies of the highly eusocial APINAE and MELIPONINAE can only survive with an egg laying queen, the strategies of gyne production are of vital significance. Colony investment into queen rearing appears reciprocal in the two groups: many young queens produced at certain times in honey bees, and single gynes continually emerging at short intervals in stingless bees. The question remains as to which of these strategies better ensures the availability of a new queen in all situations of colonial need. At first sight, the system of continuous queen production seems to minimize the risk of an irreversible queen loss. But of course the modes of caste formation as well as the behavior of the virgin and the workers also strongly influence all queen replacement.

Fig. 1 - Types of brood cell arrangement in stingless bees (MELIPONINAE). In contrast to honey bees, all stingless bees build special large storage pots for honey and pollen which are used repeatedly. **a** *Hypotrigona*-type: Cluster of single spherical cells, no involucrum between brood nest and storage area. **b** *Trigona*-type: Horizontal combs of cylindrical cells with enlarged marginal queen cells. The brood nest is separated from the storage part by several irregular layers of involucrum. **c** *Melipona*-type: Horizontal combs without special royal cells, otherwise like type **b**. - In the three types, the brood nest is built from bottom to top, and after emergence of the imaginal bees, the empty brood cells are dismantled. In types **a** and **b**, many brood cells are under construction at the same time and a series is ovipositioned in sequence. In type **c**, only a single cell is added to the brood nest, completed, and ovipositioned. The construction of the next single cell does not begin before the closure of the preceding one.

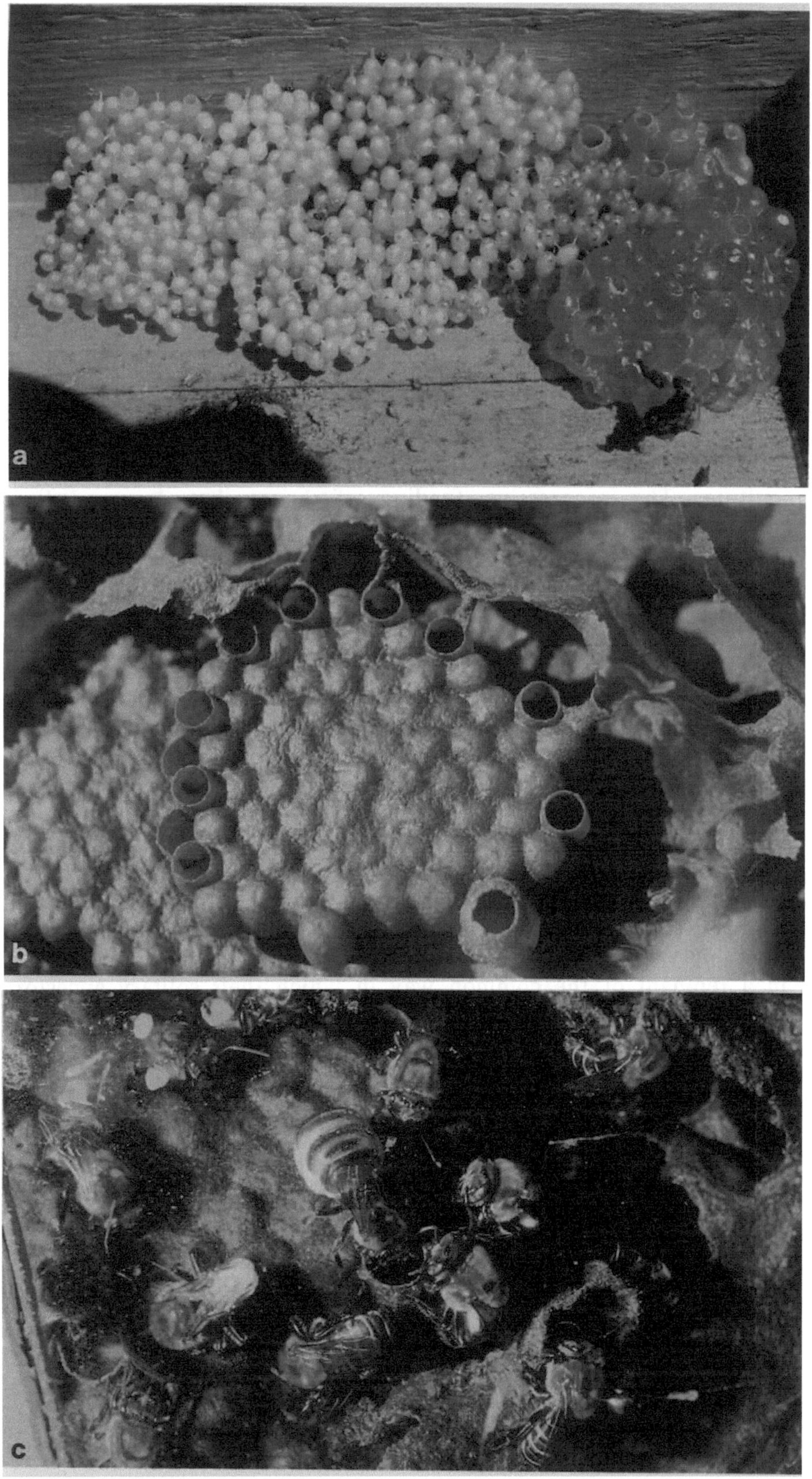
a
b
c

3 Caste Formation: Modes and Colony Requirements

The worker caste in the highly eusocial bees is distinguished by an extremely limited reproductive capacity. In bumble bees a worker sometimes can replace a queen without severe consequences for the further development of the colony (Röseler, van Honk, chapter 6). In honey bees and stingless bees, only the queen is capable of producing worker and queen progeny. Therefore, special modes of caste formation evolved.

For a long time, the queen caste was considered to be an innovation resulting from advanced socioevolution. Consequently, the great costs which the bee colony apparently invested into feeding the prospective queen larva, attained much scientific interest. This was enhanced by the significance of artificial queen breeding techniques for modern apiculture. The progressive feeding of honey bee larvae led to several presumptions one of which was the idea of a special royal diet containing a queen "determinator". This idea dominated the research concepts on caste development for more than 100 years. After the pioneer studies of Leuckart (1855), who was the first to point on the qualitative differences between royal jelly and worker food and to the queenlike differentiation of the gonad anlage in royally fed larvae, many investigations were designed to detect the chemical principle which induced queen formation (Planta 1888; Perez 1889; Butenandt 1955; Rembold and Hanser 1964; Rembold 1976). Only recently has the concept of a "nutritional balance" (Hartfelder, chapter 9) replaced this unsuccessful search. Instead of a special queen diet, the evolution of a specific nutritional program for mass rearing of workers (de Wilde 1976) has come to be regarded as the economical solution which meets the requirements of populous colonies (Wittmann and Engels 1987). Perhaps the modes of gyne production now practiced by the higly eusocial honey bees and stingless bees is nothing other than a remnant of the common way in which their solitary ancestors properly nourished their female larvae.

Nevertheless, queen formation in eusocial bee colonies is a fascinating phenomenon. The requisite worker behavior as well as the queen control of this behavior are a part of colonial homeostasis.

3.1 Nourishment of Prospective Queen Larvae

In the primitively eusocial halictines, well nourished female larvae become queens (Bohrer 1987). The same mechanism is present in the more original bumble bees. Both in HALICTINAE (Michener, chapter 4) and BOMBINAE (Röseler, van Honk, chapter 6) many caste differences appear as mere allometrically modified size differences. Larval nutrition is still of definitive importance in the highly eusocial bees. In contrast to the uniform comb structure in all recent species of honey bees, a considerably variety of brood cell arrangements and storage pots (Fig. 1) occurs in stingless bees.

Three modes of gyne production are known in the MELIPONINAE and APINAE:

1. The two-food-portion modus.
2. The large-food-portion modus.
3. The equal-food-portion modus.

Modes 1 and 2 have in common that ingestion of more larval food induces queen differentiation with the result that the emerging gyne is larger than a worker. This ecomorphic caste formation is not evident in modus 3, because apparently all female larvae receive the same amount of provisions. Caste development is probably always triggered by an endogeneous constellation (Hartfelder, chapter 9).

The consecutive consumption of two portions of provisioned larval food was observed by Terada (1974) in *Leurotrigona muelleri*, a minute Brazilian trigonine bee which constructs irregular brood cell clusters without surrounding involucrum (Fig. 2a). It can be observed mainly in periods of rapid colony growth and increased brood rearing that the pillars between the spherical cells are short or even lacking. Then the last instar larva, after consuming its own food, bites through the walls to the cell above. Since this brood cell had been built later, a young larva or even an egg will be present there. The old larva moves in, eats another food portion and becomes a giant. It is unknown whether an additional moult produces a supernumerary instar as in overfed caterpillars (Nijhout 1975b; Safranek and Williams 1984). The workers repair the two perforated brood cells thereby often forming a large one. The giant larva pupates into a queen weighing much more at emergence than a worker. Similar observations were made in *Frieseomelitta varia*. This mode of queen formation ensures that gynes are usually produced by strong colonies.

Delage-Darchen and Darchen (1970, 1977, 1982) were able to obtain adult queens from worker larvae which had been provided with additional food after having consumed the originally provisioned amount. In these experiments which have been critically discussed (Velthuis 1976a; de Wilde and Beetsma 1982), several species of African stingless bees of the tribe TRIGONINI were used. *Leurotrigona* is a neotropical genus belonging also to this tribe. Camargo et al. (1976) were unable to get comparable results when working with Brazilian species of MELIPONINI. Therefore, it remains uncertain as to whether gynes can be produced in only a few, in many, or in all stingless bee species by the two food portion modus.

Prerequisite for the second modus based on one large-food portion is the construction of special royal cells with an increased volume. Within the stingless bees, such queen cells occur in several species of the more primitive *Hypotrigona* group which build clustered brood nests, in all the numerous species of the *Trigona* group which have monolayered horizontal brood combs (Fig. 2b) and in the peculiar African species, *Dactylurina staudingeri*, the only meliponid which constructs doublelayered vertical brood combs built from top downwards similar to the honey bees (Darchen and Pain 1966; Darchen 1973). So within the tribe TRIGONINI, enlarged queen cells which often are formed at the margin of the brood combs (Fig. 2b) are the most frequent mode of producing gynes. This was already recognized by some of the first students of neotropical stingless bees, for instance by Müller (1875). Because of the mass provisioning system of larval nutrition, the future queen has a large portion of food in the sealed brood cell right from the beginning. By providing a female larva, grafted from a normal sized brood cell into an artificial queen cell, with a larger amount of food, queens could be produced in *Scaptotrigona postica* (Camargo 1972a). This shows the decisive effect of a large food portion on queen development.

In stingless bees, the larval provision is a mixture of glandular secretions, honey and pollen. These food components are already segregated partially within the vertical brood cell during the three days of embryonic development. Therefore, the freshly hatched larva feeds the lightest compounds at first while swimming on top and then, continuing to eat the stratified layers of food, comes at last to the heavy pollen on the bottom of the cell. This points to a dietary sequence which in a large queen cell results in an altered composition of the portions subsequently consumed by the different instars. Possible effects of such feeding programs on caste development have not been studied sufficiently. This is perhaps one of the special aspects of the large-food-portion modus when compared with the two-portion-modus. It resembles in certain respects the sequential larval nutrition which occurs in APINAE.

Fig. 2 - Modes of queen cells and royal nourishment in the highly eusocial bees. **a** The two food portions modus is realized in a number of cell cluster building species of the *Hypotrigona* group of the TRIGONINI, i.e. in *Leurotrigona muelleri*. One larva consumes in sequence the provisons of two closely touching brood cells. Later, the workers surround the giant pupa with a big new cell. - The large food portion modus occurs in most of the TRIGONINI and in the APINAE. **b** In most trigonine stingless bees, such as *Scaptotrigona postica*, marginal and enlarged queen cells are provisioned with 3-4 times the normal amount of larval food. **c** In *Apis mellifera* special royal cells are built at the lower border of a brood comb or, in case of emergency queen cups, brood cells containing young worker larvae are enlarged. Queen larvae are progressively fed with surplus amounts of royal jelly. - **d** The equal food portion modus is realized only in the MELIPONINI. In *Melipona quadrifasciata* queens develop in about every fourth brood cell all of which are equally sized, in between workers and drones.

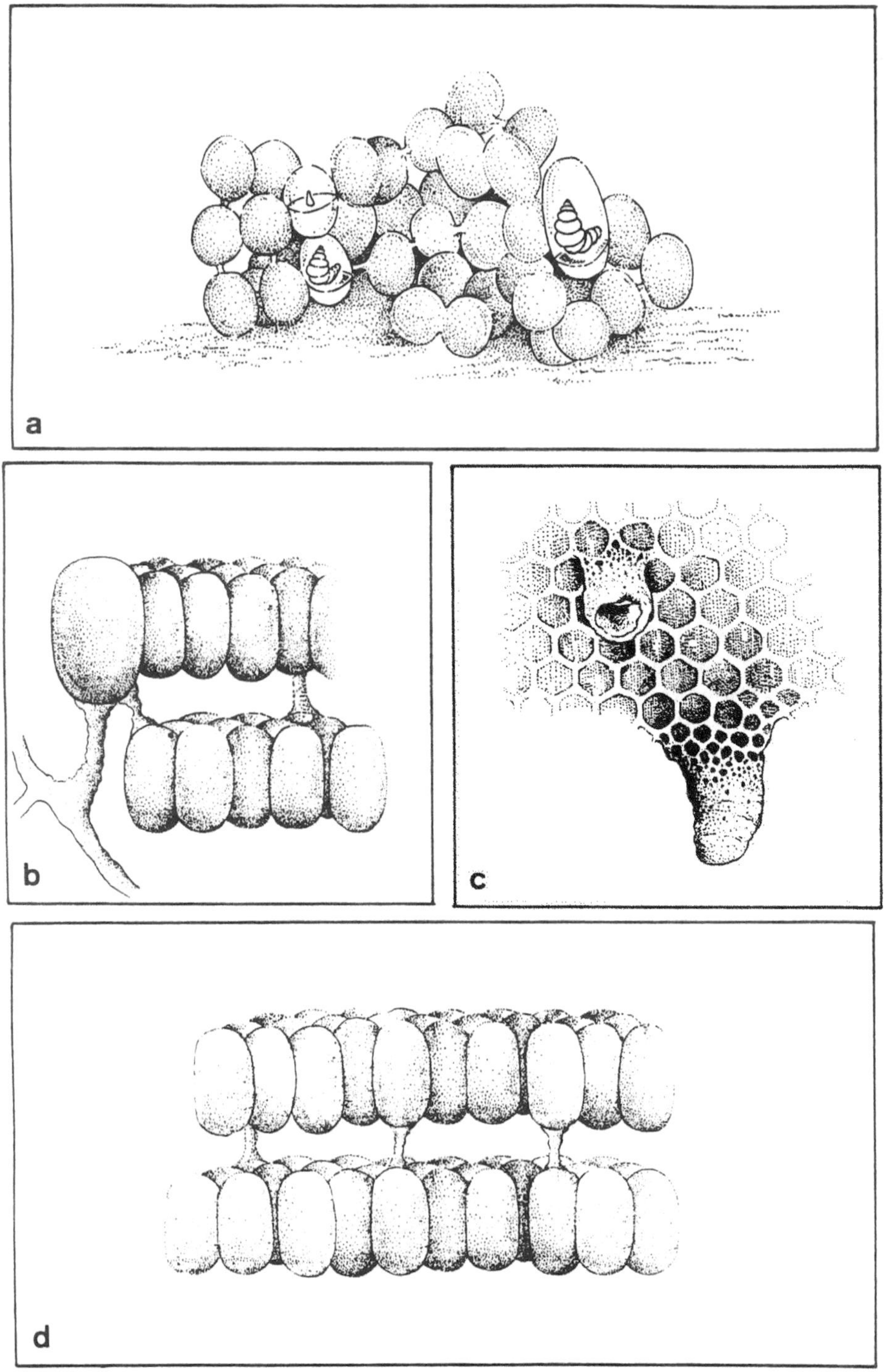
a
b
c
d

The honey bee mode of gyne formation is similar because enlarged royal cells are also built (Fig. 2c). The progressive feeding of the larvae, however, enables the workers not only to provide a future queen with more food, but also to supply a special dietary program.

3.2 Dwarf Queens, Intercastes, Giant Workers and Giant Males: Where on the Polymorphic Scale?

The quality of the royal jelly does in fact influence queen induction in honey bees (Rembold and Lackner 1981; Wittmann and Engels 1987). In *Apis mellifera carnica*, a queen larva consumes about 300 mg of food, and another 100 mg or more are left in the cell. A worker larva eats all the 175 mg of worker jelly supplied by the nurses. Under experimental conditions, but rarely in nature, dwarf queens appear. They are probably nourished with insufficient amounts of royal jelly.

A similar example can be observed in a few trigonines such as *Schwarziana quadripunctata*, a species representing an underived type of stingless bees (Zucchi 1977), where the worker-sized queens emerge from normal comb cells (Imperatriz-Fonseca 1973; Camargo 1974b). Dwarf queens have been occasionally found in a number of meliponid species, for instance in *Plebeia julianii* (Juliani 1962), *P. emerina* (Kleinert-Giovannini 1989b), *Cephalotrigona femurata* (Nogueira-Neto 1951) and *Plebeia remota* (Imperatriz-Fonseca et al. 1975).

The large-food-portion modus of queen production as observed in all the honey bee species and many stingless bees resembles the situation in the eusocial bumble bees which also feed their larvae progressively (Röseler, van Honk, chapter 6). In the BOMBINAE, however, the polymorphic scale of workers and queens is not as distinct as in APINAE and MELIPONINAE which have clearly separated castes. Rarely within colonies, but often under experimental conditions, intercastes are observed in both honey bees (Taber and Poole 1973; Beetsma 1979; Rembold and Lackner 1981; Wittmann and Engels 1987) and stingless bees (Velthuis 1976a). Intercaste individuals were obtained in the trigonine, *Scaptotrigona postica*, by artificial rearing on intermediate amounts of larval food (Camargo 1972b). Queens were produced by providing a larva removed from a worker cell with excess food as in a large royal cell (Camargo 1972a). In this species, a queen cell is provisioned with 3-4 times the quantity as is given into a normal brood cell.

Probably as an accidental event, but perhaps as a regulatory mechanism preventing excessive numbers of gynes (Imperatriz-Fonseca 1975), giant males can emerge from up to 30% of the royal cells (Nogueira-Neto 1951; Juliani 1967; Imperatriz-Fonseca 1970, 1975; Silva 1973; Imperatriz-Fonseca et al. 1975; Camillo 1977; Cortopassi-Laurino 1978; Bego and Camargo 1984). Giant drones have a longer preimaginal development (Juliani 1967), an increased number of functional spermatozoa (Camillo 1971; Bego and Camargo 1984) and genitalia which are mostly similar in shape and size to those of normal drones (Imperatriz-Fonseca 1970; Bego and Camargo 1984), but sometimes larger (Cortopassi-Laurino 1978), indicating that they are probably unfunctional. Giant males have been recorded in *Scaptotrigona postica, Nannotrigona testaceicornis, Paratrigona subnuda, Plebeia remota, P. droryana, P. emerina, P. julianii, Schwarziana quadripunctata, Geotrigona mombuca* and *Friesella schrottkyii*.

Surprisingly, giant workers were also observed in rare cases emerging from royal cells (Oliveira et al. 1973; Imperatriz-Fonseca 1975, 1976). One giant worker had worker-like ovaries, another was performing typical worker tasks during two months of observation. Therefore, the large-food-portion modus of queen induction is not without exceptions. Before not more morphogenetic data (Bego 1977; Bego and Camargo 1984) are available, no conclusive interpretation seems possible concerning either the intercaste or the exceptionally-sized individuals. Their occurrence demonstrates the existance of a broad range of possible polymorphic development in the eusocial bees. Only by social regulations do most of the colony members usually become ordinary workers, queens or drones.

3.3 Epigenetic and Genetic Control of Caste Formation

The third mode of queen formation in which no difference in the amount of larval food exists between future gynes and workers (and drones as well), is restricted to stingless bees of the tribe MELIPONINI. Queens emerge from normal-sized brood cells which are scattered all over the horizontal comb (Fig. 2d). Apparently, all the larvae are exposed to the same external conditions, suggesting an internal determination of caste formation. Von Ihering (1903a) already speculated that differences between queens and workers could result from a developmental predisposition operative from embryogenesis. Kerr (1946, 1950) counted a maximum of 25% queens within batches of hatching females in colonies of some *Melipona* species. He concluded that a two-allel constellation could be responsible for a genetic caste determination. Under this assumption, the double heterozygeous individuals become queens and all the others become workers. This appears as a 1:3 segregation.

It was later observed that environmental conditions evidently influenced queen formation in *Melipona* as well. Unfavourable circumstances resulted in less than 25% gynes. This was interpreted as an indication that part of the genetic queens could become workers (Kerr 1973, 1974). Slight differences in quantity and/or quality of larval food were thought to have such epigenetic effects. Efforts to increase the percentage of queens by manipulation of the larval food always lead to a maximum rate. A species-specific threshold was observed in 4 *Melipona* species (Camargo et al. 1976). A presumed influence of the honey bee queen substance, 9-oxo-E-2-decenoic acid, on caste induction could not be proved in *Melipona quadrifasciata* (Engel 1979).

Kerr's genetic hypothesis of caste determination in the MELIPONINI should of course be evaluated by means of appropriate cross breeding experiments (Velthuis 1976a). Since all female stingless bees can be induced to differentiate into queens by topical application of juvenile hormone to late larval instars (Velthuis and Velthuis-Kluppell 1975; Campos et al. 1975, 1983), the offspring of adult queens, which are workers according to their genetic constitution, should be analyzed. Queen/worker-ratios other than 1:3 should be expected depending on the allelic situation. Controlled mating can be realized in *Melipona* bees (Camargo 1972c). Unfortunately, the attempts to realize such experiments failed because of the high mortality of artificially raised and laboratory mated queens formed after juvenile hormone treatment (Campos 1977; Bonetti 1982). The few progeny data obtained did not allow for any statistical comparisons. Therefore, the problem of a genetic caste determination in stingless bees remains a hypothesis that still demands a definitive evaluation.

3.4 Colonial Regulation of Gyne Production

Queenlike characters of worker have often been described in stingless bees. The position of the ventral ganglia in the abdomen (Cruz-Landim and Höfling 1972; Kerr 1973) and also tergal gland structures (Cruz-Landim et al. 1980a) have been considered to indicate genetic *Melipona* queens which, under unfavorable epigenetic conditions, differentiated into adult workers. Such an interpretation implicates the assumption that morphogenesis of the structures in question is more directly under genetic control than is the overall caste differentiation. Another explanation could be related to target-specific actions of juvenile hormone as observed in larvae of non-social insects (Nijhout 1975a).

In any mode of caste formation as described for the eusocial bees, worker activities play an essential role in realizing gyne production. The building of special royal brood cells and the feeding of larvae with royal portions or with special royal jelly, all depend on colonial activities. Whether exercized by the workers regularly or at special times, these tasks reflect specific requirements of the colony. Thus, the social regulation of rearing female reproductives is evident in honey bees and in stingless bees, although the biological significance of up to 25% queens as sometimes found in *Melipona* colonies is hard to understand and appears to be a mere overproduction. Similar observations were made in laboratory colonies of *Plebeia julianii* (Juliani 1962, 1967) and *Schwarziana quadripunctata* (Imperatriz-Fonseca 1973; Camargo 1974b; Zucchi 1977). This occurrence of high percentages of small queens emerging from normal sized comb cells in some trigonine species which normally rear gynes in enlarged and marginal royal cells, is perhaps not only an interesting case of two modes of queen production within the same nest. Furthermore, this indeed exceptional intracolonial double-tracked gyne formation can be used to design new types of experiments on nutritional induction *versus* genetic determination of castes.

4 Virgin Queens: Behavior and Fates

Only in honey bees which regularly warm the royal brood, are queen cells containing pupae intensely attended by the workers. This behavior is evoked by a pheromone released by queen pupae (Koeniger and Veith 1983, 1984). Stingless bees do not excercise this function of attending the queen cells, however, if a piece of brood comb with a sealed royal cell is given to a queenless group of trigonine workers, then some bees will always inspect the queen cell. Evidently, they at least recognize the presence of a queen pupa.

4.1 Young Queens in the Monogynous Honey Bee Colony

Young honey bees stay in the cell for about a day after eclosion before they emerge. During this time the *Apis* gyne communicates with the old queen through vibrating sounds known to beekeepers as "quaking" of the royal cell which is answered by the "piping" of the mother queen or the first already emerged virgin. This dialogue stimulates the old queen to leave with a prime swarm which normally occurs within a day or two. If she is hindered in doing so, the old queen will search for the swarm cells and sting the young queens or even the pupae within them, in order to refrain her monogynous status. The worker bees often try to keep the old queen away from royal cells. In some cases, for example in hives with a queen excluder posted in front of the entrance in order to prevent successful swarming, almost all young queens are killed by the old one before or shortly after emergence. Localization of royal cells by the adult queen apparently results from pheromonal cues as well.

The newly emerged honey bee queen intends either to become head of the mother colony or to leave it with a swarm. Depending on the species and race, a few or many gynes will be born more or less at the same time. It is known only from Mediterranean bees such as *Apis mellifera syriaca, lamarckii, intermissa*, and *sicula* that the virgins stay together peacefully in their mother hive for some time (Ruttner 1988), thus forming a temporary polygynous colony during the mating period. In the other races, a young gyne tries soon after emergence to dominate the colony by killing off all other queens. In case of prime swarms, the old queen has already left before the first virgin appears. Severe fights between the young queens can be observed. If not hindered by the workers, they usually struggle and sting until only one survives. The honey bee workers never eliminate a virgin of their own nest (Tab. 1). Since the monogynous status of the colony is achieved several days before the remaining virgin is mated and ovipositioning beginning later on, there is a considerable risk of loosing the young queen in between, the chance being estimated at 20% (Ruttner 1988).

One way out of this hopeless situation for a colony, namely a thelytokous queen production by egg-laying workers through automictic diploidization (Verma and Ruttner 1983), is realized only in the Cape bee, *A. m. capensis*, a race with many queenlike peculiar features of the workers (Velthuis, Ruttner, Crewe, chapter 8). It is hard to understand why a temporary polygyny either during the swarm season or, in the case of supersedure, until one young mated queen begins to lay eggs, is not generally practiced by honey bees but rather restricted to a few Mediterranean races as mentioned. According to Ruttner (1988), this efficacious requeening system is perhaps a relatively recent aquisition with hitherto only regional distribution. It would be of particular interest to investigate the behavioral and pheromonal interactions of the coexisting young queens and their individual relationship to the workers (Tab. 1). Such a situation resembles a multi-queen afterswarm which has either already left from the mother colony with several virgin queens, or has formed through the joining of a number of small swarms having left the colony within a short time. For a few days, an afterswarm may remain polygynous, but at the latest after settling or during the period of initial comb construction, the virgins struggle with one another until only one is left who can undertake nuptial flights. If she is lost during mating, the swarm is also queenless. Again the question arises why the polygyny is not maintained a little longer to ensure the availability of a mated, egg laying queen.

Despite of the overproduction of gynes in honey bees, no reproductive strategy exists which safely protects the colony from becoming queenless. Queen competition which aims to restore monogyny as soon as possible after emergence of several young queens, seems inappropriate. Several ways of preventing this insecure colonial status of being left without an egg-laying queen without principally altering the fate of the many surplus gynes are known, but not very widespread among the some 25 subspecies of *Apis mellifera*.

Unfortunately, our knowledge about virgin queen behavior in the three Asian species of honey bees is limited. Therefore, a comparative discussion is not yet possible. In *Apis florea*, a number of small afterswarms each leaving with a virgin at intervals of a few days was recorded (Tirgari 1971; Whitcombe 1984).

On the other hand, in queenless colonies egg-laying workers develop (Velthuis, Ruttner, Crewe, chapter 8). This requires some weeks in *Apis mellifera* but only few days in *Apis cerana*. Their contribution to the reproductive performance of a bee population is also only poorly understood at this time.

4.2 Young Queens in the Multigynous Stingless Bee Nest

In stingless bees, young queen behavior is very peculiar. As mentioned before, gynes emerge from time to time, stay for a while, but normally are sooner or later eliminated by the workers (Tab. 1). Therefore, stingless bee colonies can be called multigynous, even though the occurrence of more than one physogastric old queen is restricted to a few genuine polygynous species (Nogueira-Neto 1970) like *Melipona bicolor* (Fig. 5). According to the size of the colony, here up to 5 physogastric queens can be observed, so this is better called an oligogynous status (see 4.5).

As emergency queen rearing is impossible in the system of larval mass provisioning found in all stingless bees, gynes have to be produced on stock. This requires colonial mechanisms of gyne tolerance and elimination as well. There are typical worker/gyne, gyne/gyne and gyne/queen interactions. In a number of species which have been studied with any intensity, some detailed observations concerning such interactions were made revealing tribe and group specific features.

In the MELIPONINI sometimes great numbers of gynes emerge from normal comb cells. During the first days the workers mostly take no notice of them. It was seen in colonies of *Melipona quadrifasciata* that only from the third day on the workers paid attention to the gynes. Most of them are nevertheless soon killed off, but, depending on colonial conditions, some may survive for a time, apparently the more attractive individuals. Such virgins may be tolerated or even courted by the workers.

In some species attractive gynes usually retreat to a certain area away from the central nest. In *M. marginata* several virgins were observed staying together in the exoinvolucrum or in batumen indentations and exhibiting a behavior of inter-virgin dominance (Kleinert-Giovannini and Imperatriz-Fonseca 1988). This is perhaps

Tab. 1 - Differences in worker reactions on queen pheromones in the highly eusocial colonies of MELIPONINAE and APINAE.

Queen / worker dominance	Pheromone-released intranidal caste relations in: Stingless bees	Honey bees
Releaser effects on behavior	Attraction and some aggression	Attraction only
Primer effects on fertility	Stimulation	Inhibition

due to individually varying pheromone production by the young queens depending on age or size. In addition, other refuges used by gynes were also seen in the storage area of the nest where the virgins stayed in empty or in partly dismantled pollen pots, as observed in *M. rufiventris* and *M. bicolor*. Workers often chase young queens which try to hide, but the virgins sometimes fight and defend themselves. The gynes repulse worker aggressions by swinging the abdomen rigorously towards the head of bees behind them (Zucchi 1977).

Little is known about the intermediate period in case of supersedure and replacement of the physogastric queen by a young one. A temporary multigynous status was described by Silva et al. (1972) for a colony of *M. quadrifasciata*, in which the old queen disappeared. During the next 8 days, 10 virgins were observed, 9 of which were eliminated, for the most part soon after emergence. After experimental dequeening, in other colonies 5, 10 and 12 virgins emerged. One of them was accepted after 3, 4 and 8 days, respectively. In a dequeened nest of *M. quinquefasciata*, the multigynous phase even lasted 19 days, and 31 virgins were recorded during this period. Many were killed on the day of emergence, but others survived for some time, the older gynes being courted by the workers.

In the TRIGONINI the number of gynes emerging from royal cells is relatively small as compared with the MELIPONINI. Rarely more than one young queen per day appears, and normally new gynes show up only

Fig. 3 - Virgin queen câmara in *Plebeia remota*. **a** View into a nest box with the brood combs in the left. The câmara or gyne chamber is built of clear cerumen and located in the periphery of the storage area. **b** Close-up of this chamber which is connected with numerous walkways like the brood nest. A virgin stays within the completely closed "prison" which is much larger than the surrounding pots. **c** Through a big opening in the same chamber the yellowish abdomen of the gyne is seen, visited by a few workers. (Photograph by Mauro Ramalho).

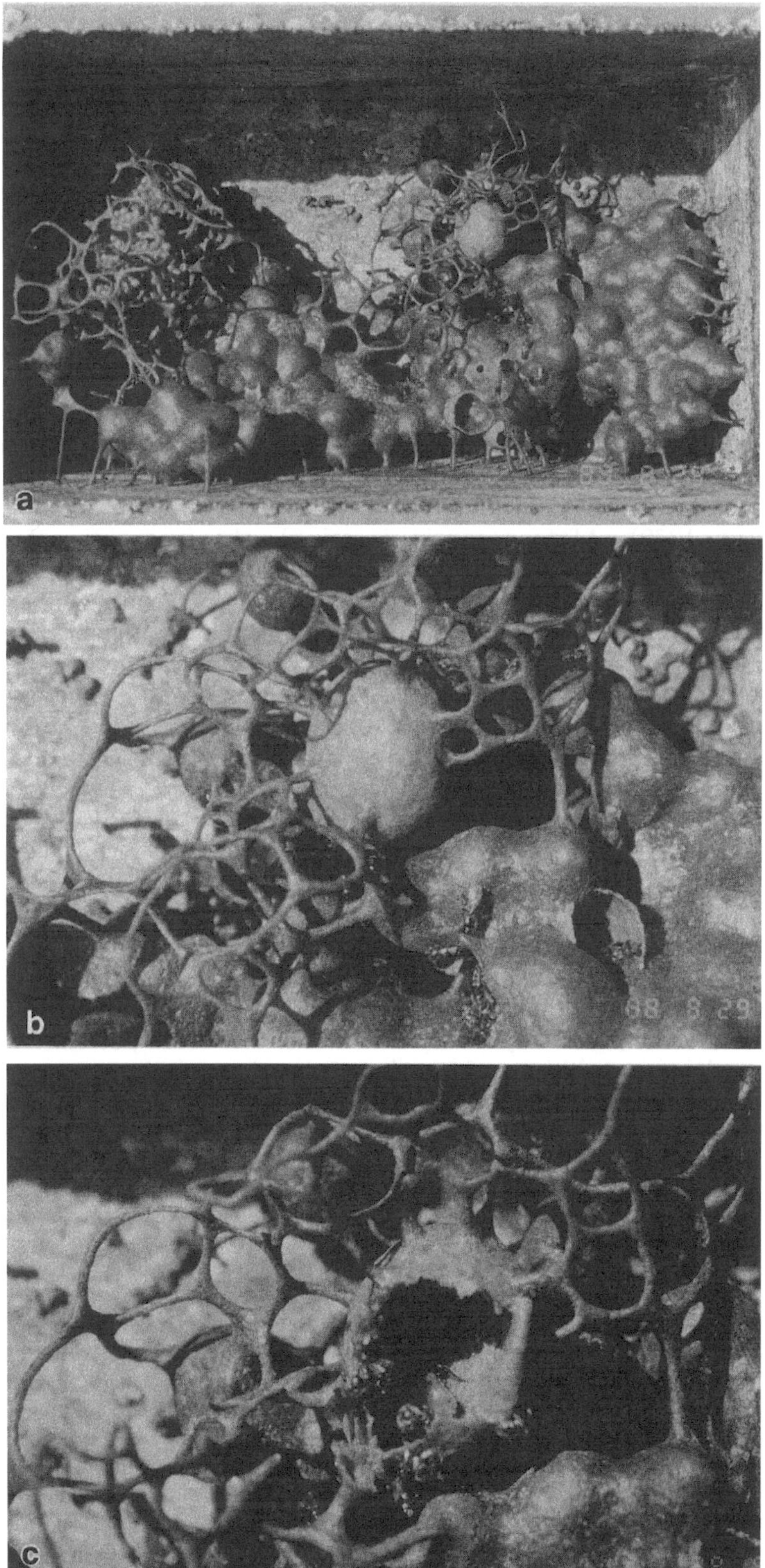

with some days in between. There are some exceptions, for instance prior to swarming quite a number of gynes can coexist within the same colony. They may be attractive or not for the workers soon after emergence, and several corresponding patterns of behavior can be distinguished. These patterns depend mainly on the beginning of gyne attractiveness and are observed in different species. A common trait is based on the ontogenetic sexual maturation of a gyne, but several routes of gradually increasing contacts can be distinguished.

In *Scaptotrigona postica* newly emerged gynes are unattractive and first join the kindergarten assemblies of young bees in the peripheral involucrum of the brood nest. After some days they hide in dark corners or in empty pots. Strong colonies usually contain a number of virgin queens. We once counted 13 gynes in a large laboratory nest. Most of them are executed by old workers (Simoes and Bego 1979). After a virgin is seized, she is rapidly plastered with sticky cerumen or propolis. She has no chance to free herself and dies after several hours. Workers do not, however, vulnerate a moriture virgin. If removed and cleaned, such a young queen continues to live without any damage.

In several other trigonine species, the gyne becomes attractive for the workers a week or more after emergence probably depending on pheromone gland development. Only then the workers begin to chase the virgin who usually retreats into an empty storage pot. The subsequent behavior was observed in *Paratrigona subnuda* (Imperatriz-Fonseca 1977) colonies containing occasionally several gynes. After becoming attractive they searched for a safe place to stay (Fig. 4). Then a gyne was sometimes working rapidly like a worker and emptied a partially filled honey pot if no other one was available. She used cerumen to close the upper opening but let small holes through which she could put out her antennae. A virgin may stay there for a while or go out, accompanied by workers or alone. The same pot can be used by several gynes within a short time because a virgin always makes periodical walks throughout the nest. During such excursions workers are attracted and trophallaxis can be seen. The gyne may return to

Fig. 4 - Scheme of virgin queen behavior in *Paratrigona subnuda*, a trigonine stingless bee. **a** Honey and pollen pots in the storage area of the nest. **b** An empty pot, sometimes with fringed or perforated top, used as a refuge for a virgin. **c** Gyne hiding in a pot. **d** Of a hiding virgin, only the antennae may be seen stretched through perforations of the closed pot. **e** Workers attending an attractive virgin in a pot. **f** The same refuge may be used by several gynes in sequence, sometimes within one day. **g** A gyne inside the pot antennates. **h** Workers forming a retinue outside. **i** An opened and empty pot indicates that the virgin became active and moves within the nest. After completion of maturation, she will try to replace the old queen or to leave with a swarm. **j** In such a phase of the colony cycle the physogastric queen may be courted by workers outside the brood nest. **k** A young queen initiates supersedure of the physogastric queen by dropping liquid cephalic secretions onto the old queen's body. This is the signal for the workers to determine which queen is more attractive and will then head the colony.

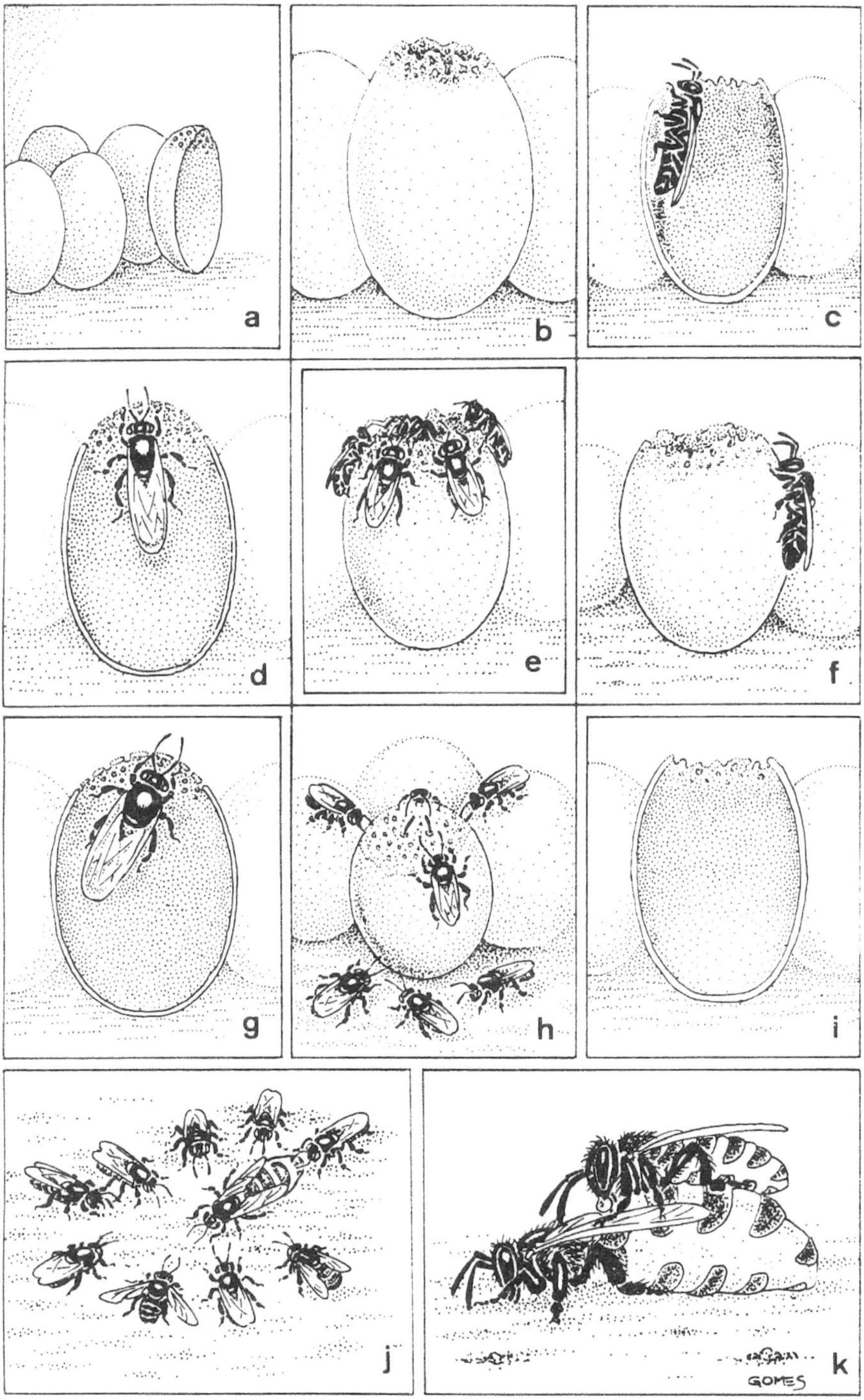
a
b
c
d
e
f
g
h
i
j
k
GOMES

the previously occupied pot or look for another one. Apparently only after sexual maturation is completed, the virgin will try a supersedure or leave with a swarm to a filial nest. There is no distinct intranidal territory for each of the gynes, but individual courts are formed always by the same workers. A gyne is only exceptionally killed by the workers before she finished her maturation phase. The young queen is eliminated if her attempts to replace the physogastric queen failed. The pot may be considered a temporal refuge, and staying there may delay such a decision.

Quite more worker/gyne conflicts seem to exist in case of virgins attractive for the workers right from emergence. This has been observed in the Asian trigonine *Tetragonula laeviceps* (Inoue et al. 1984a) and in quite a number of neotropical species. In *Plebeia* colonies the emergence of a gyne evokes strong worker excitement, and a typical smell can be recognized also by the observer. Probably according to the pheromonal performance of a virgin, the workers may decide to accept her and to discard the physogastric queen. In *Plebeia remota* once a supersedure was recorded immediately after a gyne emerged who then needed 19 days for maturation before she went on her nuptial flight. Apparently the gyne/queen interactions normally induce a spatial separation of old and young queens, effected by "emprisoning" of the virgin in a corner of the nest. This was discovered by Nogueira-Neto (1958) and confirmed by several Brazilian researchers (Moure et al. 1958; Juliani 1962; Imperatriz-Fonseca 1975). Particular and often rather large cerumen pots called câmaras or chambers (Fig. 3) are constructed for this purpose (known for instance from *Plebeia remota, P. emerina, P. droryana, P. julianii, P. schrottkyii, Tetragonisca angustula, Frieseomelitta varia, F. silvestrii, F. schrottkyii* and *Schwarziana quadripunctata*). The virgin queen is confined by the workers in a corner, and the chamber is built around her. She spends a period of days or weeks in such a câmara which apparently is a safe place. If lacking, câmaras are usually constructed when a gyne appears in the colony (Cortopassi-Laurino 1978). The gyne sometimes opens and closes again a hole in the wall of the chamber. Such a worker-like activity is never exercised by the physogastric queen. Workers may enter the chamber and feed the gyne. The virgin queen câmaras have been interpreted as prisons by the first investigators (Nogueira-Neto 1958; Moure et al. 1958; Juliani 1962) who recognized that any destruction evoked excitation and sometimes even execution of the gyne. In fact the chambers appear, however, to be rather voluntarily approached refuges. One function of such a separé is probably to avoid the diffusion of a virgin-specific pheromone within the nest, thus protecting the gyne during her temporary confinement. The virgin in some species is, at least at times, accompanied by several workers which may form a court and attend her within a câmara (Fig. 3c).

Sometimes a gyne pushes at these workers through pendular movements of her body (Imperatriz-Fonseca et al. 1975). It depends on colonial conditions how many gynes are allowed to stay for a shorter or longer period within chambers.

There are other records that a virgin was allowed by the workers to move freely, especially in medium sized colonies of *Frieseomelitta silvestrii*, which thereafter mostly replaced the old queen (Ribeiro 1989a). But normally the young queen spends most of her time within the câmara. An exception seems to be *Plebeia droryana*. At the end of the maturation phase the workers open the chamber and the attractive virgin has free interindividual contacts (Cortopassi-Laurino 1978).

Another special case is *Schwarziana quadripunctata* because of the varying gyne size. Most of the newly emerged virgins are killed. Only big gynes were seen in câmaras. But once, during the swarming cycle of a colony, also 7 small virgins were kept in chambers next door to one another. It was only after some hours that the workers began to eliminate them all, except the last staying for a few days. It seems that the individual attractiveness is correlated with the size of a virgin.

On the other hand, the elimination of surplus gynes apparently depends on the dominance status of the old physogastric queen. But in all species where virgin chambers are found, worker/gyne interactions are very important. At first the worker control is apparent. After completion of sexual maturation, there is a switch, and gyne control of worker behavior can be observed.

A perhaps less complicated and derived worker/gyne relationship was described by Terada (1974) in *Leurotrigona muelleri* and *Celetrigona longicornis*. Gynes do not hide, there are no chambers built, and the attractive virgins walk freely within all parts of the nest, accompanied by a court of workers protecting the young queen against aggression by the other workers.

The increasing attractiveness of a gyne for the workers is correlated with the secretional activities of the abdominal tergal glands as studied by Cruz-Landim et al. (1980b) and Mota (1982, 1988). The excited young queen during excursions through the nest may inflate her abdomen so as to expose the tergite pockets, obviously thereby releasing a pheromone. Massage of the abdomen and wing vibrations are also seen as common behavior of attractive virgins.

Such attractive virgins are frequently fed and this effects an increase in body weight (Engels and Engels 1977). If intensive feeding by the workers is interrupted for some minutes, the young queen rises her forelegs, opens her mandibels and offers a drop of a yellowish fluid. This droplet is quickly taken by some workers under great excitement. Immediately thereafter these and other workers begin to supply the virgin with abundant food. This observation was made in *Melipona quinquefasciata* (Silva et al. 1972), but was also reported from a number of other species, for instance *Frieseomelitta doederleini, Friesella schrottkyi, Paratrigona subnuda* (Imperatriz-Fonseca 1977). Evidently a soliciting pheromone is involved. Since this widespread behavior is exerted only by non egg-laying queens, it can be regarded as part of the various prereproductive queen/worker relationships typical found in stingless bees.

This complicated intranidal virgin queen behavior is similar in MELIPONINI and TRIGONINI, though in the latter the occurrence of câmaras and ritualized virgin/worker interactions indicates a more evolved level of young queen sociality.

Under such aspects, the queen overproduction in stingless bees appears as a limited investment of the colony. Excess virgins are allowed to survive only for a short period. Young queens are mostly eliminated by the workers before mating, while they are unable to dominate a colony. Such a queen breeding strategy, on the other hand, can minimize the risk of an irreversible queen loss which has proved to be fatal in queenless honey bee colonies without young worker brood. Ultimately, the fate of a gyne depends on her ability to attain the reproductive function either by joining a swarm which leaves to an incipient filial nest or by supersedure of the old physogastric queen, her mother. Chemocommunication (Tab. 1) seems to play a key role in the intervening conflicts during the repeated multigynous cycles in stingless bee colonies.

4.3 Queen Supersedure

Under natural conditions, a virgin honey bee queen will only undertake nuptial flights if she has become head of a colony. Such a situation may be the result of colony division in which either the old mother queen left with a prime swarm, or in which one or more virgins started together with a secondary swarm. A third case is queen replacement or supersedure. In honey bees, a reduced production of queen pheromone or the laying of unfertilized eggs into worker brood cells due to insufficient sperm content of the spermatheca leads to requeening of the colony. Normally, the old queen is killed by the workers which afterwards rear new queens. The virgins emerge when their mother has disappeared and, therefore, are not actively involved in replacing her.

Because of the multigynous situation of the stingless bee colony, much more interference between young and the old physogastric queen takes place than in honey bees. In observation hives in Brazil it was noted that an egg laying *Melipona* queen may be replaced about every half year (Silva et al. 1972). In other species like *Schwarziana quadripunctata*, *Paratrigona subnuda* and *Plebeia remota* marked queens stayed for much longer periods. In Germany, however, our laboratory colonies of *Scaptotrigona postica* were headed by the same queen even for nearly three years. Requeening after experimental removal of the old queen as well as active supersedure by a young queen, have been studied in several species.

Of the various compartments within the meliponid nest, the top of the brood nest is the main territory of the physogastric queen. In between two oviposition processes, she sometimes remains there or, as in most cases, rests on the involucrum nearby. Temporary formation of a retinue around the old queen indicates her attractiveness. Only in a few species, as in *Paratrigona lineata*, is the queen surrounded by a court of workers almost all the time (Zucchi 1977). In general, a "fixation" of the queen (Sakagami et al. 1965, 1977; Silva et al. 1972) occurs after she has arrived on the comb in the initial phase of an oviposition process (Sakagami and Zucchi 1963). The physogastric queen is encircled usually by 5-10 nurse-aged workers which in *Scaptotrigona postica* repeatedly approach the queen with ritualized aggression. The queen pheromones are dispersed during such interactions by stroking the hind legs over the abdominal terga and by repeated short wing beats which fray out the tips (Engels et al. 1979). Only in the oligogynous *Melipona bicolor* the queen does not vibrate her wings in order to distribute volatile signals, so the tips are not turn out as in many other species.

In *Melipona favosa*, it was observed that the court surrounding the queen is formed by those workers who at the same time are engaged in building brood cells (Sommeijer and de Bruijn 1983). In some species like *Schwarziana quadripunctata* court formation is also observed when, after an ovipositioning cycle, the queen has moved onto the involucrum and grooms herself.

Young queens already attractive for the workers and intending to supersede, are chased when crossing the comb. In course of their sexual maturation they become increasingly interested in the events surrounding ovipositioning. They try to eat from the provisions in unsealed larval cells and sometimes consume a worker-laid alimentary egg.

In *Plebeia remota* this indicates that a virgin will be accepted by the colony. At the same time the workers produce an increased number of alimentary eggs. Apparently, a competition can arise between a young queen and the old physogastric queen. Several observations point to a decisive role of chemocommunication in critical situations, that is circumstances under which the regular sequence of cell building, provisioning, and ovipositioning may be seriously disturbed (Silva et al. 1972; Imperatriz-Fonseca et al. 1975; Bego 1977). In *Paratrigona subnuda* the mandibular gland reaches its maximum development in young queens attempting a supersedure. If the old queen is capable of maintaining her pheromonal dominance, then the young has no chance. If not, the workers vote against the physogastric queen. In *Scaptotrigona postica*, it was seen how the workers shoved her from the combs, attacked her and covered her body with sticky cerumen (Simoes 1974; Simoes and Bego 1979). The latter behavior is essentially the same as that observed in elimination of surplus virgins.

In *Paratrigona subnuda* (Imperatriz-Fonseca 1978), the young queen initiates the process of supersedure by dropping mandibular gland secretions onto the physogastric queen (Fig. 4k) who does not react to this. If still dominant enough, she will continue to perform all normal tasks in the oviposition process, and the intended replacement will fail. However, the secretion elicits intensive licking by the workers of both the old and the young queen. In case that more than one virgin is present, the workers lick the most attractive gyne more and eliminate the others. It should be emphasized that in contrast to in honey bees, in stingless bees the workers normally do not lick the queen. They always remain at a short distance to her (Sakagami 1971). In a bioassay, licking behavior could be released by synthetic volatiles identified from worker head extracts in *Scaptotrigona postica*; 15 of 36 tested compounds evoked licking, especially 2-Undecanol (Engels et al. 1987a; Engels 1987b), which also occurs in cephalic secretions of young *Sc. postica* queens (Francke et al. 1987). Licking in stingless bees always indicates excitement, and probably serves to distribute pheromones of low volatility.

In *P. subnuda*, the superseduring young queen moves in a agitated state throughout the nest, thereby meeting the physogastric queen more and more frequently. It is apparently that during this phase the final decision is made about acceptance or rejection of the new queen. If accepted, the new queen reduces her visits to the old queen. At the same time the workers begin to attack the physogastric queen, usually after licking her body and verifying that she is less attractive than the young queen. Aggressions may begin when she is not on the comb because she already ceased egg laying. She may even avoid that part of the brood nest where cell construction is going on and where she would later oviposit under normal circumstances. She sometimes hides under the lowest comb or in the storage area of the nest. Apparently, she loses her former territory and her former subordinates begin to pay attention to the new queen. When the old queen after having been attacked again tries to dwell on the brood comb under construction, the workers kill her as observed in *Melipona marginata* (Kleinert-Giovannini 1989).

In some species, as in *Plebeia droryana*, the old queen when trying to approach the comb is not only attacked by the workers but by the young queen as well. Both queens may fight for a while, but the workers finally kill the old mother of the colony (Silva 1972). Only rarely a temporary coexistance of two egg laying queens can be observed in monogynous species. In *Paratrigona subnuda* the slaughtering of the physogastric queen begins with an excision of the abdominal tergal glands (Imperatriz-Fonseca 1978). Thereafter, the body is cut into pieces which are then carried to the garbage site in the storage region of the nest. Apparently after removal of the abdominal pheromone glands, the old queen looses her domineering immunity and can be further dismembered by the workers without hesitation.

In the oligogynous species *Melipona bicolor* an elimination of a queen not laying eggs any more was observed in colonies headed by 2-4 physogastric queens (Bego 1989). There seems to be a queen control according to the population of the nest. If there is a decrease then old queens are removed one after the other until a monogynous status is achieved.

Queen supersedure has often been observed while the colony was swarming (Imperatriz-Fonseca 1978). This coincidence probably results from the simultaneous presence of numerous virgins in a populous nest culminating in colony reproduction by fission. In such a situation the severe competition between the queens frequently ends up with the replacement of the old physogastric queen. Worker/queen chemocommunication (Tab. 1) is evidently involved in making the decision which gyne is accepted or refused and then eliminated by the maternal or a filial colony.

4.4 Nuptial Flight and Mating

In the honey bee, the virgin queen starts her mating flights either from the maternal nest or from a founding swarm. The first situation is given when the prime swarm has left with the old queen or after queen replacement. In the second case, a secondary swarm has settled only recently and the first combs are still under construction. In artificial queen breeding, such a situation is mimicked by establishing a mating nucleus, in fact a mini-colony.

In any case, the virgin queen, about 5 days old and reinforced by the workers, makes some short exploration excursions and thereafter prolonged and often repeated nuptial flights. This behavioral pattern was not influenced significantly by repeated topical application of 2 - 25 μg of juvenile hormone III (Regel 1978). On the other hand, allatectomy of newly emerged virgins prevented timely mating flights which by some queens were made up with a delay of two or three weeks with in between ovipositions of unfertilized eggs (Daerr 1978). Nuptial flights normally lead to distant mating places called drone congregation areas (Ruttner 1956, 1985). Only there does the high flying queen release the drone-attracting lure, pheromonal secretions of her mandibular (Butler et al. 1962; Gary 1962) and abdominal tergal (Renner and Vierling 1977) glands. Pursuing drones always attempt to approach the queen from downwind. Besides this, her body shape is an important optical cue (Gerig 1971) for locating her among thousands of males. Honey bee drones have large ommatidia in the upper frontal part of the compound eye for optimal visual fixation of targets against the sky like a queen flying in front and a little bit higher (van Praagh et al. 1980; Ribi et al. 1989). They obviously release a male aggregation odor while following the spur of the female sex pheromone in the air (Gerig 1972; Lensky et al. 1985). Such a male mating strategy seems to be altruistic, but probably also enhances the copulation chance of all those 50-250 drones which have already formed a chasing group, especially considering that the honey bee queen mates repeatedly, up to about 15 times during one or more nuptial flights (Adams et al. 1977). The yet unidentified male pheromone might act as an aphrodisiac on the pursuing drones, as observed in other insects (Schneider 1984). Drones immediately die after successful copulating.

The yellowish mating sign left in the sting chamber, which consists of mucus and part of the torn male endophallus, is highly attractive for the following drones (Koeniger 1988). The next one to copulate removes the mating sign of his predecessor with the hairy ventral part of his own endophallus before he is able to insert its bulb and ejaculate the semen (Koeniger 1986). After returning of the queen to the colony, the last mating sign is removed with the help of the workers.

This peculiar male cooperative behavior again appears altruistic since especially the first drone to mate has to search the queen without having been guided by a mating sign. The one he leaves helps all the subsequent males, and so on. It is yet unclear whether this special investment by the first drone is somehow compensated for, perhaps in that more of his sperms enter the spermatheca than do those of subsequent copulations.

Not nearly as many details are known about the nuptial flight in stingless bees. In the case of supersedure, the virgin queen undertakes the mating flight from the mother colony. After swarming, the virgin leaves for the nuptial flight from the incipient filial nest (Fig. 7). Up to now the complete mating flight of any stingless bee has never been observed. It is, therefore, unknown where and how the copulation occurs. In a few species kept in observation hives the mating age of the virgin, the initial steps of the mating flight and the flight duration have been recorded.

Some differences in the mating age of virgins are obvious between the MELIPONINI and the TRIGONINI. In *Melipona quadrifasciata* and *M. quinquefasciata*, the queens undertaking nuptial flights were 3-8 days old (Silva et al. 1972). In confinement, queens were capable of copulating even on the day of emergence (Camargo 1972c). In *Paratrigona subnuda*, a young queen of unknown age performed the mating flight 3 days after replacing the old queen (Imperatriz-Fonseca 1977). In *Plebeia droryana*, the nuptial flight of a virgin 18 days old was observed (Silva 1972). In *Scaptotrigona postica*, by way of a mating box bioassay the maximum attractiveness for drones was recorded in virgin queens 10-14 days after emergence. Gynes up to an age of one week were ignored by the males (Engels and Engels 1988). For this biotests, the drones were taken from an aggregation in the vicinity of colonies. Such drone aggregations also constitute nearby nuclei containing a virgin queen (Engels and Engels 1984). The males are evidently attracted by a female sex pheromone. Perhaps the respective gyne odor is dispersed also by foraging workers which had contacts with an attractive virgin, thereby acting as field messengers. Vagabonding drones might follow the airborn trails and so locate colonies. As in stingless bees many males are sons of workers, such an enhancer effect would increase their mating chance.

The analysis of pentane head extracts from individual queens revealed that secondary alcohols are predominantly present in the virgin bouquet during the second week of imaginal life (Francke et al. 1987). An appropriate mixture of the identified and synthesized compounds did in fact attract the drones of an aggregation. The source of the queen lure, probably the mandibular gland, still has to be verified, as does any contribution of the tergal and other glands to the queen pheromone.

In experimental mating nuclei of *Scaptotrigona postica* provided with a newly emerged gyne, we observed that the entrance tube was often closed for several days until the queen had reached an age of about 10 days and thereby sexual maturity. This of course is a worker function which possibly prevents a premature flight and eventual loss of the queen. As in honey bees, specific worker interactions in the initiation of a nuptial flight have been observed but only a few times (Simoes and Bego 1979). In *Melipona compressipes* the virgin ready to leave massages her abdomen with the hind legs (Kerr in litt.), probably in order to disperse a male attracting pheromone produced by the tergal glands (Mota 1982, 1988). In stingless bees, the virgin always undertakes only one mating flight. The first minutes are spent in the entrance and in the subsequent flying of short orientation circles. The nuptial flight which follows was relatively long in case of *Melipona* queens, namely 33-102 minutes with an average of 78 minutes (Silva et al. 1972). It lasted 33 minutes with a *Paratrigona* (Imperatriz-Fonseca 1977) and 15 minutes in with a *Plebeia* queen (Silva 1972).

Drone aggregations near colonies with virgin queens are known in a number of stingless bee species (Michener 1946, 1974; Kerr et al. 1962; Sakagami 1982; Engels and Engels 1984; Roubik 1989b), all belonging to the tribe TRIGONINI. Perhaps this explains why the duration of nuptial flights is long in *Melipona* queens and much shorter in those species where the drones form mating aggregations and are already waiting in the vicinity of colonies from which the matable virgins undertake their nuptial flights. The drone aggregations often include some heterospecific males. Some observations indicate that the young queen is possibly even attracted by a male assembly. A *Scaptotrigona postica* virgin, suspended from a fishing-rod or even held in the hand, is readily located by the drones when brought close to a group of resting males.

The nests of many stingless bees are found in the dense vegetation of the tropical rain forest in cavities either located not very high in the trees or in the ground (Kerr et al. 1967; Camargo 1970; Wille and Michener 1973; Michener 1974; Sakagami 1982; Wille 1983; Roubik 1979, 1983, 1989a). There is no indication that queens and drones meet and mate in the open air above the vegetation. More probable by far is a nuptial flight within the forest. Indeed the compound eyes of stingless bee drones are not specialized in number and size of the ommatidia for sighting flying targets such as a queen against the sky as are the eyes of honey bees (Ribi et al. 1989). It therefore seems unlikely that male visual allocation is important when approaching the queen in stingless bee matings.

The nuptial flights recorded in *Melipona* all occurred in the morning or in the early afternoon, suggesting that sun compass orientation plays a role in homing of the mated queens which is difficult in the tropics at hours around noon (Silva et al. 1972). Although a systematic study is still lacking, many occasional observations indicate that the time and duration of the nuptial flight are species-specific features.

Stingless bees defecate within the nest. Worker feces are deposited at intranidal garbage heaps (Engels et al. 1979). Feces of a dominant, egg-laying queen and likewise of an accepted queen are attractive for the workers and are always eaten up. Surprisingly the same was seen in a recently mated young *Melipona* queen already five minutes after she had returned from the nuptial flight (Kerr in litt.). The compound making the excrements only of dominant queens so desirable for the workers remains to be identified.

Spermathecal sperm counts both of open mated *Melipona* queens (Kerr et al. 1962) and of queens who copulated once in confinement (Camargo 1972c, 1984) indicate single mating. In the *Scaptotrigona postica* mating box bioassay, we observed repeated copulations, but it was not controlled whether all drones transferred sperma (Engels and Engels 1988). However, this multiple mating could of course be induced by the artificial conditions, since sometimes physogastric queens were also mated. A similar observation was first reported by Sakagami and Laroca (1963) after disturbance of a nest of *Lestrimelitta ehrhardti*. The copulation of a pair at rest was described for *Melipona quadrifasciata* (Camargo 1972c) and photographed in *Scaptotrigona postica* (Engels and Engels 1988). Whether copulation is normally carried out in this position or during flight is not yet known. The mating sign observed in queens who recently copulated both under natural (Kerr and Krause 1950; Kerr et al. 1962) and laboratory conditions (Camargo 1972c; Engels and Engels 1988) was removed after returning to the colony by the queen herself by pressing the abdominal tip against the waxy cerumen (Silva et al. 1972; Camargo 1972c). This occurred without any worker help. The remains of the torn male genitalia of which the mating sign consists would probably hinder a second copulation. In contrast to the honey bees, the drones do not die soon after mating (Engels and Engels 1988).

The evolution of a multiple mating system in honey bees has been discussed under aspects of individual *versus* group selection (Page 1980; Moritz 1985c). Multiple copulations seem to be limited in number according to a minimal risk during one nuptial flight. The worker/queen conflict in reproductive investment ratio in each sex is likewise reduced by multiple matings which result in a low average relatedness between workers and drones of a colony. Evidently, selection has not favoured multiple mating in stingless bees as much as in honey bees.

4.5 Accession to the Lead

Since in honey bees only queens which already lead a colony undertake mating flights, the problem of queen acceptance is already solved by this time. It could be demonstrated in a bioassay that a colony is able to recognize the own queen according to her individual blend of tergal gland volatiles (Moritz and Crewe 1988). Therefore, a mated young queen coming back from nuptial flights just enters her hive. A few days later she begins to lay eggs. Only under experimental conditions the recognition and acceptance of an alien virgin by the workers has been tested and compared with reactions towards close kin (Page and Ericson 1986). Of course this worker/queen interactions (Tab. 1) are of special interest for techniques of queen introduction in beekeeping.

In stingless bees, a virgin or, in many cases, for example in *Plebeia* and in *Scaptotrigona*, a young mated queen still has to be accepted by the colony (Tab. 1). Two factors probably provide the workers with the necessary informations: A pheromonal message and a specific behavior. Both are offered together by the young queen who needs to establish her dominance. In *Scaptotrigona postica* a ritualized acceptance or inthronisation dance is performed by the debuting queen on top of the brood nest. The queen runs rapidly and restlessly around with frequent wing beats. Apparently, her pheromones from the queen-specific abdominal tergite glands are distributed and dispersed through this activity. When the young, attractive queen stops for a moment, she is often fed by the workers. If not, she may offer a drop of liquid held between the mandibles (Simoes and Bego 1979). This evokes intensive trophallactic interactions of the workers, including repeated feeding of the gyne. In intranidal biotesting of synthetic volatiles, we found that only very few components of the workers' cephalic secretions released any trophallaxis (Engels et al. 1987a). 1-hexanol and hexylpentanoate are most effective in making young workers exchange food (Engels 1987b). The acceptance dance can last one or two days and is accompanied by much worker exitement. When both cease the queen has assumed her dominant position. In *Plebeia remota* she is then allowed to eat alimentary eggs which at that time are frequently laid. Thereafter the young queen usually begins to lay the first eggs within a few days.

In a rare case, it was observed for instance in *Schwarziana quadripunctata* and *Melipona rufiventris* how two physogastric queens headed a colony for a short while. The young queen produced more eggs and eventually replaced the old one. This has been also described by Silva (1972) for *Plebeia droryana*. Such exceptions perhaps are due to a balanced attractiveness of both queens, but the final result is the same as in a supersedure taking its normal course.

A true and permanent oligogynous status is realized in *Melipona bicolor* (Nogueira-Neto 1970; Silva 1977; Bego 1989). In the latter species, the virgins were observed participating in the court and remaining on the brood comb during ovipositioning by the physogastric queens (Bego 1988). Sometimes only one of the old queens was replaced after she did not lay any more eggs (Bego 1989). Obviously, in a oligogynous colony, a gyne moving forward into the brood nest is not regarded an intruder into the territory of the dominant queen with the intention of replacing her. The colonial homeostasis is apparently not affected if an additional queen is adopted since the oligogynous status requires a tolerant interqueen behavior (Fig. 5).

5 Drone Production: Worker Chance to participate in Reproduction

In honey bees, the production of drone brood is seasonal and linked to swarming (Free and Williams 1975). The proportion of male to female progeny is a race-specific character (Rinderer et al. 1987). The principal decision rests with the workers which build the necessary drone combs. This occurs about three weeks before the first swarm cells are constructed. The queen lays unfertilized eggs (Dzierzon 1845) into cleaned comb brood cells with a large diameter. She measures them with her forelegs (Koeniger 1969, 1970) during an inspection of the cell. When laying an egg into a worker-sized comb

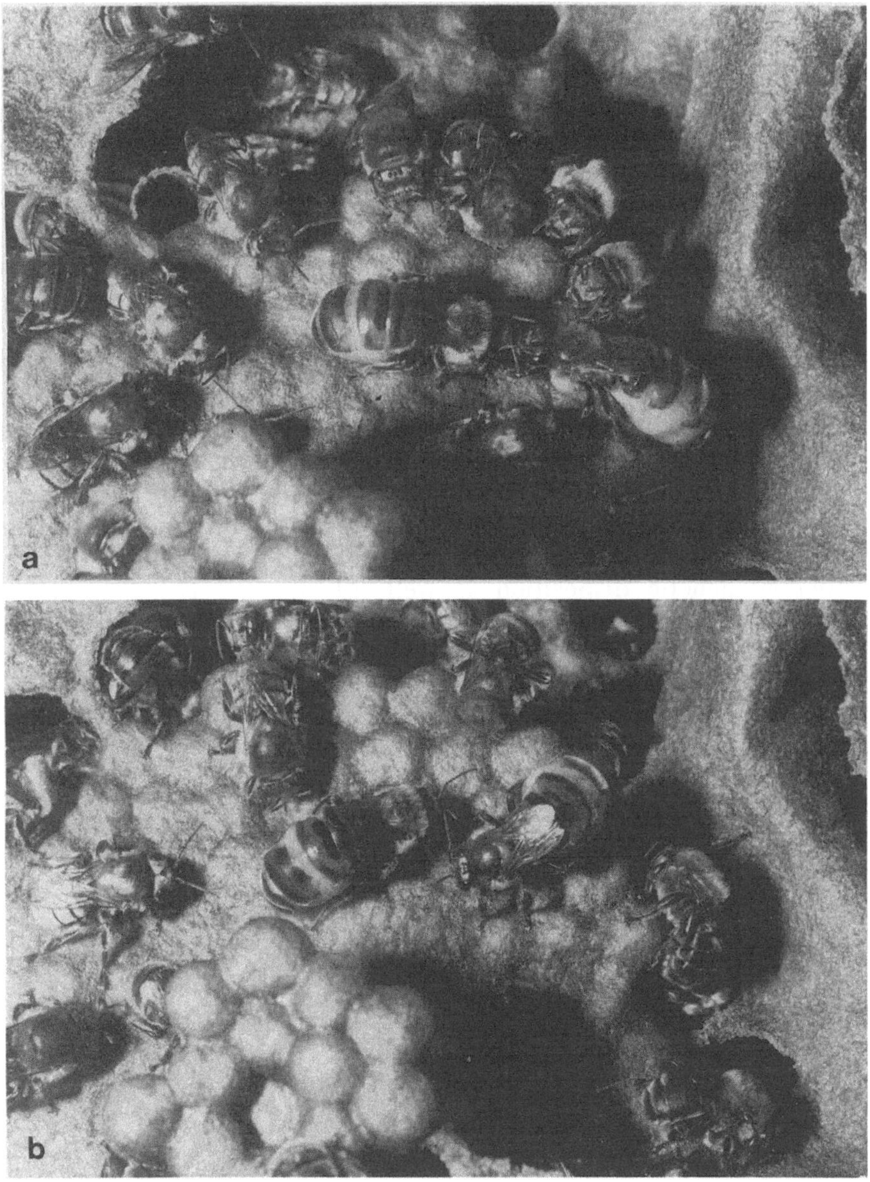

Fig. 5 - In the oligogynous stingless bee, *Melipona bicolor*, usually 2, but up to 5 physogastric queens simultaneously lead a colony. Two old queens are seen on top of the brood nest together with nurse-aged workers. **a** The physogastric queens meet and antennate without any aggression. They are surrounded by a common court. Often both queens participate in the same ovipositioning process, but sometimes only one of them lays eggs at a time. **b** After oviposition always a ritualized queen dance is performed. The workers around seem to be more interested in the sealing and construction of brood cells. (Photograph by Luci Rolandi Bego).

cell or into a queen cup, a droplet of spermathecal fluid is added by Bresslau's semen pump to the micropylar region of the anterior pole of the egg whilst passing the common oviduct (Bresslau 1905). By this mechanism of fertilization, the production of male or female progeny can be controlled by all hymenopteran females. In addition, unmated females can only produce male offspring. In the eusocial bees, the workers remain unmated and can lay unfertilized eggs which develop into drones, but only under certain colonial conditions (Tab. 2).

In the queenright honey bee colony, a genuine feminine monarchy (Butler 1634), reproduction is a royal privilege. The queen is the mother of all queens, workers and drones. Worker fertility is completely suppressed by the primer functions of the queen's pheromone (Tab. 1). It has been known since Aristotle that in orphan colonies the workers lay eggs into normal-sized comb cells and rear drones which are a little smaller than those emerging from drone combs. Under queenless conditions, not all the workers become queenlike in that they reproduce. As a result of intranidal dominance conflicts (Engels 1986, 1988) and probably depending on the individual genetic disposition as well (Moritz and Hillesheim 1985), only a few worker bees will to a limited extent become fertile (Tab. 2). Therefore, the chance of a honey bee worker reproducing is small but not negligible. Data on the occurrence of queenless colonies in natural populations of honey bees are still lacking. Orphan colonies are probably an important source of drones outside of the swarming

Tab. 2 - Caste specific reproductive features in MELIPONINAE and APINAE.

Colony	Caste	Age	Fertility status in: Stingless bees	Honey bees
queenright	queen	young	infertile	subfertile
		old	***fertile***, physogastric or potentially fertile	***fertile***, non-physogastric or potentially fertile
	worker	young	infertile	infertile
		nursing	***fertile***	subfertile
		old	infertile	infertile
queenless	worker	young	infertile	infertile
		nursing	infertile or ***fertile***	subfertile or ***fertile***
		old	infertile	infertile

season when replacement queens search for mates. Even when competing with queen-born drones, the sons of workers succeeded in mating (Martinho and Gonçalves 1980). The significance of worker-derived drones for the population dynamics in honey bees still has to be investigated.

In queenright stingless bee colonies, not only the physogastric queen, but also the workers produce eggs (Sakagami and Zucchi 1963). Worker fertility is a normal phenomenon in meliponids (Tab. 2). It has been studied mainly in *Scaptotrigona postica*, a species in which all the workers are capable of laying eggs during the nurse age (Engels and Engels 1977). Drones are produced throughout the year, the rates depending on the season (Bego 1982). They originate from queen and worker eggs. The latter are laid mainly by the youngest nurse bees (Bego et al. 1983), after aggressive actions towards the queen (Bego 1977), into cells already ovipositioned by her (Sakagami et al. 1965). In addition workers produce special alimentary eggs (Akahira et al. 1970; Engels and Engels 1977) mainly serving as a royal food. The percentage of queen and worker born males varies between colonies and is probably a species-specific character as well. In *Melipona quadrifasciata* (Silva 1977), most of the drones are sons of the queens, whereas in *Scaptotrigona postica* the workers are considered to produce the bigger fraction of males (Beig 1972; Bego 1977, 1982). In fact, which ever worker contributes the last provisions into a brood cell, waits until immediately before the queen lays an egg and may then deposite an alimentary egg at the cell rim (Fig. 6c) or later during sealing the cell may lay a slender egg beside the queen egg (Fig. 6i). This worker is actively defending its temporary mini-territory during the respective phase of the ovipositioning process for the purpose of his own reproduction. Later on, the young male larva kills the cohabitant so that from a brood cell containing two eggs, always a drone emerges (Beig 1972). In rare cases, the worker oviposits instead of the queen directly on the provisions (Fig. 6h), then closing the cell. This was in particular observed in overpopulated colonies of *Melipona quadrifasciata, Scaptotrigona bipunctata* and *Schwarziana quadripunctata* and may be interpreted as an indication of diminished dominance of the physogastric queen. In *Friesella schrottkyi* an abridged operculation occurs after a worker has laid a functional egg, but sometimes several eggs may be deposited into the same cell, resulting in an inter-male larval competition (Imperatriz-Fonseca and Kleinert-Giovannini 1988).

Therefore, in stingless bees at least those workers which do not only produce alimentary eggs participate regularly in reproduction (Tab. 2). Even meliponid workers in queenright colonies are not mere serving helpers of the queen, contributing to the inclusive fitness of the society. We have to expect that they are more interested in

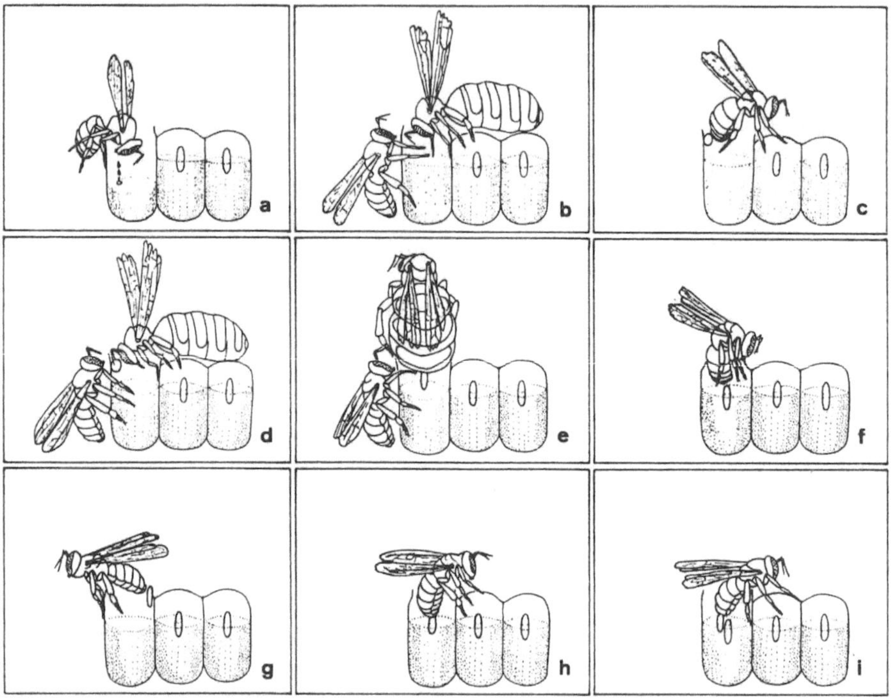

Fig. 6 - Types of eggs and ovipositioning by the queen and the workers in queenright stingless bee colonies as observed in *Scaptotrigona postica*. **a** Provisioning of a newly built brood cell by a nurse-aged worker. **b** The queen eats a little bit of the provisions at the beginning of an ovipositioning cycle while a worker waits alongside. **c** A worker lays an alimentary egg at the rim of the provisioned cell. **d** The queen comes and eats the alimentary egg. **e** The queen lays an egg on top of the provisions. **f** The previously guarding worker closes this cell. **a** - **f** corresponds to a normal queen ovipositioning cycle. **g** A worker lays a slender functional egg at the rim of an unoviposited cell; later, the queen will eat this like an alimentary egg. A normal ovipositioning may follow. **h** A worker lays a functional egg on top of the provisions and then closes the cell before the queen comes. **i** A worker, while operculating a cell previously ovipositioned by the queen, also lays a functional egg beside the queen egg, and then closes the cell. **h** and **i** result in developing of a drone originating from a worker egg.

their own descendants and invest energy in reproductive activities for producing their sons.

The tricky and discrete worker which oviposits during closure of a brood cell after the queen has laid her egg (Fig. 6i) is circumventing an oophagous loss. Worker/queen aggression at this time can be interpreted as a rudimentary defense of the worker's egg. The evolution of special alimentary eggs as an offering for the queen who is coming to a provisioned cell is something like an appeasement in a critical moment, distracting the queen's attention. This complicated part of the fertile workers' egg production which on the surface favours royal nutrition is in fact protecting the worker progeny and, therefore, increases the individual worker fitness. It requires different modes of worker oogenesis, of course a special investment of particular reproductive value.

6 Mating Systems: Insurance Against Inbreeding

In all HYMENOPTERA the appearance of diploid drones as a consequence of high degrees of homozygosity (Page and Laidlaw 1985) strongly favours outbreeding strategies. Only by experimental inbreeding were diploid drones produced at high rates in honey bees (Woyke 1963) and stingless bees (Camargo 1974a, 1984) as well. The omission of copulation within the nest is already generally known in solitary bees. For instance, in the primitively eusocial halictine bee, *Lasioglossum malachurum*, matings are brought about by male choice based on pheromonal cues (Ayasse 1987). As the males prefer alien females, matings between siblings or nestmates have attained a low level of probability (Smith and Ayasse 1987). In the highly eusocial bees long distance mating flights have a panmictic effect. In honey bees and stingless bees, the sexuals evolved different behavioral patterns as a reasult of outbreeding mating strategies.

In the honey bee, both queens and drones from many colonies gather at places called drone congregation areas (Ruttner 1956). Such areas are approached time and again. Some are known to local beekeepers for centuries. Drones commonly fly more than 3 km to reach these mating areas whereas queens fly shorter distances. Therefore, males and females meeting at a congregation area rarely come from the same apiary or even same hive (Ruttner and Ruttner 1966, 1972).

The mating distance was estimated at a maximum of 17 km (Currie 1987), but of course the spatial distribution of drones and queens depends on many factors like weather, topography, and population density. Furthermore, inbreeding is minimized by the repetitive copulations with 10-20 drones (Adams et al. 1977; Page 1980), which normally carry several sex allels. In this polyandrous mating system of the honey bee (Fig. 7a), the random distribution of both males and females within a huge mating range contributes to a highly effective outbreeding result, enhanced by the existence of multiple sex alleles, the number of which has been determined to exceed 10 in a number of bee populations (Mackensen 1955; Woyke 1976, 1988). Among diploid nestmates, both full and half siblings occur, though at a given moment the different patrilines are present in unequal percentages (Martinho 1979; Moritz 1985a).

Since the queen pheromone attracting the drones does not differ much in the 4 recent species of the genus *Apis* (Shearer et al. 1970;

Sannasi et al. 1971), the question is what type of interspecific mating barrier exists. In Asian regions with a sympatric occurrence of *Apis cerana*, *A. florea* and *A. dorsata*, the periods of daily drone flight are temporally separated (Koeniger and Wijayagunasekera 1976; Koeniger et al. 1988). Consequently, conspecific matings are highly probable.

Stingless bees evolved similar mating systems, since drone aggregations occur here as well (Engels and Engels 1984). But there are some important differences and peculiarities. First of all, the drone aggregations are formed in the vicinity of colonies containing virgin queens. Secondly, the males wait there all the day and may remain for long periods exceeding a week in the same assembly. Thirdly, the panmictic effect results only from the migration of the drones, as the mature males abandon their maternal nest, vagabond, and occasionally join a drone aggregation which therefore consists of mostly unrelated males. The virgins are probably mated by a drone coming out of the assembly next to the nest from where the nuptial flight is undertaken. Fourthly, there is no evidence for multiple matings, so that in the monandrous mating system of stingless bees (Fig. 7b) all diploid nestmates are full siblings. Hence the meliponid colony fits the conception of kin selection (Hamilton 1964) quite well.

In honey bees and stingless bees, huge numbers of drones gather at the encounter sites where the sex ratio is in the range of 1,000 : 1. As in most APOIDEA, also in the highly eusocial bees there is no female choice. Nothing like a courtship behavior of the drones happens. There is perhaps a slight male choice when more than one virgin is available. The experimental proof of varying queen attractiveness (Engels and Engels 1988) suggests that pheromonal performance could play a role, although there is no evidence that a queen remains unmated when drones are present. Male competition during mating at congregation and aggregation sites appears as a scramble in which the mating chance of an individual drone seems unpredictable.

Selection pressure may favour males which are able in locating a queen and are fast and persevering flyers. The haploid hymenopteran males probably serve in general as tryouts for optimized genomic constellations because no allelic balance is possible. This explains the remarkably low degree of protein polymorphism as proved by several studies of population genetics in bees (Mestriner and Contel 1972; Brückner 1973; Wagner and Briscoe 1983; Del Lama et al. 1985; Moritz 1985b). It would also explain why drones in highly eusocial colonies are seemingly produced in excess. Much of the colonial performance perhaps is inherited from those drones which are fit enough to mate.

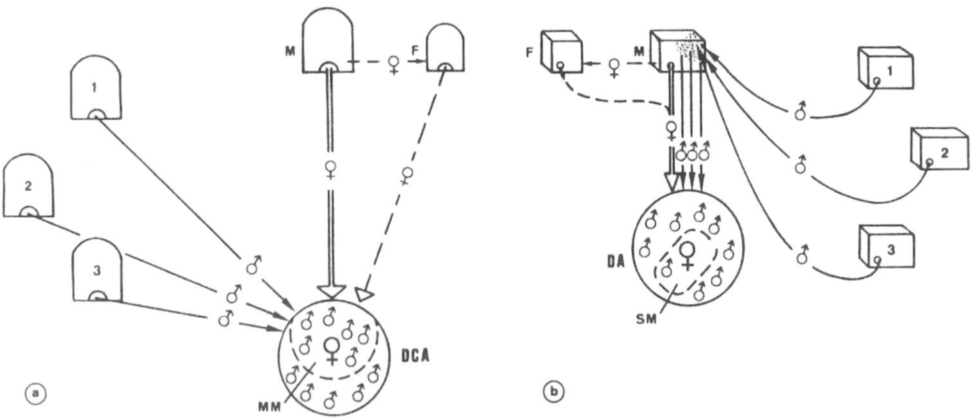

Fig. 7 - Mating systems in **a** honey bees and **b** stingless bees. Male migrations and male assemblies are part of either system. **a** In the honey bee, *Apis mellifera*, the drones from many hives approach drone congregation areas (DCA) to which also virgin queens fly from mother colonies (M) or nest foundations (F) settled by swarms. Queens copulate repeatedly with 10-20 drones. Multiple mating (MM) and long distance flights by both sexes to the congregation areas result in panmixis and ensure outbreeding, since matings do not occur within the hive. - **b** In the stingless bee, *Scaptotrigona postica*, vagabonding males from many colonies are attracted to nests containing virgins and form aggregations of waiting drones (DA) in their vicinity. Virgins from the maternal colony (M) or a filial nest (F) undertake a short nuptial flight during which a single mating (SM) occurs. Only drone migrations have a panmictive effect and avoid inbreeding. - 1, 2, 3 = distant colonies from which drones approach the male assemblies.

7 Colony Reproduction: Filial Nest or Swarming at first?

In the highly eusocial bees, all new nest foundations result from a fission of the colony. In contrast to independent founding of nests by queens in bumble bees and in other groups of primitively eusocial bees, in honey bees and in stingless bees the queen is not involved in the selection of a new nesting site. This decision is made only by the workers. The time at which they do this and the manner in which this is done, especially in relation to swarming, is distinctly different in the APINAE and MELIPONINAE. The corresponding strategies of colony reproduction in honey bees seem to involve more risks.

Division of a honey bee colony during the swarming season depends exclusively on its state and not at all on the availability of a suitable nesting site. Although scout bees probably already look for possible nest cavities before the swarm leaves, the search becomes intense only from the bivouac where the swarm rests either for several hours or for one or two days. Prime swarms often fly farther to bivouac sites than do secondary swarms, perhaps because of their size. In Brazil migrating swarms many of which may have absconded enabled the Africanized bees to spread some hundred kilometers per year and to colonize rapidly all tropical and subtropical zones of South and Central America. Scout bees, dancing on the surface of the clustering swarm, compete in indicating the best lodging (Lindauer 1955). The scouts carefully inspect cavities and measure their volume (Seeley 1977; Seeley and Morse 1978) its suitability depending on the race-specific requirements (Winston et al. 1981; Seeley 1988) for a honey bee nest (Seeley and Morse 1976). Comb building begins after the swarm with the queen moved into the new home which may be located pretty far off the maternal colony (Fig. 7a). Up to that moment, the alimentation of the swarm bees is completely supplied with honey engorged by the workers immediately before the swarm left the mother colony (Otis et al. 1981). Foraging activities are observed as soon as combs to store nectar and pollen are available. At the same time, the queen begins ovipositioning even before the new comb is fully drawn out into brood cells. In a secondary swarm, the virgin first has to go on nuptial flights. If she is lost, the queenless swarm either disperses or workers begin laying eggs.

Thus, in honey bees swarming precedes the foundation of a new nest which is a sudden event. In an *Apis mellifera* prime swarm, colony foundation occurs with the old queen. The advantage of this most probably derived strategy is obvious under temperate climate conditions (Seeley 1978). The earlier a swarm issues, the better the survial chance in the next winter (Seeley and Visscher 1985; Ruttner 1988). Practically the same founding process is observed after absconding (Smith 1961; Winston et al. 1979) or in migratory swarms as known from the Asian species of honey bees, especially from *Apis dorsata* (Hadorn 1946; Reddy 1980). Prime swarming with the old queen mated in the previous year is perhaps an adaptation to non-tropical environments and evolved from nest foundation after long distance seasonal migrations.

Colony reproduction in stingless bees (Fig. 8) is a slow and long lasting process in contrast to honey bees, a difference already recognized by Spinola (1840). Again scout bees (Fig. 8a) look for suitable nesting sites (Kerr 1951) in the foraging range of the colony, but often very nearby (Fig. 7b). In *Partamona*, a genus nesting in termitories on trees (Kerr et al. 1967), quite a number of entrance funnels with whitish shields can commonly be seen on the dark surface of a big termitory (Camargo 1980) indicating that filial nests were founded in the immediate vicinity of the maternal colony. In *Scaptotrigona postica* we have similarily observed 7 neighbouring colonies in a large hollow tree. In the same species, when kept in an indoor flight room, an incipient nest was found in a piece of bamboo lying directly on top of the nest box (Engels and Engels 1987). In this case, the abandoned nest of another meliponid species was

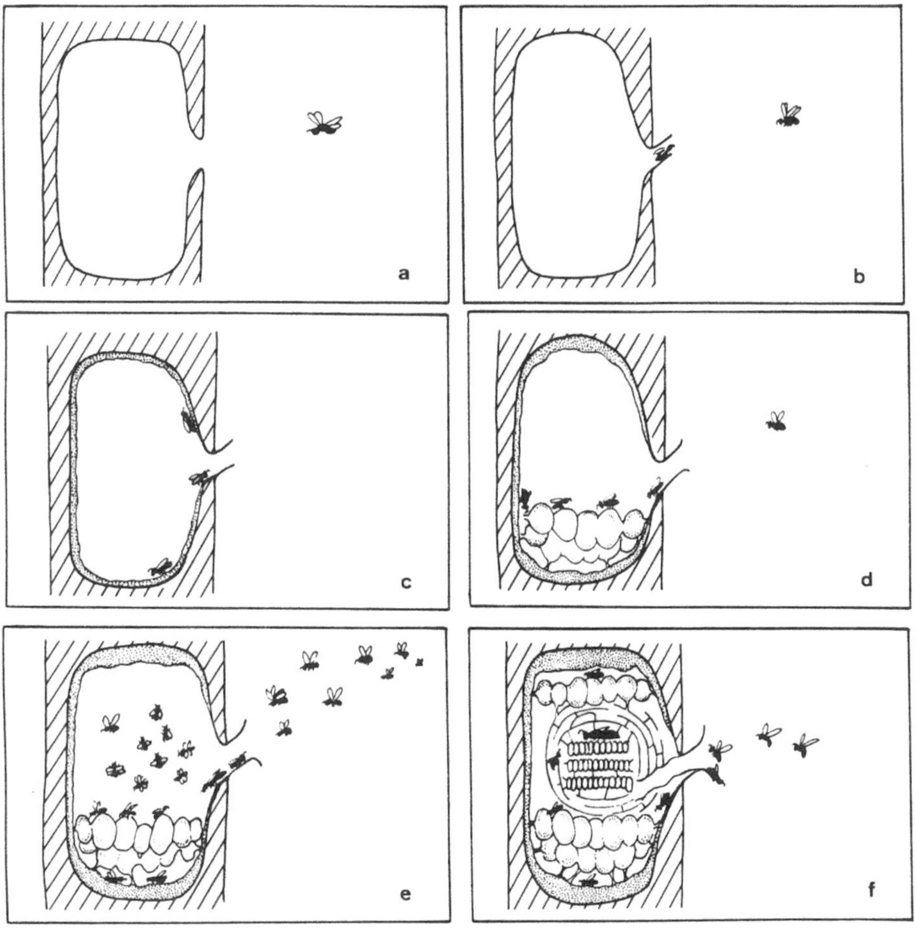

Fig. 8 - Foundation of a filial nest and swarming in stingless bees. **a** A scout bee finds a suitable nesting site, for instance a cavity in a tree. **b** Workers from the maternal nest bring cerumen and construct an entrance tube. **c** Workers are cleaning the hole by removing waste material and plastering the walls with laminated batumen. **d** Storage pots are contructed. **e** A swarm with a virgin queen moves in. Later, the virgin will undertake a nuptial flight. **f** After returning from the mating flight, the inseminated queen will stay in this nest for the rest of her life. The workers contruct a brood nest which is surrounded by involucrum, and the queen oviposits into brood cells. Foraging makes the incipient colony become more and more independent from the maternal nest to which worker traffic is continued over long periods. - This scheme is a combination of characteristics of several species.

occupied, a proven strategy of nest site selection which has also been observed repeatedly under natural conditions (Schwarz 1948; Nogueira-Neto 1954; Wille and Orozco 1974). Some species like *Tetragonisca angustula* are also known to occupy weak heterospecific nests.

After taking posession of a nesting site, the first activity is usually to build the species-specific nest entrance tube (Fig. 8b). Later clefts and cracks are closed and in some species the inner wall is smoothed out with laminated batumen (Fig. 8c). To separate the nest room in large cavities, again batumen in thick horizontal plates may be used. Thereafter, cerumen is carried in and storage pots are constructed and filled (Fig. 8d). All the material is brought from the maternal colony. The work in an incipient nest may go on some days or weeks before a swarm arrives (Nogueira-Neto 1954; Juliani 1967; Terada 1972, 1974; Darchen 1977; Inoue et al. 1984a).

Workers in the mother nest prepare for swarming by gathering large amounts of cerumen from the involucrum, from pots and even from brood cells which is carried in the corbiculae. Frequently, trophallaxis is observed between these workers and the virgin queen. The departure of the swarm happens very quickly. Many worker bees of all ages together with an attractive virgin and possibly some drones leave the mother colony and fly directly to the filial nest (Fig. 8e). The swarm can issue with more than one virgin, and several young queens may follow later. After the departure of the bulk of the swarm, still many more workers may leave for the filial nest during the next hours carrying honey and cerumen. In the incipient nest, the construction of brood cells soon begins. At first, only worker eggs are laid which are then eaten by the queen. If more than one virgin is present in the filial colony, the workers eliminate the surplus queens. The most attractive virgin is probably the one accepted. She undertakes the nuptial flight and begins to lay eggs about 5 days later. The first batches of brood cells are small, but the comb grows rapidly (Fig. 8f). Within a week the young fertile and egg laying queen is already becoming physogastric (Tab. 2).

During this period and sometimes for weeks or even months afterwards, the connection with the maternal nest continues and worker traffic can still be observed, by way of which building material and food is transported to the new settlement from which little foraging is carried out (Nogueira-Neto 1954; Wille and Orozco 1974; Imperatriz-Fonseca 1975; Inoue et al. 1984a). This dependency usually ends when the first batches of workers emerge. In fact, stingless bee colonies in this phase of the reproductive cycle are temporarily polydomus. Because of the persisting exchange between maternal and filial nest, the latter has to be located at close range. Large colonies may initiate the construction more than one incipient nest at the same time. Depending on the climate, swarming may occur seasonal, but less pronounced than in honey bees.

The strategy of colony reproduction through subnests which often separate with remarkable delay certainly includes the advantage of ensuring the filial nest against shortage of supplies in the initial phase of low population and, furthermore, allows for replacement of a lost queen. The diminished risk of this mode of colony foundation, on the other hand, is probably compensated by a possible overcrowding of the nesting territory and the slow rate of expansion into free areas. The rapid spread after introduction into a new environment as was the case with the Africanized honey bees in Brazil is therefore not imaginable for meliponids.

In stingless bees, the old physogastric queen is uncapable of flying and will normally never leave her home. Only young queens participate in swarming and colony foundation. In case of absconding, the old queen remains alone with the remnants of the nest. A contradictory report of successful absconding with an old queen in *Tetragonula laeviceps* from Sumatra (Inoue et al. 1984b) is really surprising. We once found a *Melipona* nest with a single physogastric queen who had opened some brood cells and fed on the larval provisions. Of course she had no chance of surviving. Similar observations were made in *Frieseomelitta languida* (Ribeiro 1989b) where in a meliponary sometimes all the workers left one queenright colony and joined others.

The stingless bee strategy of colony foundation, which includes a polydomus phase, appears to be more original and safer than the honey bee mode of first swarming and then settling. Since honey bee colonies under favourable conditions and depending on the genetic background can cast numerous swarms and afterswarms within a year (Winston 1979, 1980; Winston et al. 1981; Otis 1982; Seeley and Visscher 1985), dispersal strategies contribute to the avoidance of overpopulation. Perhaps bee communities in the diverse tropical forest and savanna environments are subject to different selection pressures resulting in specific features of colonial reproduction.

8 Queen Fertility: Optimized but Dependent Fecundity

Vitellogenin is a proven parameter (Engels 1974, 1981, 1985, 1987a,b) for measuring bee fertility (Tab. 3). Since insects eggs contain large quantities of proteineous yolk or vitellin, the rate of synthesis of the precursor molecule vitellogenin and its incorporation into growing oocytes reflect the intensity and extent of egg formation very well (Hagedorn and Kunkel 1979). In the honey bee, vitellogenin is a typically high molecular serum protein of about 180 Kd, occurring not only in queens, but also in workers (Engels 1972a, 1974) and less in drones (Trenczek and Engels 1986). In the queen hemolymph, the first traces of vitellogenin appear about a day before eclosion. At this time, the imaginal fat body becomes competent to synthesize vitellogenin as a result of metamorphic differentiation (Kaatz 1988). The rate of vitellogenin synthesis and the dependent vitellogenin hemolymph titer then increase rapidly

Tab. 3 - Relation of vitellogenin metabolism and oviposition in females of highly eusocial bees, according to the actual individual level of fertility.

Physiological status	Vitellogenin: synthesis	titer	incorporation	Egg laying
infertile	---	---	---	-
subfertile	low	low	---	-
fertile	***high***	***high***	***high***	+
potentially fertile	low	very high	low	-

after emergence. At this time, long before vitellogenesis begins, vitellogenin already amounts to about 20-30% of the hemolymph proteins in the young and still infertile virgin. A further stimulation of the vitellogenin metabolism is observed subsequent to mating. A first physiological response might be the release of neurohormones from the medial neurosecretory cells (Herrmann 1969), through which the endocrine system could in turn initiate oogenesis. This can be phenocopied by double CO_2 treatment of unmated gynes, but is only observed in queens maintained in colonies. Mated or CO_2 treated queens caged in an incubator even show a decrease in the rate of vitellogenin synthesis (Engels et al. 1976). The CO_2 technique has proved useful in the artificial insemination procedure and has become important in honey bee genetics and breeding programs.

Under normal conditions, a maximum vitellogenin production is attained about three days after the last nuptial flight. One day later, egg laying begins and within another few days the young honey bee queen develops the full capacity of a fertile reproductive female, ovipositing quasi continously with a daily production of up to some thousand eggs. Then the vitellogenin portion of all hemolymph protein synthesis can exceed 90% (Engels 1974), a maximum unsurpassed by other insects. Probably a neuropeptide is involved in the modulation of vitellogenin production (Kaatz 1988). Initiation and maintenance of vitellogenin synthesis (Engels 1980) do not depend on *corpora allata* functions, since allatectomy in queen pupae, pharate adults, newly emerged or mature egg laying queens does not prevent vitellogenin production and egg formation (Engels and Ramamurty 1976; Ramamurty and Engels 1977; Daerr 1978; Kaatz 1988). Accordingly application of juvenile hormone has no decisive influence on vitellogenin synthesis by the queen either *in vivo* (Engels 1978; Regel 1978) or, in the case of cultured fat body, *in vitro* (Kaatz 1988). In the fertile queen the rate of vitellogenin synthesis will later on never be reduced to a zero-level (Engels 1974; Kaatz 1987).

ven under winter conditions or during swarming, a considerable tellogenin production is observed, enabling the potentially fertile mey bee queen to increase oogenesis and to lay eggs within a short ne if necessary (Tab. 3). Such a rapid regulation of actual rtility according to colonial needs and potentialities is not known om any non-social insect.

The unique fertility of the honey bee queen (Tab. 3) enables her to oduce a daily egg-mass equivalent to two or three times her body eight. Of course this depends on an adequate nutrition which is ovided constantly by the worker bees, especially by those which rm the court. They feed the queen glandular secretions, especially od sap from the hypopharyngeal glands which are enlarged in the irse phase. So both the larvae and the queen are fed by the workers. ie reproductive output of the queen depends essentially on such lper functions of her daughters.

In stingless bees, the vitellogenin metabolism has only been udied to some extent in *Scaptotrigona postica*. The exact time of itial vitellogenin synthesis in young infertile queens has not yet en determined. Considerable amounts of vitellogenin do not appear the hemolymph of virgins before the tenth day after emergence. At e same age, a sharp increase in vitellogenin synthesis can be easured (Engels and Engels 1977). Furthermore, virgins of this age come highly attractive for the drones (Engels and Engels 1988). The condary alcohols, which probably are important constituents of the male sex pheromone, disappear from the cephalic volatile spectrum the queen soon after mating, normally within approximately 5 days ngels et al. 1987b). It has been assumed that the synchronous duction of pheromone production and vitellogenin synthesis is rhaps mediated by a common control factor (Engels 1987b). Both the te of vitellogenin synthesis and the hemolymph titer rise after the iptial flight and later during the first ovipositioning activity of e fertile queen who within a few days becomes physogastric (Engels id Engels 1977). In this respect, the vitellogenin metabolism in ingless bee queens reflects the situation described for honey bee ieens (Tab. 3). The difference concerns the first week after nergence. In stingless bee queens, the initiation of fertility curs later in comparison with honey bees, and apparently does not clude a preceding subfertile phase (Tab. 2). Perhaps this is lated to the quite uncertain situation of the virgin, which lasts least 10 to 15 days in stingless bees, whereas a young honey bee ieen who survived the first or second day after emergence has a good iance of heading a colony for the rest of her life with egg oduction as the main duty.

In the honey bee queen oviposition behavior seems to be quite mple if compared with that of stingless bees. Surrounded by her urt, the *Apis* queen walks on the brood nest combs. She inspects cleaned empty cells and thereby determines worker and drone cell size. For oviposition, the abdomen is dipped into the cell and the egg is stuck on the bottom. The queen looks downwards during the procedure, so the egg remains in an upright-oblique position (Fleig 1987). The queen can lay one hundred or more eggs in sequence which takes about one hour. Ovipositioning is performed day and night with intervening breaks.

In stingless bees, ovipositioning has become an extremely complicated behavioral complex with much species-specific diversification in which numerous nurse-aged workers are involved (Zucchi 1977; Sakagami 1982). The many comparative studies of the ritualized worker/queen interactions have revealed a sophisticated sequence of behavioral steps. Two reasons can be given for these complications: First, in all meliponids the brood cells are dismantled after emergence of the imago and, therefore, have to be constructed cell by cell again for every reproductive cycle. Second, stingless bees are mass-provisioners (Fig. 6a) like most solitary bees. Thus, the brood cell has to be filled up with all the larval food before the queen can oviposit. Furthermore, this cannot happen too early and, consequently, requires behavioral regulations. The queen feeds on a small amount of the provisions just before laying an egg (Fig. 6b). In addition she eats worker-laid eggs, in particular the large (Fig. 6d) and special alimentary eggs (Sakagami and Zucchi 1963, 1966; Sakagami et al. 1963; Akahira et al. 1970; Bego 1977; Engels et al. 1979). Although this is known from many species, it has to be mentioned that there are some like *Frieseomelitta languida* in which the workers never lay any eggs (Tab. 2), and others like *Friesella schrottkyi* in which the queen does not eat the worker-laid eggs. Direct feeding of the physogastric queen by the workers is rarely seen. So the royal nutrition consists, as in honey bees, of larval food and, in addition, of worker eggs. As both are products of nurse-aged bees, the nutritional dependence of the old queen on the workers is essentially the same in both groups of highly eusocial bees.

Besides this, the workers have to maintain the appropriate microclimate within the brood nest. This has been the subject of detailed studies in *Apis mellifera* (Seeley and Heinrich 1981; Heinrich 1985; Rosenkranz 1988). In the open-nesting tropical species *A. dorsata* and *A. florea*, a high percentage of house bees forms a curtain protecting and insulating the brood area of the single comb. In stingless bees, thermoregulation depends on the nest structure (Zucchi and Sakagami 1972; Darchen 1973). In *Scaptotrigona postica*, the brood nest temperature is effectively controlled by warming (Engels et al. 1979; Rosenkranz et al. 1987), the bees, however, being uncapable of cooling. In nests free of involucrum, such as in *Plebeia wittmanni*, a recently discovered Brazilian species inhabiting cravices in granite rocks, the temperature of the ambient mass determines the nest conditions (Wittmann 1989).

The maximum daily rate of egg production by a physogastric queen in stingless bees varies according to the species. It does not reach, however, the frequency seen in queen honey bees. The latter has an ovary containing 150-200 ovarioles. Thus, in a large *Apis mellifera* queen, normally about 350 functional ovarioles are found. Each is capable of producing 5-10 eggs per day (Engels 1973b). In stingless bees, only 4 ovarioles per ovary are normal for both queens and workers. A few species like *Lestrimelitta limao* have queens with some more, but always less than 15 ovarioles (Sakagami 1982). In the physogastric queen, the 8 ovarioles are extremely enlongated and folded in a S-shape. In each ovariole, 10-50 follicles grow up and ripen per day, a unique capacity among all insect orders. Thus, the enormous output of eggs produced by the queen in the monogynous, highly eusocial bees is attained in two different ways in the APINAE and MELIPONINAE, both having populous colonies headed by only one reproductive female of extreme fecundity (Tab. 2), and depending on a very specialized worker-derived diet. The hypothesis of a dietary transmission of vitellogenin from the workers to the queen (Engels 1971, 1972b) could not be confirmed (Engels 1973b, 1974; Rutz and Lüscher 1974), but the mode of queen nutrition undoubtedly includes biochemical adaptations for the production of large numbers of eggs.

9 Worker Fertility: Rudimentary or Emergency Reproduction?

In the process of social insect evolution, completely different reproductive capacities evolved in the worker caste of the ISOPTERA and the HYMENOPTERA. Whereas in the hemimetabolous termites the preadult and therefore sexually immature stages became workers, female larvae as well as male (Noirot, chapter 1), in the holometabolous ants (Buschinger, chapter 2), social wasps (Strambi, chapter 3) and social bees only female adults were differentiated into workers and queens. Apparently this late ontogenetic bias did not allow for the development of an exclusively fertile caste. The optimized fecundity of the queen could probably not evolve without maintaining worker fertility. Only by physiological inhibition is the worker bee forced to be more or less infertile. Remarkable differences in the extent and mode of control of worker fertility (Tab. 1, 2) are realized in the APINAE and MELIPONINAE.

An important divergency concerns the development of the ovarian structure. In the honey bee, one of the first steps in worker caste formation is the degeneration of most of the ovariole anlagen. This process begins as early as in the fourth larval instar under low juvenile hormone titer conditions (de Wilde and Beetsma 1982) and continues into the pupal phase. The enhanced synthetic activity of the *corpora allata* in a future queen larva (Rachinsky and Hartfelder 1989) resulting in an increased juvenile hormone titer (Rachinsky et al. 1989) protects the ovary and inhibits its rudimentation. Consequently, in the imago, the caste difference in ovariole numbers figures to be in the range of 1 : 50. In *Apis mellifera*, worker ovaries normally contain only 3-5 ovarioles (von Siebold 1856; Velthuis 1970). It is interesting to note that in *Apis dorsata*, the honey bee species with the least pronounced caste dimorphism, workers have significantly more ovarioles, normally about 20-30 (Velthuis et al. 1971). Ovariole numbers in the workers of the Cape bee with 10-20 (Velthuis, Ruttner, Crewe, chapter 8) resemble the numbers of intercaste individuals (Weiss 1978). In contrast to this, nearly no caste differences in ovariole numbers occur in stingless bees. In most species, 4 ovarioles are found, in both queens and workers.

Vitellogenin as a prerequisite for egg development is also produced by worker bees as first discovered in *Apis mellifera* (Engels 1971, 1972a) and later confirmed in *Scaptotrigona postica* and *Melipona quadrifasciata* (Engels and Engels 1977). This is therefore a common worker feature in the APINAE and in both tribes of the MELIPONINAE. Since honey bee workers do not produce eggs under queenright conditions (Tab. 2), vitellogenin synthesis can be regarded a rudimentary character. On the other hand, the vitellogenin titer in workers is regulated according to a strict ontogenetic pattern (Engels and Fahrenhorst 1974). In comparing the vitellogenin titer of workers in the four recent honey bee species, a higher level was found in *Apis florea* and especially in *A. dorsata* than in *A. cerana* and *A. mellifera* (Engels 1973a). Since high vitellogenin titer is a typical queen feature (Engels 1974), this was interpreted as pointing to less pronounced differences in caste characteristics including fertility (Engels 1973a,c). The maximum vitellogenin titer is reached during the nurse age of adult honey bee workers (Engels and Fahrenhorst 1974) which can be classified as subfertile (Tab. 2, 3). In old foragers, vitellogenin is in the process of dissapearing from the hemolymph. It is interesting to note that a similar ontogenetic pattern of vitellogenin production is realized in drones, although much less is synthesized (Trenczek et al. 1989). In a queenless colony, egg-laying workers (Velthuis, Ruttner, Crewe, chapter 8) develop from subfertile nurse bees and thereby increase their vitellogenin titer (Engels 1974) and become fertile (Tab. 2). In queenless worker groups, a considerable individual variation in vitellogenin titer was found (Velthuis 1976b) in *Apis mellifera*. This has not been studied in the other species of honey bees. Since in *Apis cerana* egg laying by worker bees is observed already a few days after dequeening of a colony, probably even under queenright conditions some workers retain a more queenlike status of fertility.

On the other hand, all the workers become fertile to a limited extent in stingless bees, also under queenright conditions (Tab. 2). Due to the slower changes in age-dependent tasks, which is also reflected in alterations of the cephalic volatile blends (Francke et al. 1983), and the greater adult lifespan, vitellogenin in *Scaptotrigona postica* occurs in the worker hemolymph up to 5 weeks after emergence (Engels and Engels 1977). Within this period of nurse functions, slender reproductive eggs (Fig. 6g) are formed at first (Bego 1977) in small numbers by both the right and left ovary (Staurengo da Cunha 1977, 1979; Bego et al. 1983). Since by this egg-laying the workers regularly contribute to reproduction in form of drone offsprings, worker fecundity has to be regarded a normal event in the MELIPONINAE. No inhibition of worker fertility by the queen occurs as in honey bees.

Also, the situation in orphan colonies is different in honey bees when compared to stingless bees (Tab. 2). Without the inhibiting primer effects of the queen pheromone (Tab. 1), a honey bee worker can become fertile. Just how many bees of a queenless colony will finally lay eggs and how many they will lay, depends on factors like group size, age of the bees, season, and also on race-specific properties (Velthuis, Ruttner, Crewe, chapter 8). Egg-laying is observed within 1-4 weeks after loss or removal of the queen in *Apis mellifera* (Ruttner and Hesse 1979), but occurs much sooner in *Apis cerana*. Unless except a false queen as a replacement reproductive develops (Sakagami 1958), irregular multiple ovipositioning often occurs. The easiest transition into the fertile state (Tab. 2) is achieved in the nurse phase under previously queenright conditions. Yolk uptake into oocytes and follicle growth are seen within 7 days in all individuals of a caged group of *A. m. carnica* workers set up with former queenright nurse bees one week old (Trube 1988). In workers kept queenless since emergence, egg follicle development is a little delayed and not uniform.

Because of the immense attractiveness of a queen, a worker will never voluntarily escape a queenright honey bee colony. Once liberated from the reign of the domineering royal pheromonal, however, the workers immediately take the chance to reproduce and at least some of them can have drone progeny (Tab. 1). Since a queenless group will survive only a few months, thereby showing signs of disorganisation, worker reproduction seems to be an emergency response, especially considering that - except for the Cape bee - no workers or queens are reared. From the viewpoint of population genetics, this form of worker reproduction of drones is even more effective. Drones are sexuals and potential transmitters of 50% of the worker's genome into the filial generation, the female offsprings of a queen. Worker reproduction may therefore contribute directly to the genetic maintenance of a given bee population. This perhaps explains as well why worker fertility was not erased in the course of socioevolution. Indeed, worker oviposition in queenless colonies is widespread in social HYMENOPTERA. In wasps the reconstruction of a lost nest and rearing of drones by a surviving worker group was described (Fahrenhorst and Engels 1976).

Ovipositioning in orphan colonies has also been recorded for most genera of stingless bees (Sakagami et al. 1964). Only in *Frieseomelitta* and *Duckeola* (Sakagami 1982) does this not occur. In queenless honey bee colonies combs are normally retained which is making an oviposition site available for egg laying workers. In contrast to this stand the orphan stingless bee colonies where the irregularities begin with the contruction and provisioning of brood cells. In *Scaptotrigona postica*, single cells, intermingled with storage pots, are constructed, filled, emptied, filled again, ovipositioned or not, closed and reopened, and so on. As in honey bees, several eggs are laid often into the same cell and the eggs may be eaten by the workers. Sometimes cells are overfilled with a sticky white mass instead of larval food. Seldom is a male larvae found and adult drones are normally not produced in significant numbers. But this can be observed in other species like *Hypotrigona* (Bassindale 1955) and *Leurotrigona* (Sakagami and Zucchi 1974). Furthermore, it seems that orphan colonies in stingless bees may be requeened by a virgin entering the nest. An alien queen perhaps is more readily accepted by a queenless group of stingless bees than by queenless honey bees. Worker bees apparently inspect neighbouring nests and possibly recognize queenless colonies. By odor cues this information could be transmitted to sister virgin queens. Consequently, worker fertility (Tab. 1) is not enhanced in orphan meliponid nests as it is in honey bees colonies (Tab. 2), and drone production by such nests is probably without significance in most stingless bee species. Nevertheless, there are some exceptions like *Plebeia saiqui* (Imperatriz-Fonseca and Oliveira 1974) and *Paratrigona subnuda*; after a period without any physogastric or virgin queen, some workers can become dominant and fertile (Tab. 2), and again a regular comb is built and worker eggs are deposited.

This does have all signs of a rudimentary reproduction and could be misinterpreted as a diminished contribution of meliponid workers to drone production. But, as described above, the stingless bee worker already has a chance to reproduce under queenright conditions and in many species a remarkable portion of the drones is regularly derived from worker eggs. Thus, in the APINAE and MELIPONINAE, different strategies have evolved to ensure the workers' participation in reproduction (Tab. 2). A varying worker contribution to drone production under "normal" colonial circumstances as seen in stingless bees appears to be original and reflects the permanent reproductive competition of the female castes. The more casual chance to reproduce which remains for the queenless workers in honey bees seems more derived since the royal privilege of reproduction is thereby better protected. This, of course, is a question of colonial homeostasis.

10 Colony Homeostasis: Releaser and Primer Functions of Domineering Queen Pheromones

The honey bee colony is more strictly monogynous than meliponid nests. It is possible to manage two queen hives, but only with two distinct brood nests. Such bigynous hives consist, therefore, in fact of two neighbouring colonies with some fluctuation of the worker populations, but without any contact betwen the two egg-laying queens. If the queens meet at all, they immediately begin to fight until only one remains. Releaser effects of the queen pheromones evidently also include the obligatory instigation of aggressive queen/queen interactions. The maintenance of the monogynous state is an essential feature of colonial homeostasis in the APINAE.

In comparison, the stingless bee colony is virtually never completely monogynous, although in nearly all species only one old and physogastric queen is fertile and is capable of laying fertilized eggs. The permanent emergence of gynes, most of which eventually disappear, results in a multigynous nest situation which includes some competition between the old and young queens without, however, bringing into question the reproductive privilege of the physogastric queen. Interestingly enough, there are a few exceptions in the form of polygynous species. Since the maximum number of old queens does not exceed 5 or 6, this colonial status is better named oligogynous. The best known oligogynous species is the Brazilian *Melipona bicolor* (Silva 1977; Bego 1988, 1989) with usually two, but sometimes 3, 4 or 5 physogastric queens present in a colony (see 4.2, 4.3, 4.5). They do not have separate subterritories and meet in the top brood comb whenever participating in a common ovipositioning cycle (Fig. 5). No interqueen aggression occurs and all the queens are apparently to the same degree attractive for the workers. Clearly, colonial homeostasis in *Melipona bicolor* is based on the absence of any competition between old queens, probably due to an altered behavioral pattern in the intranidal queen/queen interactions. Unfortunately, it is yet unknown how the accession of an additional dominating queen takes place. Furthermore, there must be an royal equilibrium system preventing the establishment of more than 2-5 old queens. Special pheromone actions are presumably involved in this unique oligogynous homeostasis.

Releaser and primer functions of the domineering queen pheromones (Tab. 1) have up to now been extensively studied only in honey bees (Free 1987). The main differences between stingless bees and honey bees in the colonial dominance hierarchy evidently concerns the interqueen relations. In the MELIPONINAE, the reproductive superiority of the physogastric queen is less pronounced than in *Apis*. The more permissive behavior of the old queen towards young virgins is thought to have made feasible the evolution of an oligogynous colonial organization as realized in *Melipona bicolor*.

In the honey bee colony two inhibitory effects of the queen pheromone relating to worker functions in reproduction are well documented (Tab. 1): the suppression of the building of queen cells and of worker fertility (Velthuis 1985; Engels 1986). The former is one of the typical releaser functions, since after removal of the queen within a few hours the workers start to feed a few female larvae in worker-sized brood cells amply and later to construct emergency queen cells around them. The latter is a primer effect since an increase in the rate of vitellogenin synthesis and vitellogenic growth of egg follicles in worker bees is observed only after some days under queenless conditions.

Comparable pheromone-mediated functions of the dominant female, the physogastric queen, are obviously not present in stingless bees (Tab. 1). Under queenright conditions neither the construction of royal cells in the TRIGONINI nor the worker fertility is inhibited in any way whatsoever. On the other hand, in orphan colonies many irregularities in the reproductive behavior of the workers occur, an indication that a controlling stimulus of the physogastric queen is lacking.

It must be concluded that in the course of socioevolution in the APINAE and the MELIPONINAE the dependance of colonial homeostasis on domineering queen pheromones developed in divergent directions. In honey bees a royal suppression and inhibition of the reproductive capacities of the workers evolved, whereas in stingless bees a stimulation of worker reproduction by the queen developed. It still has to be evaluated whether any primer effects of queen pheromones are involved in meliponid queen/worker relations. Intercaste competition and worker control seem to be weaker in queenright colonies of stingless bees than in those of honey bees. Perhaps the multigynous state of meliponid societies can be regarded as more original than the strictly monogynous organization (von Ihering 1903b).

The persisting royal dominance in the highly eusocial bee colonies requires effective instruments. In honey bees, the female sex pheromone has attained an extensive and lasting function in domineering the workers all during the long postmating phase of a queen's life. Although exact data on the amounts of pheromonal components synthesized in the imaginal ontogenesis of a queen are still lacking, it is obvious that at least the main compound of the mandibular gland secretions, 9-oxo-E-2-decenoic acid, is produced more by old egg-laying queens than by virgins. It can be considered that in the course of socioevolution in *Apis*, the sex pheromone became a social pheromone as well and established the dominant colonial position of the fertile queen (Engels 1988).

Our knowledge on queen pheromones in stingless bees is still in an initial stage. The first analyses conducted on cephalic volatiles of *Scaptotrigona postica* queens revealed a striking difference in composition between the secretions in virgin and those in physogastric queens (Engels et al. 1987b; Francke et al. 1987). A possible interpretation is that stingless bees have two complexes of queen substances. One acts as a female sex pheromone, which attracts the drones (Engels and Engels 1988), and a second maintains social dominance within the nest. This would mean that, in stingless bee evolution, the queen developed another sentence of pheromonal signals for the secondary functions of keeping colonial homeostasis. This sytem differed from the one which probably already existed in a similar form in the solitary ancestors, the primary function being that of a female mating lure. This assumption of course must be substantiated by further investigations.

11 Fitness Control: Individual *versus* Colonial Contribution

Socioevolution in the haplo-diploid HYMENOPTERA (Trivers and Hare 1976) is presumably a consequence of kin selection (Hamilton 1964) resulting in inclusive fitness. The different interpretations of presumed altruistic and domineering queen/worker interactions, especially during initial phases of socioevolution (Michener and Brothers 1974) in bees, have recently been reviewed by Michener (1985) and Dietz (1986). In all the social groups of APOIDEA, a substantial number of reproductive strategies evolved (Eickwort and Ginsberg 1980; Thornhill and Alcock 1983; Engels 1988). The question is to what degree individual fitness remained and is now still operating in honey bee and stingless bee male and female reproduction.

The well known interpretation of altruistic sib worker behavior has often been stressed and needs no further discussion. But does the regular production of drones by queenright stingless bee workers not contradict the inclusive fitness model? This question becomes especially acute when we consider that the two larvae developing in one brood cell, one from a queen-laid and one from a worker-laid egg, compete for the same portion of the provisions. Since apparently the worker-born male larva nearly always wins and since, at least in some species, most of the drones originate from workers, such a bias towards individual worker reproduction instead of the carrying out of mere helper functions of nursing queen offspring cannot be regarded as a minor and accidental development in brood rearing. On the other hand, the production of specific alimentary eggs by stingless bee workers, which are important nutrients for the queen, must be considered a costly investment into colonial reproduction. All this contributes to the inclusive fitness and is somehow controlled by the queen, as well as the enhanced secretion of glandular provisions, which in part also serve as a royal diet and not only as larval food. Therefore, in meliponids the workers' contributions to reproduction consist of behavior and performance relevant for both individual and inclusive fitness.

In the highly eusocial bee colony, the old and dominant queen is an egg-laying machine. Her output depends completely on the nutrition and attendance provided by the workers, which is consequently a colonial contribution, based of course on physiological functions and gonadal structures which are optimized for the main royal task of reproduction. Colonial fitness can be measured both in terms of colony survival which exceeds a queen's lifespan and in the establishment of successful filial nests. Queen production is a prerequisite for both, whereas drone rearing only indirectly contributes to the fitness of a given population, since outbreeding strategies normally prevent a direct patrilineal continuation within the same nest. Perhaps these circumstances may contribute to the maintenance of a dual origin of the males, which can either be sons of the queen or sons of the workers, i.e. grandsons of the queen. The fatherless males in case of queen sons transmit exclusively maternal queen genes. In case of grandsons, only 50% of their genome and accordingly of their gametic DNA is queen-derived. This difference explains the caste competition in producing drones and presumably stabilizes the individual contribution of successfully reproducing workers, as observed in stingless bees mainly under queenright and in honey bees exclusively under queenless conditions. Accordingly, fitness control acts more strictly on the colonial level if concerning female progeny and depends more on individual contributions in case of male offspring. Recompensation for such alternative male production strategies is probably provided in the rigorous struggle which takes place in the scrambling competition of the supernumerous drones during mating. Physical fitness may be decisively for selection here and control male performance. In the monandrous mating system of stingless bees, this appears to be of much more consequence than in the polyandrous reproductive strategy of honey bees.

12 Honey Bees and Stingless Bees: Parallel Evolution Towards high Eusociality

In most of the older literature, honey bees and stingless bees were regarded as close sister groups, mainly because of the highly evolved level of social organization (von Ihering 1903a). Originally, some stingless bees were even described as a species of the genus *Apis*. Up to recently the term "stingless honey bees" was occasionally used. The quite separate phylogenetic position of the subfamilies MELIPONINAE and APINAE became obvious through comparative studies of external morphology (Schwarz 1948; Winston and Michener 1977). The results indicated a dual origin of the highly eusocial organization among bees.

We have presented here several examples of differences between honey bees and stingless bees which touch nearly all areas of caste development and reproductive biology, including the intranidal dominance hierarchy. This justifies the following conclusions. Socioevolution in bees is linked with finding solutions for problems which automatically arise with increasing reproductive specialization. The latter originates from the advantage provided by populous and perennial colonies in helping establish more stable conditions and allowing for an extensive exploitation of environmental resources. The amplification of the caste syndrome includes optimizations and restrictions as well, but cannot exclude intercaste and interindividual conflicts. Honey bees and stingless bees often went different ways only to end up with rather similar social regulations. Evidently, not a common but a parallel evolution towards high eusociality took place. At first glance, much appears similar in APINAE and MELIPONINAE. Both groups have large colonies, build nests with extensive use of worker-produced wax, store honey and pollen, have female castes and have workers which perform age-related tasks and exhibit a remarkable polyethism. A closer second look and intensified research have shown the many differences in general organization of social regulation in honey bees and in stingless bees, particularly in regard to reproduction, fertility and colonial homeostasis. It could be that if in time other groups of bees were to reach a comparable standard of social life, then the result would resemble the current solutions evolved by the two existing unique groups of highly eusocial bees. They are fascinating and amazing subjects, in particularly for those who study their reproduction and sociobiology.

Acknowledgements

Supported by DFG, CNPq, DAAD and FAPESP grants. The technical assistance of Rejane Rosa, Nelson Gomez and Isabel Schlak in preparing the figures is appreciated. Figures 3 and 6 were kindly provided by M. Ramalho and L.R. Bego, respectively.

13 References

Adams J, Rothman ED, Kerr WE, Paulino ZL (1977) Estimation of the number of sex alleles and queen matings from diploid male frequencies in a population of *Apis mellifera*. Genetics 86:583-596

Akahira Y, Sakagami SF, Zucchi R (1970) Die Nähreier von den Arbeiterinnen einer stachellosen Biene, *Trigona (Scaptotrigona) postica*, die von der Königin kurz vor der eigenen Eiablage gefressen werden. Zool Anz 185:85-93

Ayasse M (1987) Die Rolle von Duftstoffen bei der Partnerwahl von Furchenbienen (Hymenoptera: Halictidae). Apidol 18:371-373

Bassindale R (1955) The biology of the stingless bee *Trigona (Hypotrigona) gribodoi* Magretti (Meliponidae). Proc Zool Soc Lond 125:49-6

Beetsma J (1979) The process of queen-worker differentiation in the honeybee. Bee World 60:24-39

Bego LR (1977) Aspectos da regulaçao social em *Nannotrogona (Scaptotrigona) postica* Latreille (Hymenoptera, Apidae, Meliponinae). Tese Dout Univ Ribeirao Preto, 181 p

Bego LR (1982) On social regulation in *Nannotrigona (Scaptotrigona) postica*, with special reference to male production cycles (Hym., Apidae, Meliponinae). Bolm Zool Univ Sao Paulo 7:181-196

Bego LR (1988) Padrao geral de oviposiçao em *Melipona bicolor bicolor* Lep. (Hym., Meliponinae). Ciência e Cultura 40:920-921

Bego LR (1989) Behavioral interactions among queens of the polygynic stingless bee, *Melipona bicolor bicolor* Lepeletier (Hymenoptera, Apidae). Bras J Med Biol Res 22:587-596

Bego LR, Camargo CA de (1984) The occurrence of giant males in *Nannotrigona (Scaptotrigona) postica* Latreille (Hymenoptera, Apidae, Meliponinae). Bolm Zool Univ Sao Paulo 8:11-6

Bego LR, Simoes D, Zucchi R (1983) On some structures in *Nannotrigona (Scaptotrigona) postica* Latreille (Hymenoptera, Meliponinae). Bolm Zool Univ Sao Paulo 6:79-88

Beig D (1972) The production of males in queenright colonies of *Trigona (Scaptotrigona) postica*. J Apicult Res 11:33-39

Bohrer K (1987) Do sweat bees also have a trophogenic basis of caste development? Apidol 18:369-371

Bonetti AM (1982) Açao do hormônio juvenil sobre a expressao gênica em *Melipona* (Hymenoptera, Apidae, Meliponinae). Tese Mestr Univ Ribeirao Preto, 140 p

Bresslau E (1905) Der Samenblasengang der Bienenkönigin. Zool Anz 29:299-323

Brückner D (1973) Reduction of biochemical polymorphisms in honeybees (*Apis mellifica*). Experientia 30:618-619

Butenandt A (1959) Wirkstoffe des Insektenreiches. Naturwiss 46:461-471

Butler C (1634) Feminin monarchi or the histori of bees. Oxford

Butler GG, Callow RK, Johnston FRS, Johnston NC (1962) The isolation and synthesis of queen substance, 9-oxodec-trans-2-enoic acid, a honeybee pheromone. Proc Roy Soc B 155:417-432

Camargo CA de (1972a) Determinaçao de castas em *Scaptotrigona postica* Latreille (Hymenoptera, Apidae). Rev Bras Biol 32:133-138

Camargo CA de (1972b) Produçao *in vitro* de intercastas em *Scaptotrigona postica* Latreille (Hym.: Apidae). In: Cruz-Landim CL da , Hebling NI, Lello de E, Takahashi CS (eds) Homenagem à Warwick E. Kerr, Rio Claro, pp 37-45

Camargo CA de (1972c) Mating of the social *Melipona quadrifasciata* under controlled conditions (Hymenoptera, Apidae). J Kansas Entomol Soc 45:520-523

Camargo CA de (1974a) Produçao de machos diplóides de *Melipona quadrifasciata* Lep. (Hymenoptera, Apidae). Ciência e Cultura 26:267

Camargo CA de (1984) Spermatozoa number and migration to the seminal vesicles in haploid and diploid males of *Melipona quadrifasciata* Lep. J Apicult Res 23:15-17

Camargo CA de, Almeida MG de, Parra MGN, Kerr WE (1976) Genetics of sex determination in bees. IX. Frequencies of queens and workers from larvae under controlled conditions (Hymenoptera: Apoidea). J Kansas Entomol Soc 49:120-125

Camargo JMF de (1970) Ninhos e biología de algumas espécies de Meliponídeos (Hymenoptera: Apidae) da regiao de Pôrto Velho, Território de Rondônia, Brasil. Rev Biol Trop 16:207-239

Camargo JMF de (1974b) Notas sobre a morfologia e a biologia de *Plebeia (Schwarziana) quadripunctata quadripunctata* (Hymenoptera, Apidae: Meliponinae). Stud Entomol 17:433-470

Camargo JMF de (1980) O grupo *Partamona (Partamona) testacea* (Klug): suas espécies, distribuiçao e diferenciaçao geográfica (Meliponinae, Apidae, Hymenoptera). Acta Amazonica 4, Suppl, 175 p

Camillo C (1971) Estudos adicionais sobre zangoes de *Trigona (Friesella) schrottkyi* Friese (Hym., Apidae). Ciência e Cultura 23:273

Camillo-Atique C (1977) Estudo da variabilidade etológica de *Friesella schrottkyii* incluindo a caracterizaçao de espêcies cripticas (Hymenoptera, Meliponinae). Tese Dout Univ Ribeirao Preto, 203 p

Campos LAO (1977) O hormônio juvenil nas abelhas: seu papel na diferenciaçao das castas e nos aspectos do controle social. Tese Dout Univ Ribeirao Preto, 67 p

Campos LAO, Velthuis HHW, Velthuis-Kluppell FM (1975) Juvenile hormone and caste determination in a stingless bee. Naturwiss 62:98-99

Campos LAO, Drummond MS, Lacerda L de M (1983) Sex determination in bees. XVIII: The role of juvenile hormones I, II and III in caste differentiation in *Scaptotrigona xanthotricha*. Ciência e Cultura 35:209-211

Cortopassi-Laurino M (1978) Observaçoes sôbre atividades de machos de *Plebeia droryana* Friese (Hymenoptera, Apidae, Meliponinae). Rev Bras Entomol 23:177-191

Cruz-Landim C da, Höfling JF (1972) The post-embryonic changes in *Melipona quadrifasciata anthidioides* Lep. (Hym., Apoidea). V. Development of the nervous system. Papéis Avulsos Zool 26:137-147

Cruz-Landim C da, Santos SMF dos, Höfling MCA (1980a) Sex determination in bees. XV. Identification of queens of *Melipona quadrifasciata anthidioides* (Apidae) with the worker phenotype by a study of the tergal glands. Rev Brasil Genet 3:295-302

Cruz-Landim C da, Höfling MCA, Imperatriz-Fonseca VL (1980b) Tergal and mandibular glands in queens of *Paratrigona subnuda* (Moure) (Hymenoptera: Apidae). Morphology and associated behaviour. Naturalia 5:121-133

Currie RW (1987) The biology and behaviour of drones. Bee World 68:129-143

Daerr E (1978) Effekte einer Allatektomie frischgeschlüpfter Bienenköniginnen auf Hochzeitsflug und Eiablage. Mitt dtsch Ges allg angew Entomol 1:317-319

Del Lama MA, Mestriner MA, Paiva JCA (1985) Est-5 and PGM_1: new polymorphisms in *Apis mellifera*. Rev Bras Genet 8:17-27

Darchen R (1973) La thermorégulation et l'écologie de quelques espècies d'abeilles sociales d'Afrique (Apidae, Trigonini et *Apis mellifica* var. *adansonii*). Apidol 4:341-370

Darchen R (1977) L'essaimage chez les *Hypotrigona* au Gabon - Dynamique de quelques populations. Apidol 8:33-59

Delage R, Delage-Darchen B (1970) Facteur déterminant les castes chez les trigones (Hym., Apidae). CR Acad Sci Paris D 270:1372-1373

Delage R, Delage-Darchen B (1977) Sur le déterminisme des castes chez les mélipones (Hyménoptères Apidés). Bull Biol France Belgique 111:91-109

Delage-Darchen B, Delage R (1982) Déterminisme des castes chez mélipones et trigones. In: Jaisson P (ed) Social insects in the tropics. Univ Paris-Nord, Vol 1, pp 31-40

Darchen R, Pain J (1966) Le nid de *Trigona (Dactylurina) staudingeri* Gribodoi (sic). Biol Gabonica 2:25-35

Dietz A (1986) Evolution. In: Rinderer TE (ed) Bee genetics and breeding. Academic Press, London, pp 3-21

Dzierzon J (1845) Gutachten über die von Herrn Direktor Stöhr im ersten und zweiten Kapitel des General-Gutachtens aufgestellten Fragen. Eichstädt Bienenz 1:109-113, 119-121

Eickwort GC, Ginsberg HS (1980) Foraging and mating behavior in *Apoidea.* Annu Rev Entomol 25:421-446

Engel MS (1979) Is caste determination in *Melipona quadrifasciata*, a stingless bee, influenced by 9-oxo-decenoic acid? Insect Soc 26:273-278

Engels E, Engels W (1984) Drohnen-Ansammlungen bei Nestern der Stachellosen Bienen *Scaptotrigona postica.* Apidol 15:315-328

Engels E, Engels W (1988) Age-dependent queen attractiveness for drones and mating in the stingless bee, *Scaptotrigona postica.* J Apicult Res 27:3-8

Engels E, Engels W, Schröder W, Francke W (1987a) Intranidal worker reactions to volatile compounds identified from cephalic secretions in the stingless bee, *Scaptotrigona postica* (Hymenoptera, Meliponinae). J Chem Ecol 13:371-386

Engels E, Engels W, Francke W (1987b) Do stingless bees have two complexes of queen substances? In: Eder J, Rembold H (eds) Chemistry and biology of social insects. Peperny, München, pp 275-281

Engels W (1971) Herkunft von Dotterproteinen bei *Apis mellifera.* Proc Int Apimondia Congr, Moskau, 23:439-441

Engels W (1972a) Quantitative Untersuchungen zum Dotterprotein-Haushalt der Honigbiene (*Apis mellifica*). Roux' Arch 171:55-86

Engels W (1972b) Der intra- und interindividuelle Protein-Stoffwechsel bei *Apis mellifica* unter besonderer Berücksichtigung der Oogenese. Insect Soc 19:415

Engels W (1973a) Die Evolution der Haemolymph-Protein-Spektren als entwicklungsphysiologisches Merkmal bei den weiblichen Kasten der 4 rezenten *Apis*-Arten *florea, dorsata, cerana* und *mellifica.* Proc Int Apimondia Congr, Buenos Aires, 24:327-332

Engels W (1973b) Das zeitliche und räumliche Muster der Dottereinlagerung in die Oocyte bei *Apis mellifica.* Z Zellforsch 142:409-430

Engels W (1973c) Vorkommen und Synthese von Dotter-Proteinen. Beziehungen zur Oogenese bei den weiblichen Kasten der rezenten *Apis*-Arten im Hinblick auf die Evolution monogyner Staaten. Proc Int IUSSI Congr, London, 7:92-103

Engels W (1974) Occurrence and significance of vitellogenins in female castes of social hymenoptera. Amer Zool 14:1229-1237

Engels W (1978) Der Einfluß verschiedener Juvenilhormone und Dosierungen auf den Vitellogenin-Stoffwechsel eierlegender Bienenköniginnen (*Apis mellifica*). Mitt dtsch Ges allg angew Entomol 1:308-312

Engels W (1980) Initiation and control of vitellogenin synthesis and yolk incorporation in social bees. Adv Invertebr Reprod 2:345

Engels W (1981) Control of fertility in bees. In: Sehnal F, Zabza A, Menn JJ, Cymborowsky B (eds) Regulation of insect development and behaviour, Vol II. Wroclaw Techn Univ Press, Bresslau, pp 655-660

Engels W (1985) Fertilitätskontrolle bei Honigbienen. Int Apimondia Congr, Nagoya, 30:98-100

Engels W (1986) The concept of chemical communication in arthropods as realized in social bee reproduction. Adv Invertebr Reprod 4:285-296

Engels W (1987a) Reproduction and caste development in social bees. In: Eder J, Rembold H (eds) Chemistry and biology of social insects. Peperny, München, pp 275-281

Engels W (1987b) Pheromones and reproduction in Brazilian stingless bees. Int Symp Insect, Mem Inst Oswaldo Cruz 82, Suppl III:35-45

Engels W (1988) Fortpflanzungsstrategien bei Bienen. Reproductive strategies in bees. Verh Dtsch Zool Ges 81:155-167

Engels W, Engels E (1977) Vitellogenin und Fertilität bei Stachellosen Bienen. Insect Soc 24:71-94

Engels W, Engels E (1987) Maintaining and overwintering of stingless bee colonies in a flight room. In: Eder J, Rembold H (eds) Chemistry and biology of social insects. Peperny, München, pp 575-576

Engels W, Fahrenhorst H (1974) Alters- und kastenspezifische Veränderungen der Haemolymph-Protein-Spektren bei *Apis mellifica*. Roux' Arch 174:285-296

Engels W, Gonçalves LS, Engels E (1976) Effects of carbon dioxide on vitellogenin metabolism in unmated queen honeybees. J Apicult Res 15:3-10

Engels W, Ramamurty PS (1976) Initiation of oogenesis in allatectomized virgin honey bee queens by carbon dioxide treatment. J Insect Physiol 22:1427-1432

Engels W, Engels E, Lotz G, Dröscher R, Wittmann H (1979) Nestbiologie der Stachellosen Biene *Scaptotrigona postica*. Hochschulunterrichts-Farb-Ton-Film 16 mm, IWF-Nr C 1351, Göttingen

Fahrenhorst H, Engels W (1976) Bau eines Ersatznestes durch ein weiselloses Restvolk von *Vespa (Paravespula) germanica*. Natur und Heimat 36:40-41

Fleig R (1987) Posture of the queen honeybee during egg deposition. In: Eder J, Rembold H (eds) Chemistry and biology of social insects. Peperny, München, p 322

Francke W, Rosengren R, Engels W, Engels E (1987) News from ants and bees. In: Eder J, Rembold H (eds) Chemistry and biology of social insects. Peperny, München, pp 430-431

Francke W, Schröder W, Engels E, Engels W (1983) Variation in cephalic volatile substances in relation to worker age and behaviour in the stingless bee, *Scaptotrigona postica*. Z Naturforsch 38c:1066-1068

Free JB (1987) Pheromones of social bees. Chapman and Hall, London, 218 p

Free JB, Williams IH (1975) Factors determining the rearing and rejection of drones by the honeybee colony. Anim Behav 23:650-675

Gary NE (1962) Chemical mating attractants in the honeybee. Science 136:773-774

Gerig L (1971) Wie Drohnen auf Königinnenattrappen reagieren. Schweiz Bienenztg 94:558-562

Gerig L (1972) Ein weiterer Duftstoff zur Anlockung der Drohnen von *Apis mellifica*. Z Angew Entomol 70:286-289

Hadorn H (1946) Betrachtungen über wilde Bienen in Sumatra. Schweiz Bienenztg 69:309-314

Hagedorn HH, Kunkel JG (1979) Vitellogenin and vitellin in insects. Annu Rev Entomol 24:475-505

Hamilton WD (1964) The genetical theory of social behaviour. Theoret Biol 7:1-52

Hartfelder K, Engels W (1989) The composition of larval food in stingless bees: evaluating nutritional balance by chemosystematic methods. Insect Soc 36:1-14

Heinrich B (1985) The social physiology of temperature regulation in honeybees. Fortschr Zool 31:393-406

Herrmann H (1969) Neurohormonale Kontrolle der Paarungsflüge und Eilegetätigkeit bei der Bienenkönigin (*Apis mellifica*). Z Bienenforsch 9:509-544

Hubbell SP, Johnson LK (1977) Competition and nest spacing in a tropical stingless bee community. Ecology 58:949-963

Ihering von H (1903a) Biologie der stachellosen Honigbienen Brasiliens. Zool Jb Syst Oekol Geogr Tiere 19:179-287

Ihering von R (1903b) Zur Frage nach dem Ursprung der Staatenbildung bei den sozialen Hymenopteren. Zool Anz 27:113-118

Imperatriz VL (1970) Aparecimento de supermachos em *Friesella schrottkyi* (Apoidea, Apidae, Meliponinae). Ciência e Cultura 22:291

Imperatriz-Fonseca VL (1973) Miscellaneous observations on the behavior of *Schwarziana quadripunctata* (Hym., Apidae, Meliponinae). Biol Zool Biol Mar NS 30:633-640

Imperatriz-Fonseca VL (1975) Estudos bionômicos sobre a rainha virgem de *Paratrigona subnuda* (Moure). Tese Dout Univ Sao Paulo, 217 p

Imperatriz-Fonseca VL (1976) Studies on *Paratrigona subnuda* (Moure) (Hymenoptera, Apidae, Meliponinae). I. Members of the colony. Rev Bras Ent 20:101-112

Imperatriz-Fonseca VL (1977) Studies on *Paratrigona subnuda* (Moure) (Hymenoptera, Apidae, Meliponinae). II. Behaviour of the virgin queen. Bolm Zool Univ Sao Paulo 2:169-182

Imperatriz-Fonseca VL (1978) Studies on *Paratrigona subnuda* (Moure) (Hymenoptera, Apidae, Meliponinae). III. Queen supersedure. Bolm Zool Univ Sao Paulo 3:153-162

Imperatriz-Fonseca VL, Godoi CBM (1978) Estudo comparativo do comportamento de abelhas Meliponinae (Hymenoptera, Apidae). Ciência e Cultura 30:604

Imperatriz-Fonseca VL, Kleinert-Giovannini A (1988) Relaçoes de dominância entre rainhas e operárias de *Plebeia (Friesella) schrottkyii*. Ciência e Cultura 40:655

Imperatriz-Fonseca VL, Oliveira MAC de (1974) Observations in a queenless colony of *Plebeia saiqui* (Friese) (Apidae, Meliponinae). Bol Zool Univ Sao Paulo 1:299-312

Imperatriz-Fonseca VL, Oliveira MAC de, Iwama S (1975) Notas sobre o compartamento de rainhas virgens de *Plebeia (Plebeia) remota* Holmberg (Apidae, Meliponinae). Ciência e Cultura 27:665-669

Inoue T, Sakagami SF, Salmah S, Yamane S (1984a) The process of colony multiplication in the Sumatran stingless bee, *Trigona (Tetragonula) laeviceps*. Biotropica 16:100-111

Inoue T, Sakagami SF, Salmah S, Nukmal N (1984b) Discovery of successful absconding in the stingless bee *Trigona (Tetragonula) laeviceps*. J Apicult Res 23:136-142

Juliani L (1962) O aprisionamento de rainhas em colônias de Trigonini. Bolm Univ Paraná 20:1-11

Juliani L (1967) A descriçao do ninho e alguns dados biológicos sobre a abelha *Plebeia juliani* Moure. Rev Bras Ent Sao Paulo 12:31-58

Kaatz HH (1987) Regulation of vitellogenin synthesis in honey bee queens. In: Eder J, Rembold H (eds) Chemistry and biology of social insects. Peperny, München, pp 317-318

Kaatz HH (1988) Juvenile hormone independent initiation and regulation of vitellogenin synthesis in the honey bee queen (*Apis mellifera* L.). Verh Dtsch Zool Ges 81:272-273

Kerr WE (1946) Formaçao das castas no gênero *Melipna* (Illinger, 1806). An Esc Sup Agric Luiz de Queiroz USP 3:300-312

Kerr WE (1950) Genetic determination of castes in the genus *Melipona*. Genetics 35:143-152

Kerr WE (1951) Bases para o estudo de genética de populaçoes dos Hymenoptera em geral e dos Apinae sociais em particular. Anais Esc Sup Agric Luiz de Queiroz Univ Sao Paulo 8:221-354

Kerr WE (1973) Genética e biologia de abelhas. Ciência e Cultura 25:927-934

Kerr WE (1974) Sex determination in bees. III. Caste determination and genetic control in *Melipona*. Insect Soc 21:357-368

Kerr WE, Krause W (1950) Contribuiçao para o conhecimento da bionomia dos *Meliponini*. Dusenia 1:275-282

Kerr WE, Zucchi R, Nakadaira JT, Butolo JE (1962) Reproduction in the social bees (Hymenoptera, Apidae). J NY Entomol Soc 70:265-276

Kerr WE, Sakagami SF, Zucchi R, Portugal Aráujo V de, Camargo JMF de (1967) Observaçoes sôbre a arquitetura dos ninhos e comportamento de algumas espécies de abelhas sem ferrao das vizinhanças de Manaus, Amazonas (Hymenoptera, Apoidea). Atas Simpósio Biota Amazônica Vol 5 (Zoologia):255-309

Kleinert-Giovannini A (1989a) Mecanismos de controle reprodutivo em colônias de *Melipona marginata* Lepeletier (Apidae, Meliponinae). Tese Dout Univ Sao Paulo

Kleinert-Giovannini A (1989b) Swarming activity in *Plebeia emerina*. In prep

Kleinert-Giovannini A, Imperatriz-Fonseca VL (1988) A existência de refúgios de rainhas virgens em colônias de *Melipona marginata* Lepeletier (Apidae, Meliponinae). Ciência e Cultura 40:

Kleinert-Giovannini A, Imperatriz-Fonseca VL (1989) Estratègias reprodutivas em *Scaptotrigona bipunctata* (Apidae, Meliponinae) como consequência da seleçao de parentesco. Ciência e Cultura 41:739-740
Koeniger G (1986) Reproduction and mating behavior. In: Rinderer TE (ed) Bee breeding and genetics. Academic Press, Orlando FA, pp 255-288
Koeniger G (1988) Mating behavior of honey bees. In: Needham GR, Page RE, Delfinado-Baker M, Bowman CE (eds) Africanized honey bees and bee mites. Ellis Horwood, Chichester, pp 167-172
Koeniger N (1969) Experiments concerning the ability of the queen (*Apis mellifera* L.) to distinguish between drone and worker cells. Proc Int Apimondia Congr, 22:138
Koeniger N (1970) Über die Fähigkeit der Bienenkönigin (*Apis mellifica* L.) zwischen Arbeiterinnen- und Drohnenzellen zu unterscheiden. Apidol 1:115-142
Koeniger N, Veith HJ (1983) Identification of a triglyceride (glyceryl-1,2-dioleate-3-palmitate) as a brood pheromone of the honey bee (*Apis mellifera* L.). Apidol 14:59
Koeniger N, Veith HJ (1984) Spezifität eines Brutpheromons und Bruterkennung bei der Honigbiene (*Apis mellifera* L.). Apidol 15:205-210
Koeniger N, Wijayagunasekera HNP (1976) Time of drone flight in the three Asiatic honeybee species (*Apis cerana, Apis florea, Apis dorsata*). J Apicult Res 15:67-71
Koeniger N, Koeniger G, Tingek S, Mardan M, Rinderer TE (1988) Reproductive isolation by different time of drone flight between *Apis cerana* Fabricius, 1793 and *Apis vechti* (MAA, 1953). Apidol 19:103-106
Lensky Y, Cassier P, Nutkin M, Delorme-Joulie M, Levinson M (1985) Pheromonal activity and fine structure of the mandibular glands of the honeybee drones. (*Apis mellifera* L.) (Insecta, Hymenoptera, Apidae). J Insect Physiol 31:265-276
Leuckart R (1855) Seebacher Studien. Eichstädt Bienenz 11:199-211
Lindauer M (1955) Schwarmbienen auf Wohnungssuche. Z vergl Physiol 37:263-324
Mackensen O (1955) Further studies on a lethal series in the honey bee. J Heredity 46:72-74
Martinho MR (1979) Competiçao reprodutiva entre machos de *Apis mellifera* L. e migraçao de espermatozóides para a espermateca de rainhas. Tese Dout Univ Ribeirao Preto, 134 p
Martinho MR, Gonçalves LS (1980) Valor competitivo de zangoes filhos de rainhas e de operarias de *Apis mellifera ligustica* e *Apis mellifera adansonii* (Hymenoptera, Apidae). Anais Congr Bras Apicult, Viçosa/Brasil, 5:76-89
Mestriner MA, Contel EBP (1972) The P-3 and Est loci in the honeybee *Apis mellifer.* Genetics 72:733-738
Michener CD (1946) Notes on the habits of some Panamanian stingless bees (Hymenoptera, Apidae). J NY Entomol Soc 54:179-197
Michener CD (1974) The social behavior of the bees. Harvard Univ Press, Cambridge, 404 p
Michener CD (1985) From solitary to eusocial: need there be a series of intervening species? Fortschr Zool 31:293-305
Michener CD, Brothers DJ (1974) Were workers of eusocial Hymenoptera initially altruistic or oppressed? Proc Nat Acad Sci USA 71:671-674
Moritz RFA (1985a) Intracolonial worker relationship and sperm competition in the honeybee (*Apis mellifera* L.). Experientia 42:445-448
Moritz RFA (1985b) Inzuchteffekte bei Drohnen (*Apis mellifera* L.). Insect Soc 32:104-105
Moritz RFA (1985c) The effects of multiple mating on the worker-queen conflict in *Apis mellifera* L. Behav Ecol Sociobiol 16:375-377
Moritz RFA, Crewe RM (1988) Chemical signals of queens in kin recognition of honeybees, *Apis mellifera* L. J Comp Physiol A 164:83-90
Moritz RFA, Hillesheim E (1985) Inheritance of dominance in honeybees (*Apis mellifera capensis* Esch.). Behav Ecol Sociobiol 17:87-89
Mota MHVB (1982) Estudo comparativo das glândulas abdominais relacionadas com o comportamento social em Apidae (Apinae e Meliponinae). Tese Mestr Univ Rio Claro, 174 p

Mota MHVB (1988) Estudo comparativo das glândulas epidérmicas do abdômen dos Meliponineos (Hymenoptera, Apidae, Meliponinae). Tese Dout Univ Rio Claro, 170 p
Moure JS, Nogueira-Neto P, Kerr WE (1958) Evolutionary problems among Meliponinae. Proc Int Congr Entomol 10, Vol 2:481-494
Müller H (1875) Stachellose brasilianische Honigbienen zur Einführung in zoologischen Gärten empfohlen. Zool Garten 26:41-55
Nijhout HF (1975a) Dynamics of juvenile hormone action in larvae of the tobacco hornworm, *Manduca sexta* (L.). Biol Bull 149:568-579
Nijhout HF (1975b) A threshold size for metamorphosis in the tobacco hornworm, *Manduca sexta* (L.). Biol Bull 149:214-225
Nogueira-Neto P (1951) Stingless bees and their study. Bee World 32:73-76
Nogueira-Neto P (1954) Notas bionômicas sôbre meliponíneos. III. Sôbre a enxameagem. Arq Mus Nac Rio de Janeiro 42:419-452
Nogueira-Neto P (1958) Rainhas prisoneiras. Chácaras e Quintais, Sao Paulo 97:477
Nogueira-Neto P (1970) A criaçao de abelhas indígenas sem ferrao. Chácaras e Quintais, Sao Paulo, 365 p
Oliveira MAC de, Imperatriz-Fonseca VL (1973) Observaçoes sobre o comportamento de uma colônia mista de *Plebeia saiqui* - *Plebeia droryana* (Hymenoptera, Apidae, Meliponinae). Ciência e Cultura 25:460-462
Otis GW (1982) Population biology of the Africanized honey bee. In: Jaisson P (ed) Social insects in the tropics. Univ Paris-Nord, Vol 1, pp 209-219
Otis GW, Winston ML, Taylor OR, Jr (1981) Engorgement and dispersal of africanized honeybee swarms. J Apicult Res 20:3-12
Page RE (1980) The evolution of multiple mating behaviour by honey bee queens (*Apis mellifera* L.). Genetics 96:263-273
Page RE, Erickson EH (1986) Kin recognition and virgin queen acceptance by worker honey bees (*Apis mellifera* L.). Anim Behav 34:1061-1069
Page RE, Laidlaw HH (1985) Closed population honeybee breeding. Bee World 66:63-72
Perez J (1889) Les abeilles. Paris, Hachette, 348 p
Planta AV (1888) Über den Futtersaft der Bienen. Hoppe Seyl Z Physiol Chemie 12:327-354 and 13:552-561
Praagh JP van, Ribi WA, Wehrhahn C, Wittmann D (1980) Drone bees fixate the queen with the dorsal frontal part of their compound eyes. J Comp Physiol 136:263-266
Rachinsky A, Hartfelder K (1989) Corpora allata activity controls caste-specific juvenile hormone titers in last instar honey bee larvae. J Insect Physiol 35:
Rachinsky A, Strambi C, Strambi A, Hartfelder K (1989) Caste and metamorphosis: haemolymph titers of juvenile hormone and ecdysteroids in last instar honey bee larvae. Gen Comp Endocrinol 74:
Ramamurty PS, Engels W (1977) Allatektomie- und Juvenilhormon-Wirkungen auf Synthese und Einlagerung von Vitellogenin bei der Bienenkönigin (*Apis mellifica*). Zool Jb Physiol 81:165-176
Reddy Ch (1980) Studies on the nesting behavior of *Apis dorsata F.* Int Conf Apicult Trop Climate 2:391-397 New Delhi Inst Agr Res
Regel R (1978) Auswirkungen einer Langzeitbehandlung mit unterschiedlichen Dosen von Juvenilhormon III auf Verhalten und Fertilität von Jungköniginnen (*Apis mellifica*). Mitt dtsch Ges allg angew Entomol 1:313-316
Rembold H (1976) The role of determination in caste formation in the honey bee. In: Lüscher M (ed) Phase and caste determination in insects. Pergamon Press, Oxford, pp 21-34
Rembold H, Hanser H (1964) Über den Weiselzellenfuttersaft der Honigbiene. VIII: Nachweis des determinierenden Prinzips im Futtersaft der Königinnenlarve. Hoppe Seyler's Z Physiol Chem 316:206-212
Rembold H, Lackner B (1981) Rearing of honeybee larvae *in vitro*: Effect of yeast extract on queen differentiation. J Apicult Res 20:165-171
Renner M, Vierling G (1977) Die Rolle des Taschendrüsenpheromons beim Hochzeitsflug der Bienenkönigin. Behav Ecol Sociobiol 2:329-338

Rhein W von (1933) Ueber die Entstehung des weiblichen Dimorphismus im Bienenstaate. Roux's Arch 129:601-655

Ribeiro MF (1989a) Etologia de rainhas virgens de *Frieseomelitta silvestrii* (Friese) (Hymenoptera, Apidae, Meliponinae). Ciência e Cultura 41:

Ribeiro MF (1989b) Estudos etoecológicos em *Frieseomelitta languida* Moure & Camargo, in litt (Hymenoptera, Apidae, Meliponinae). Tese Mestr Univ Sao Paulo, 129 p

Ribi WA, Engels E, Engels W (1989) Sex and caste specific eye structures in stingless bees and honey bees (Hymenoptera: Trigonidae, Apidae). Entomol Gener 14:233-242

Rinderer T, Collins AM, Hellmich II RL, Danka RG (1987) Differential drone production by africanized and european honey bee colonies. Apidol 18:61-68

Rosenkranz P (1988) Temperturpräferenz der *Varroa*-Milbe und Stocktemperaturen in Bienenvölkern an Tropenstandorten (Acarina: Varroidae/Hymenoptera: Apidae). Entomol Gener 14:123-132

Rosenkranz P, Engels W, Engels E, Kläger S (1987) Reaction of *Scaptotrigona depilis* colonies to a change in external temperature. In: Eder J, Rembold H (eds) Chemistry and biology of social insects. Peperny, München, p 721

Roubik DW (1979) Nest and colony characteristics of stingless bees from French Guiana (Hymenoptera: Apidae). J Kansas Entomol Soc 52:443-470

Roubik DW (1983) Nest and colony characteristics of stingless bees from Panamá (Hymenoptera: Apidae). J Kansas Entomol Soc 56:327-355

Roubik DW (1989a) Ecology and natural history of tropical bees. Cambridge Univ Press, New York, 248 p

Roubik DW (1989b) Reproductive ecology of stingless bees: observations on *Tetragona* and mate competition (Hymenoptera: Apidae: Meliponinae). Entomol Gener 15:

Ruttner F (1956) The mating of the honeybee. Bee World 37:2-15, 23-24

Ruttner F (1985) Reproductive behaviour in honeybees. Fortschr Zool 31:225-236

Ruttner F (1988) Biogeography and taxonomy of honeybees. Springer-Verlag, Berlin, Heidelberg, New York, 284 p

Ruttner F, Ruttner H (1966) Untersuchungen über die Flugaktivität und das Paarungsverhalten der Drohnen. 3. Flugweite und Flugrichtung der Drohnen. Z Bienenforsch 8:332-354

Ruttner H, Ruttner F (1972) Untersuchungen über die Flugaktivität und das Paarungsverhalten der Drohnen. Apidol 3:303-232

Ruttner F, Hesse B (1979) Rassenspezifische Unterschiede in Ovarentwicklung und Eiablage von weisellosen Arbeiterinnen der Honigbiene *Apis mellifera* L. Apidol 12:159-183

Rutz W, Lüscher M (1974) The occurrence of vitellogenin in workers and queens of *Apis mellifica* and the possibility of its transmission to the queen. J Insect Physiol 20:897-909

Safranek L, Williams CM (1984) Critical weights for metamorphosis in the tobacco hornworm, *Manduca sexta*. Biol Bull 167:555-567

Sakagami SF (1958) The false queen: fourth adjustive response in dequeened honey bee colonies. Behav 8:280-296

Sakagami SF (1971) Ethosoziologischer Vergleich zwischen Honigbienen und stachellosen Bienen. Z Tierpsychol 28:337-350

Sakagami SF (1982) Stingless bees. In: Hermann HR (ed) Social insects. Academic Press, New York, Vol 3, pp 361-423

Sakagami SF, Laroca S (1963) Additional observations on the habits of the cleptobiotic stingless bees, the genus *Lestrimelitta* Friese (Hymenoptera, Apoidea). J Fac Sci Hokkaido Univ, Ser 6 Zool 15:319-339

Sakagami SF, Zucchi R (1963) Oviposition process in a stingless bee, *Trigona (Scaptotrigona) postica* Latreille (Hymenoptera). Studia Entomol 6:497-510

Sakagami SF, Zucchi R (1966) Estudo comparativo do comportamento de várias espécies de abelhas sem ferrao, com especial referência ao processo de aprovisionamento e postura das células. Ciência e Cultura 18:283-296

Sakagami SF, Zucchi R (1974) Oviposition behavior of two dwarf stingless bees, *Hypotrigona (Leurotrigona) muelleri* and *H. (Trigonisca) duckei*, with notes to the temporal articulation of oviposition process in stingless bees. J Fac Sci Hokkaido Univ Ser VI Zool 19:361-421

Sakagami SF, Beig G, Zucchi R, Akahira Y (1963) Occurrence of ovary-developed workers in queenright colonies of stingless bees. Rev Brasil Biol 23:115-129

Sakagami SF, Beig G, Akahira Y (1964) Behavior studies of the stingless bee with special reference to the oviposition process. III. Appearance of laying workers in an orphan colony of *Partamona (Partamona) testacea testacea* (Klug). Jap J Ecol 14:50-57

Sakagami SF, Montenegro MJ, Kerr WE (1965) Behavior studies of the stingless bee, with special reference to the oviposition process. V. *Melipona quadrifasciata anthidioides* Lepeletier. J Fac Sci Hokkaido Univ Ser VI Zool 15:578-607

Sakagami SF, Zucchi R, Araujo VP (1977) Oviposition behavior of *Meliponula bocandei*, an aberrant African stingless bee, with notes to the mechanism and evolution of oviposition behavior in stingless bees. J Fac Sci Hokkaido Univ Ser VI Zool 20:647-690

Sannasi A, Ratalu GS, Sundara G (1971) 9-oxo-trans-2-decenoic acid in the Indian honeybees. Life Science 10:195-201

Schneider D (1984) Pheromones and insect reproduction. Adv Invertebr Reprod 3:435-439

Schwarz HF (1948) Stingless bees (Meliponinae) of the western hemisphere. Bull Am Mus Nat Hist 90:1-546

Seeley TD (1977) Measurement of nest cavity volume by the honey bee (*Apis mellifera*). Behav Ecol Sociobiol 2:201-227

Seeley TD (1978) Life history strategy of the honey bee, *Apis mellifera*. Oecologia 32:109-118

Seeley TD (1988) What we do and do not know about the nest-site preferences of African honey bees. In: Needham GR, Page RE, Delfinado-Baker M, Bowman CE (eds) Africanized honey bees and bee mites. Ellis Horwood, Chichester, pp 87-90

Seeley TD, Heinrich B (1981) Regulation of temperature in the nests of social insects. In: Heinrich B (ed) Insect thermoregulation. Wiley, New York, pp 159-234

Seeley TD, Morse RA (1976) The nest of the honey bee. Insect Soc 23:495-512

Seeley TD, Morse RA (1978) Nest site selection by the honey bee, *Apis mellifera*. Insect Soc 25:323-337

Seeley TD, Visscher PK (1985) Survival of honeybees in cold climes: the critical timing of colony growth and reproduction. Ecol Entomol 10:81-88

Shearer DA, Boch R, Morse RA, Laigo FM (1970) Occurrence of 9-oxodec-trans-2-enoic acid in queens of *Apis dorsata*, *Apis cerana* and *Apis mellifera*. J Insect Physiol 16:1437- 1441

Siebold CT E von (1856) Wahre Parthenogenesis bei Schmetterlingen und Bienen. W. Engelmann, Leipzig, 149 p

Silva DLN da (1972) Consideraçoes em torno de um caso de substituçao de rainha em *Plebeia (P.) droryana* (Friese, 1900) (Hym., Apidae). In: Cruz-Landim C da, Hebling NI, Lello E de, Takahashi CS (eds) Homenagem à Warwick E. Kerr. Rio Claro, SP, pp 267-273

Silva DLN da (1973) Estudos bionômicos em colônias mistas de Meliponinae (Hymenoptera, Apidae). Tese Dout Univ Ribeirao Preto, 114 p

Silva DLN da (1977) Estudos bionômicos em colônias mistas de Meliponinae (Hymenoptera, Apoidea). Bolm Zool Univ Sao Paulo 2:7-106

Silva DLN da, Zucchi R, Kerr WE (1972) Biological and behavioural aspects of the reproduction in some species of *Melipona* (Hymenoptera, Apidae, Meliponinae). Anim Behav 20:123-132

Simoes D (1974) Estudos sobre a regulçao social em *Nannotrigona (Scaptotrigona) postica* Latreille, com especial referência a aspectos comportamentais (Hymenoptera, Apidae, Meliponinae). Tese Mestr Univ Ribeirao Preto, 92 p

Simoes D, Bego LR (1979) Estudo da regulaçao social em *Nannotrigona (Scaptotrigona) postica* Latreille, em duas colônias (normal e com rainhas virgens) com especial referência ao polietismo etário (Hym., Apidae, Meliponinae). Bolm Zool Univ Sao Paulo 4:89-98

Smith FG (1961) Races of honeybees in East Africa. Bee World 42:255-260

Smith BH, Ayasse M (1987) Kin-based male mating preferences in two species of halictine bee. Behav Ecol Sociobiol 20:313-318

Sommeijer MJ, Bruijn LLM de (1983) Social behavior of stingless bees: "Bee-dances" by workers of the royal court and the rhythmicity of brood cell provisioning and oviposition behavior. In: Sommeijer MJ (ed) Social mechanisms in stingless bees. Utrecht, Krips Repro Meppel, pp 47-68

Spinola M (1840) Observations sur les apiaries meliponides. Ann Sci Nat Zool 13:116-140

Staurengo da Cunha MA (1977) Ovarian development in *Scaptotrigona postica* Latr. (Hymenoptera - Apidae). I. Morphological study. Dusenia 10:205-215

Staurengo da Cunha MA (1979) Ovarian development in *Scaptotrigona postica* Latr. 1807 (Hym.: Apidae). II. A quantitative study. Insect Soc 26:196-203

Taber S III, Poole KH (1973) Caste determination in honey bees. The production of queen-worker intermediates. J Apicult Res 12:111-116

Terada Y (1972) Enxameagem em *Frieseomelitta varia* Lep. (Hymenoptera, Apidae). In: Cruz-Landim C da, Hebling NI, Lello E de, Takahashi CS (eds) Homenagem à Warwick E. Kerr. Rio Claro, SP: pp 293-299

Terada Y (1974) Contribuiçao ao estudo de regulaçao social em *Leurotrigona muelleri* e *Friesomelitta varia* (Hymenoptera, Apidae). Tese Mestr Univ Ribeirao Preto, 63 p

Thornhill R, Alcock J (1983) The evolution of insect mating systems. Harvard Univ Press, Cambridge, Mass

Tirgari S (1971) Biology and behavioral characteristics of the Iranian dwarf honeybee (*Apis florea*). Proc Int Apimondia Congr 23:344-345

Trenczek T, Engels W (1986) Occurrence of vitellogenin in drone honeybees (*Apis mellifica*). Int J Invertebr Reprod Developm 10:307-311

Trenczek T, Zillikens A, Engels W (1989) Developmental patterns of vitellogenin haemolymph titer and rate of synthesis in adult drone honey bees (*Apis mellifera*). J Insect Physiol 35:475-481

Trivers RL, Hare H (1976) Haplodiploidy and the evolution of the social insects. Science 191:249-263

Trube A (1988) Aufbau eines Biotests für Wirkungen von Komponenten des Königinnenpheromons der Honigbiene (*Apis mellifera carnica*) auf Follikelentwicklung bei Arbeiterinnen. Diplomarbeit Univ Tübingen

Velthuis HHW (1970) Ovarian development in *Apis mellifera* worker bees. Ent exp appl 13:377-394

Velthuis HHW (1976a) Environmental, genetic and endocrine influences in stingless bee caste determination. In: Lüscher M (ed) Phase and caste determination in insects. Pergamon Press, Oxford, pp 63-70

Velthuis HHW (1976b) Egg laying, aggression and dominance in bees. Proc Int Congr Entomol, Washington, 25:436-449

Velthuis HHW (1985) The honeybee queen and the social organization of her colony. Fortschr Zool 31:343-357

Velthuis HHW, Velthuis-Kluppell FM (1975) Caste differentiation in a stingless bee, *Melipona quadrifasciata* Lep., influenced by juvenile hormone application. Proc Kon Neder Akad Wetenschapen Ser C 78:81-94

Velthuis HHW, Clement J, Morse RA, Laigo FM (1971) The ovaries of *Apis dorsata* workers and queens from the Philipines. J Apicult Res 10:63-66

Verma S, Ruttner F (1983) Cytological analysis of the thelytokous parthenogenesis in the Cape honeybee (*Apis mellifera capensis* Escholtz). Apidol 14:41-58

Wagner AE, Briscoe DA (1983) An absence of enzyme variability within two species of *Trigona* (Hymenoptera). Heredity 50: 97-103

Weiss K (1978) Zur Mechanik der Kastenentstehung bei der Honigbiene (*Apis mellifica* L.). Apidol 9:223-258

Whitcombe RP (1984) The biology of *Apis* spp. in Oman with special reference to *Apis florea* Fab. Doct Thesis Univ Durham, 621 p

Wilde J de (1975) An endocrine view of metamorphosis, polymorphism and diapause in insects. Am Zool 15:13-28

Wilde J de (1976) Juvenile hormone and caste differentiation in the honey bee (*Apis mellifera* L.). In: Lüscher M (ed) Phase and caste determination in insects. Pergamon Press, Oxford, pp 5-20

Wilde J de, Beetsma J (1982) The physiology of caste development in social insects. Adv Insect Physiol 16:167-146

Wilde J de, Wirtz P (1974) Hormone und Polymorphismus der Insekten. Fortschr Zool 22:93-98

Wille A (1983) Biology of the stingless bees. Annu Rev Entomol 28:41-64

Wille A, Michener CD (1973) The nest architecture of stingless bees with special reference to those of Costa Rica (Hym., Apidae). Rev Biol Trop 2:1-278

Wille A, Orozco E (1974) Observations on the founding of a new colony of *Trigona cupira* (Hymenoptera, Apidae) in Costa Rica. Rev Biol Trop 22:253-287

Winston ML (1979) Intracolony demograpy and reproductive rate of the Africanized honeybee in South America. Behav Ecol Socioiol 4:279-292

Winston ML (1980) Swarming, afterswarming and reproductive rate of unmanaged honeybee colonies (*Apis mellifera*). Insect Soc 27:391-398

Winston ML, Michener CD (1977) Dual origin of highly social behavior among bees. Proc Nat Acad Sci USA 74:1135-1137

Winston ML, Otis GW, Taylor OR (1979) Absconding behaviour of the Africanized honeybee in South America. J Apicult Res 18:85-94

Winston ML, Dropkin JA, Taylor OR (1981) Demography and life history characteristics of two honey bee races. Oecologia 48:407-413

Wittmann D (1989) Nest architecture, nest site preferences and distribution of *Plebeia wittmanni* (Moure & Camargo, 1989) in Rio Grande do Sul, Brazil (Apidae: Meliponinae). Stud Neotrop Fauna Environm 24:17-23

Wittmann D, Engels W (1987) Welche Diät ergibt Arbeiterinnen bei *in vitro*-Aufzucht von Honigbienen? Apidol 18:279-288

Woyke J (1963) Drone larvae from fertilized eggs of the honeybee. J Apicult Res 2:19-24

Woyke J (1976) Population genetic studies in sex alleles in the honeybee using the example of the Kangaroo Island bee sanctuary. J Apicult Res 15:105-123

Woyke J (1988) Brood survival in productive bee apiaries in Australia as a test for breeding honeybees in closed populations. J Apicult Res 27:30-34

Zucchi R (1977) Aspectos etológico-evolutivos da bionomia dos Meliponinae (Hymenoptera. Apidae). Tese Doc.-Livre Univ Ribeirao Preto, 204 p

Zucchi R, Sakagami SF (1972) Thermoregulatory capacity of *Trigona spinipes* and some other species of stingless bees. In: Cruz-Landim C da, Hebling NI, Lello E de, Takahashi CS (eds) Homenagem à Warwick E. Kerr. Rio Claro, SP: pp 301-309

Chapter 8

Differentiation in Reproductive Physiology and Behaviour During the Development of Laying Worker Honey Bees

HAYO H.W. VELTHUIS[1], FRIEDRICH RUTTNER[2], ROBIN M. CREWE[3]

CONTENTS

1 Introduction

Reproduction in the social insects is governed by physiological factors within the individual organism and by behavioural factors acting at the colony level. The interaction of these two types of factors is the topic of this chapter.

---

[1] *Laboratorium voor Vergelijkende Fysiologie, Rijksuniversiteit Utrecht, Nederlands*
[2] *Institut für Bienenkunde, Universität Frankfurt, Oberursel, BRD*
[3] *Department of Zoology, University of the Witwatersrand, Johannesburg, South Africa*

2 Castes in Honey Bees

The egg producing female is, by definition, the queen, and the other females are called workers. In eusocial bees there is generally only one queen per colony. In the more primitive social bees, like the Halictids (Michener, chapter 4), it is often impossible to distinguish workers from the queen on morphological grounds and their functional differentiation occurs only during adulthood, as a result of competitive interactions. In higher bees, the differentiation into worker or queen takes place during larval and pupal development; here it is mainly the consequence of different food quantities ingested during the larval stages. This earlier initiation of the development of caste specific characters leads to two extreme types of females, there being no longer a continuum of female morphs and functions within the colony.

Although there are distinct castes for the female sex, worker honey bees may nevertheless enter the no man's land between the two castes. This happens for instance temporarily when a bee, by being in the court around the queen, attains a small amount of the queens' pheromones. In the following 10-30 minutes, such a worker bee experiences some of the favours normally shown to the queen, such as a high rate of being offered food. Her biological function is to make known to the colony at large that the queen is still present. Such workers have been called messenger bees (Seeley 1979) or, when they are introduced into groups of queenless workers that will react more intensely, substitute queens (Velthuis 1972).

3 The Laying Worker

Another way towards an intermediate position is the development of the laying worker. Although all worker bees possess ovaries, they normally are unable to produce eggs because the queen inhibits the activation of the hormonal system required for egg formation. The rare exception occurs when the queen has disappeared and is not replaced by a new one. Then a very limited number of bees is able to achieve reproductive potency. Since these individuals do not have a number of other queen specific characters, we still call them workers, although, in a strict sense, they meet the requirements of the definition of a queen given above. However, along with the activation of the ovary, they develop a means to suppress such activation in their companion worker bees, as does the queen.

There are rare occasions where such a laying worker crosses the opposite border of no man's land. She then also evokes court behaviour by the other workers and is treated as if she were a queen. Apparently, in her relation to the other workers, she produces the necessary stimuli to regulate their behaviour. At the same time she adopts behavioural characteristics of real queens. Therefore, the denomination 'false queen' (Sakagami 1958) seems appropriate.

The scarcity of false queens has prevented a systematic study of the causation of the transitional caste development in the adult stage. However, by exploiting the difference between honey bee races, this can be overcome. In such a study the Cape honey bee, *Apis mellifera capensis*, plays a decisive role.

4 The Cape Honey Bee

Apis mellifera capensis is a race of honey bee that occurs in the Cape province of South Africa (Ruttner 1976, 1988). Its most peculiar characteristics, the presence in the worker of a functional spermatheca and the production of female offspring by laying workers that are not inseminated, were reported by Onions (1912) and Jack (1916) and confirmed later by Anderson (1963).

The biological significance of this thelytokous parthenogenesis of the worker is that from eggs of laying workers a new queen may develop. This is particularly important where the geographic range of *capensis* borders that of *scutellata* (better known as the *adansonii* of Zimbabwe and South Africa, see Ruttner 1988). In this border area *capensis* workers may penetrate queenless colonies of *scutellata*, where they develop into laying workers. Their offspring gradually replace those of the host race. In these areas we may consider *capensis* as a facultative parasite.

In comparison to other *Apis mellifera* races, the *capensis* worker has characteristics of an intermediate caste (Table 1). This is not only true for her reproductive organs, as we will see later on.

Once the queen of a *capensis* colony is removed, laying workers are rather quickly produced. In comparison with other races this interval was estimated to be 6.5 days, whereas European races needed 22-30 days in this experiment. Only *Apis mellifera intermissa*, needing only

Table 1 - Some characters of a *A. m. capensis* worker in comparison to the queen and the worker of *A. m. mellifera* :

	A. m. mellifera queen	*A. m. capensis* worker	*A. m. mellifera* worker
rate of development (days)	16	19	21
number of ovarioles / ovary	160-180	10-25	3-10
diameter of spermatheca (mm)	1.2-1.5	0.4-0.9	0.0-0.1
maximum rate of egg laying / day	1.000-2.000	50-200	5-15
duration of reproductive period	3-6 years	2-5 months	up to 10 days

5.6 days on the average, was quicker. However, *intermissa* laying workers produce male offspring as is the case in other races (Ruttner 1988).

Within a colony of a different race, the development of a laying worker of *capensis* may be even more rapid, especially when only one young *capensis* is present. Egg laying may start after 5 days. Her most favourable age for introduction is at about one week; then egg laying may even start after only 3 days. However, after having reached an age of 2-3 weeks she becomes incapable of developing into a laying worker. When introduced into the host colony upon emergence, or when the worker emerges within the host colony, the reproductive stage is reached in 7-10 days. Apparently, there is a period of optimal sensitivity to social factors permitting activation of reproductivity.

The laying *capensis* worker is immediately recognizable by the presence of a court of host bees. This is related to the production of queen pheromones (Ruttner, Koeniger and Veith 1976; Velthuis 1976; Hemmling, Koeniger and Ruttner 1979; Crewe and Velthuis 1980). Such production of queen pheromones is not exclusive to the mandibular glands; also the tergal glands, so far being considered to be absent in workers, may develop if conditions are favorable. Especially in the *capensis* workers these glands may become very active (Billen, Dumortier and Velthuis 1986). As mentioned before, such queenlike laying workers are to be called false queens. They are reported to appear naturally in *Apis cerana* (Sakagami 1958) and also occasionally in other *Apis mellifera* races (Velthuis 1970). After introduction of several *capensis* pupae or young adult workers into *Apis mellifera mellifera* or *Apis mellifera carnica* colonies, one or more workers may develop into false queens. The number seems to depend on the size of the host colony, there being generally but one false queen in a 2-frame observation hive. The other *capensis* workers are evicted. If several *capensis* false queens are present simultaneously, there is no mutual aggressiveness between them as is generally the case when two queens meet inside the same colony. When the *capensis* false queens meet, they have a common court.

False queens of *capensis* are observed to reach an age of 3-5 months and are capable of reproducing during this entire period. However, they seldom reach that age, for the appearance of their first offspring causes much aggressiveness by the host workers. Many *capensis* workers are thrown out of the colony. In this period also the false queen may be attacked and killed. However, provided enough *capensis* brood is emerging, bees of this race become the most numerous. Several laying workers may develop at the same time in the colony, which now contains a majority of *capensis* workers. Peculiarly, the laying workers now show all the characteristics known of laying workers from other races: absence of a court, irregular

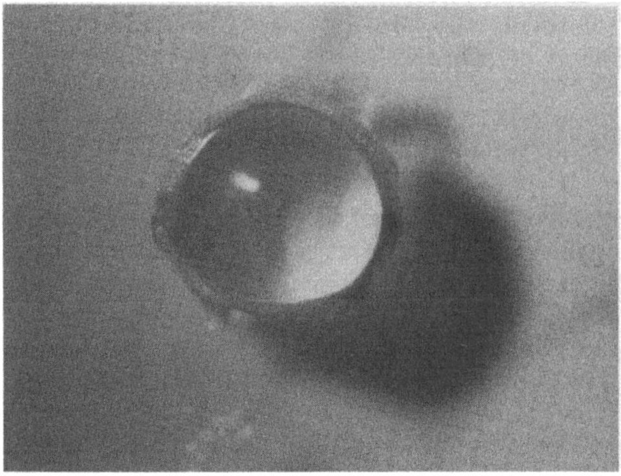

Fig. 1 - Spermatheca from a queen that produced mainly female offspring. The tracheal envelope has been removed, disclosing the spermatheca being well filled with liquid but devoid of sperms.

deposition of eggs in a cell, many eggs in the same cell, generally submissive behaviour towards companions. The colony is now apparently in a "queenless condition" and queens are therefore reared in emergency cells from the eggs being laid. In other instances the emergence of the *capensis* bees leads to swarming, in which the false queen may keep the swarm together while it is moving.

5 Rearing Methods for *Capensis*

The subtropical *capensis* has been reared in temperate Europe since its introduction in Oberursel, Germany, in 1973. From there it reached Utrecht, Holland, in 1975. In Oberursel the race is maintained by means of artificially inseminated queens, that are kept in whole colonies, either pure *capensis* or mixed with a European race. The latter is especially favourable for overwintering.

The *capensis* bee can also be kept in the indoor flight rooms. In Utrecht we used exclusively the thelytokous reproduction of the workers and got 27 generations without a single insemination. After the 18th worker generation, we discovered that queens reared from the worker eggs also reproduced thelytokously (Fig. 1), a feature not present or only to some extent in the original *capensis*.

Workers have been artificially inseminated by Ruttner and by Woyke (pers. comm.) with the result that more than 600.000 living spermatozoa have been found in the spermatheca and that in some cases fertilized eggs were laid.

Normally they do not mate and their spermatheca therefore is then devoid of sperms.

6 The Ontogeny of the False Queens

By recording interactions between a *capensis* worker and individuals of the host race, *Apis mellifera mellifera*, it became clear that at an early stage the *capensis* can be distinguished by its behaviour. Most important seem to be the behavioural elements belonging to the trophallaxis system: food offering and food asking. The *capensis* starts to make a gain in food exchanges long before she is able to elicit court behaviour and to lay eggs. Therefore, a detailed analysis was made of the role of trophallaxis (Korst and Velthuis 1982). Food exchanges in an observation hive may be initiated by food offering and by food begging, and the exchange may be broken off by either of the two bees. Some workers may predominantly offer while others beg more frequently. These observations led to the conclusion that the individuals were far from identical, both with respect to their initiatives and responses.

In order to eliminate age differences and the effects of the division of labour within the worker caste, these interactions were studied in small groups of 2-5 workers, which permitted us to obtain sufficient data on all the interacting individuals. These groups were of either *mellifera* or *capensis* workers or they contained one *capensis* and a few *mellifera* worker bees. They were studied for 2-3 weeks after emergence and establishment of the group.

Trophallaxis was generally not of a random nature, but resulted in distinct gains or losses for certain individuals. Also, it became clear that true exchange of food occurs in a low percentage of the contacts (see for an example Table 2).

Generally, such gains or losses are more or less characteristic for a given animal although in the course of 2-3 weeks important changes within the group may occur (Fig. 2). These gains and losses are correlated with certain physiological differences that develop among the bees. Those bees that are known to be losers show, upon dissection, a notably distended rectum, due to the enormous amount of pollen. Where bees from a cage were allowed to fly out daily and defecate, some use this opportunity eagerly, while others, especially the ones that gain the most in food exchanges, stay behind on the comb. Perhaps due to the burden placed on their digestive apparatus, losers are apt to die sooner. On the other hand the gainers may show activated ovaries at the end of the observations. Together these observations suggest a causal relationship between the dominance relations and future physiological features.

Table 2 - Trophallactic contacts between two *A. m. mellifera* worker bees

number of times of:	workers marked red	green
begging for food	29	252
responding to begging by presenting	58	6
offering food	109	26
responding to offering by taking	1	38
net gain or loss in food transmissions	-89	+89

Table 3 - The occurrence of mandibular gland substances in worker bees out of a group of 50 bees, of which 2 were egg layers (*A. m. mellifera*)

10-HDA = (E)-10-hydroxy-2-decenoic acid
10-HDDA = 10-hydroxy-decanoic acid
9-HDA = (E)-9-hydroxy-2-decenoic acid

	total acids (µg/head)	percentage of each component		
		10-HDA	10-HDDA	9-HDA
4 individuals	1.5	100.0		
10 individuals	3.0	87.0	13.0	
1 individual	4.3	78.4	15.7	5.9
2 egg layers	4.5	81.4	13.6	5.1

Table 4 - The occurrence of mandibular gland substances in laying worker bees (*A. m. mellifera*) taken from an observation hive

9-ODA = (E)-9-oxo-decenoic acid
8-HOA = 8-hydroxyoctanoic acid
For further legend see Table 3

	total acids (µg/head)	percentage of each component				
		10-HDA	10-HDDA	9-HDA	9-ODA	8-HOA
3 laying workers	5.4	65.2	34.8			
7 laying workers	6.0	80.3	12.2	7.5		
1 laying worker	22.5	52.5	7.8	12.0	18.3	9.4

7 Mandibular Gland Secretions and Reproduction

In the case of a mixed group, one *capensis* with a few *mellifera* workers, the *capensis* is a distinct gainer, as is to be expected. One is inclined to ascribe these behavioural and physiological differentiations to the different physiological abilities of the mandibular gland, the *capensis* worker being able to synthesize a queen pheromone compound, the 9-oxo-E-2-decenoic acid (9-ODA).

This picture has to be refined however. Chemical analysis of mandibular glands of bees for which behavioural data had been sampled show a distinct differentiation of the mandibular gland secretion in correlation with the functioning of the ovary. Some examples will be given.

Table 3 shows the components found in the mandibular glands of *mellifera* workers that had been together with 50 bees in a cage for 2 weeks. Two of them had become egg layers by that time. Of the others 15 were taken randomly. The two egg layers produced a greater range of acids than did the workers with underdeveloped ovaries, with one exception. This one produced a range of acids similar to that of the laying workers. This suggests that the production of appropriate mandibular gland signals may precede ovarial development.

Table 4 shows the composition patterns of the mandibular glands of 11 laying workers taken from a queenless *mellifera* observation hive. Their mandibular gland secretions are not identical and there is even one worker that produces 9-ODA, a constituent of queen mandibular gland secretion.

Table 5 presents the data obtained from a group of *mellifera* workers that were kept in a cage for 3 weeks. Their mandibular gland contents are again correlated with the degree of ovary development, but this experiment is also noteworthy for the fact that here almost all individuals produce 9-ODA, and the mandibular glands of all individuals are rich in substances. This finding has been repeated in the case of *intermissa*, where queenless workers produce mandibular gland secretions that are complex and contain 9-ODA (Crewe 1988; Crewe and Moritz 1988). Even workers in queenright colonies of this race produce small quantities of 9-ODA (Crewe and Moritz 1988).

Table 6 gives a comparison among *capensis* workers that were kept in groups of 4, either consisting of only *capensis* or of one *capensis* and three *mellifera* workers. Both groups of *capensis* workers have 5 different acids but the relative amounts differ. This is related to the fact that in the groups with *mellifera*, the *capensis* develops her ovaries, while this does not occur in the pure *capensis* groups.

From these data we conclude that development towards egg laying is not correlated with the presence or amount of a certain specific substance but with having a relative advantage over the other members of the group in the range of fatty acids produced and the ratio of

Table 5 - The occurrence of mandibular gland secretions in *A. m. mellifera* worker bees kept in one group for three weeks, in relation to the activation of their ovaries

MHPE = 2-methoxy-4-hydroxy phenylethanol
For further legend see preceding Tables 3 and 4

number of bees	degree of ovary activity	total acids (µg/head)	percentage of each compound 10-HDA	10-HDDA	9-HDA	MHPE	9-ODA	8-HOA
5	slight	4.4	70.5	13.6	2.1	-	8.6	5.2
2	intermediate	8.3	73.4	18.5	2.8	-	2.3	3.0
1	intermediate	3.9	36.4	9.7	23.1	5.9	14.4	10.5
3	high	8.9	38.5	15.1	19.9	5.5	16.3	4.7

Table 6 - The occurrence of mandibular gland substances in *A. m. capensis* worker bees in relation to their social environment

	total acids (µg/head)	percentage of each component 10-HDA	10-HDDA	9-HDA	9-ODA	8-HOA
17 *capensis* with *capensis* conditions	61.5	42.4	3.0	6.8	33.9	14.0
3 *capensis* with *mellifera* conditions	22.4	5.5	0.5	7.7	76.2	10.0

the substances involved. Apparently, the workers of both races are capable of synthesizing these substances, but *capensis* does so more readily than *mellifera.*

Finally, there is a correlation between behavioural data related to trophallaxis sampled over a period of 2-3 weeks and the chemical composition of the mandibular gland. This is examplified by data obtained from groups of 4-5 *capensis* workers. If their behavioural characteristics are pooled and subsequently expressed in a single rank number, and at the same time the composition of their mandibular gland material is expressed as a ratio of the most workerlike substance, 10-HDA, and the most queenlike substance, 9-ODA, the two sets of data are correlated ($r=0.342$, $p<2.5\%$) (Fig. 3).

From this figure we see that a number of individuals fit on the diagonal representing the ideal correlation but that others, having a biochemical rank number 1 or a behavioural rank number 1, appear at the margins of the diagram. There are apparently other factors beside

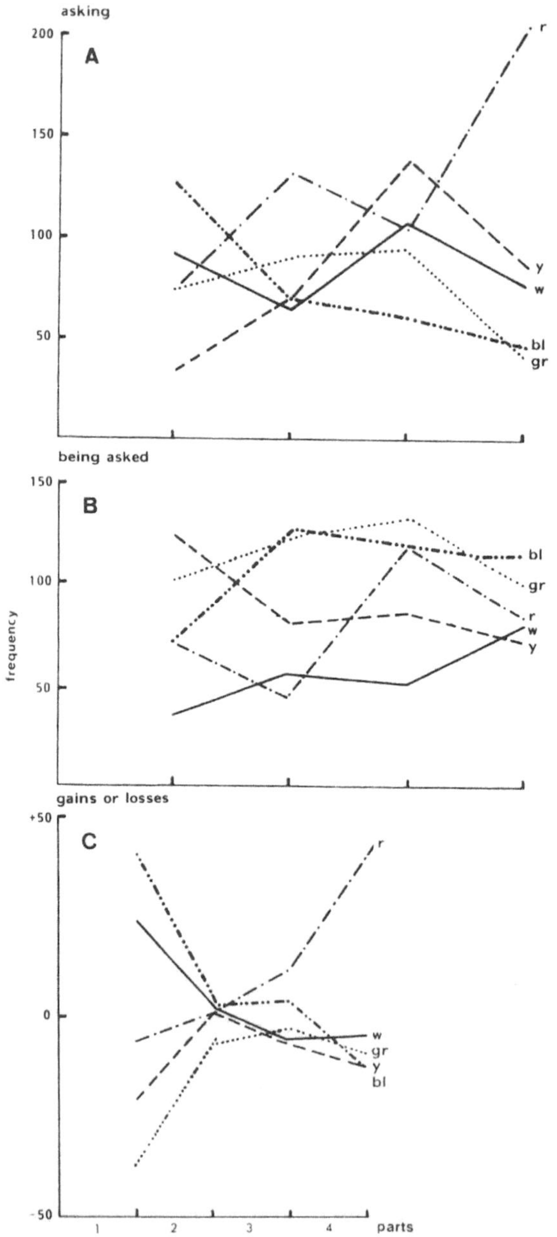

Fig. 2 - Differences among individuals from a group of 5 *A. m. capensis* workers. A Food begging. **B** Being solicited for food. **C** Gains or losses in food transmission. The observation period, 28. Feb. - 19. March 1979, has been devided in 4 parts with about an equal number of interactions. Animal "red" is characterized by distinct higher scores for food begging, overall activity and success in food transmissions. "Yellow" is the only bee with a notable score for food offering (not shown), the other three are more or less the same for the analysed characters. Their rank numbers (compare Fig. 3) for behaviour and biochemistry resp. are: red: 1, 1; green: 2, 1; blue: 3, 3; white: 4, 4; yellow: 2, biochemical data insufficient for ranking.

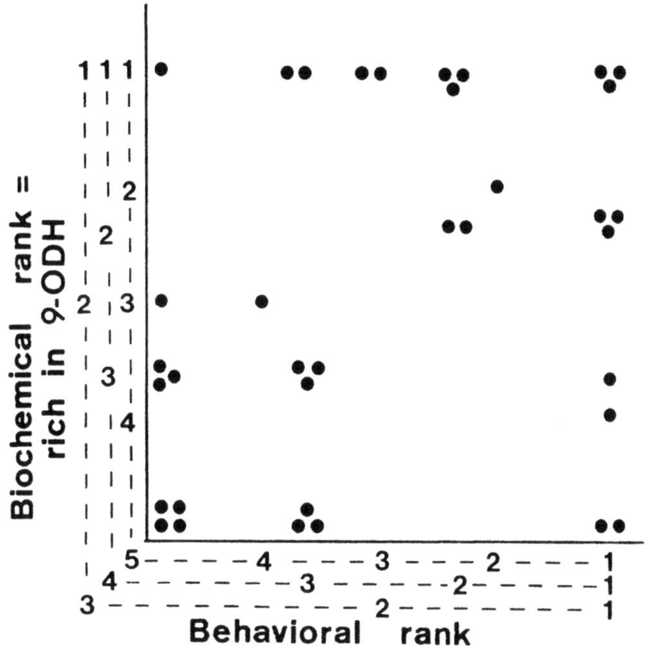

Fig. 3 - Correlation between overall-behavioural rank number and biochemical rank number based on the relative amounts of 10-HDA and 9-ODA in the mandibular glands of *A. m. capensis* workers that were kept together in groups of 4-5 bees. Correlation coefficient r = 0.342, p < 2.5%.

the relative amount of 9-ODA that determine the relation between signal and status. One factor might be that the behavioural data reflect only the situation at the end of that period, therefore not covering the dynamics within the group. Another factor might be, that several characteristics of a glandular secretion may develop independently, and therefore may independently reflect the physiological condition of the messenger. A combination between any given property of the secretion and the social status would then reveal only a slight statistical correlation, but nonetheless this secretion, in one way or other, would still reflect the physiological condition. The complexity of the mandibular gland secretion in relation to physiological differentiation of reproductive properties and age has been studied by Velthuis and Van der Kerk (1988). A further analysis of these factors is still to be made.

It is tempting to suggest that the mandibular gland secretion is related to the success in trophallactic contacts and physiological differentiation within the worker caste. This would help us to understand why worker bees perform as they do within the colony and how, at the colony level, the distribution of labour is maintained under different population structures.

However, we have to realize that these data leave open the important question whether the correlation between enrichment of the mandibular gland secretion and ovarial differentiation is a direct or an indirect one. Both processes may be more or less correlated with age and with physiological potencies that are different for the individuals of the same group. Evaluation of the pheromonal properties of individual components of these secretions either in a metabolic bioassay (Moritz and Crewe 1988) or in a behavioural bioassay (Slessor et al. 1988) has indicated low levels of activity. Mixtures of components or total extracts produce the highest responses in these assays indicating that the secretions cannot be decomposed into their individual components for behavioural purposes. The challenge is to define the quantitative variability in the mixtures necessary to effect behavioural responses.

What is clearly evident in *capensis*, and to lesser extent in *mellifera* and *intermissa*, is a change in the composition of the mandibular gland secretion of workers when reared under queenright and queenless conditions. Such changes may be translated into signals of different qualities that affect the behaviour of the other workers (e.g. producing retinue behaviour). Indeed, the range of different signals that may be produced by appropriately reared queenless workers of different honey bee races, may provide a very suitable set of mixtures of components that could be used in behavioural tests to determine the relationship between composition and signal.

8 References

Anderson RH (1963) The laying worker of the Cape honey bee, *Apis mellifera capensis*. J Apicult Res 2:85-92

Billen JPJ, Dumortier KTM, Velthuis HHW (1986) Plasticity of honeybee castes, occurrence of tergal glands in workers. Naturwiss 73:332

Crewe RM, Velthuis HHW (1980) False queens: a consequence of mandibular gland signals in worker honey bees. Naturwiss 67:467-469

Crewe RM (1988) Natural history of honey bee mandibular gland secretions: development of analytical techniques and the emergence of complexity. In: Needham GR, Page RE, Delfinado-Baker M, Bowner CE (eds) Africanized honey bees and bee mites. Ellis Horwood, Chichester, pp 149-158

Crewe RM, Moritz RFA (1988) Mandibular gland fatty acid variation in workers and queens of *Apis mellifera intermissa* Buttel-Reepen.

Hemmling C, Koeniger N, Ruttner F (1979) Quantitative Bestimmung der 9-Oxodecensäure im Lebenszyklus der Kapbiene (*Apis mellifera capensis* Escholtz). Apidol 10:227-240

Jack RW (1916) Parthenogenesis amongst the workers of the Cape honey bee: Mr GW Onion's experiments. Trans Ent Soc Lond 64:396-403

Korst PJAM, Velthuis HHW (1982) The nature of trophallaxis in honeybees. Insect Soc 29:209-221

Moritz RFA, Crewe RM (1988) Reaction of honey bee workers (*Apis mellifera* L.) to fatty acids in queen signals. Apidol 19:333-342

Onions GW (1912) South African "fertile" worker bees. Agric J S Afr 7:44-46

Ruttner F (1976) Die Bienenrassen Afrikas. Verh Int Apimondia Kongr, 25:334-364

Ruttner F (1988) Biogeography and taxonomy of honeybees. Springer-Verlag, Berlin, Heidelberg

Ruttner F, Koeniger N, Veith HJ (1967) Queen substance bei eierlegenden Arbeiterinnen der Honigbiene (*Apis mellifera* L.). Naturwiss 63:434

Sakagami SF (1958) The false queen: fourth adjustive response in dequeened honey bee colonies. Behav 13:280-296

Seeley TD (1979) Queen substance dispersal by messenger workers in honey bee colonies. Behav Ecol Sociobiol 5:391-415

Slessor KN, Kaminski L-A, King GGS, Borden JH, Winston ML (1988) Semiochemical basis of the retinue response to queen honey bees. Nature 332:354-356

Velthuis HHW (1970) Ovarian development in *Apis mellifera* worker bees. Ent Exp & Appl 13:377-394

Velthuis HHW (1972) Observations on the transmission of queen substances in the honey bee colony by the attendents of the queen. Behav 41:105-129

Velthuis HHW (1976) Egg laying, aggression and dominance in bees. Proc Int Congr Entomol, Washington, 15:436-449

Velthuis HHW, Van der Kerk A (1988) Age, environment, and genes in relation to the mandibular gland secretion of pure and hybrid *Apis mellifera capensis* worker bees. In: Needham GR, Page RE, Delfino-Baker M, Bowner CE (eds) Africanized honey bees and bee mites. Ellis Horwood, Chichester, pp 80-86

Chapter 9

Regulatory Steps in Caste Development of Eusocial Bees

KLAUS HARTFELDER[1]

CONTENTS

1 Introduction

With the availability of highly sensitive techniques for the analysis and quantification of the two morphogenetic hormones, ecdysone and juvenile hormone (JH), a "quantum evolution" has occurred in research on endocrine regulatory mechanisms of phase and caste polymorphisms in insects. With respect to the phenomenon of caste development, a large number of research activities has concentrated on the honey bee, revealing the highly complex nature of caste differentiation as a multilevel interaction of trophogenic and endocrine regulatory steps. During the last years it has become obvious that a comprehensive understanding of evolutionary and ecological aspects and constraints of caste development calls for a comparative view.

Outside of the family APIDAE, eusocial bee species with distinct castes have only been described in the HALICTIDAE (Sakagami and

---

[1] *Zoologisches Institut der Universität Tübingen, BRD*

Fukushima 1961; Knerer and Atwood 1966; Knerer and Plateaux-Quénu 1966; Brooks and Roubik 1983). Unfortunately, however, no data are available so far for an endocrine mechanism regulating caste differentiation in any of these primitively eusocial halictine species (Michener, chapter 5). For the three most important groups of eusocial bees, the genera *Bombus* and *Apis*, and the species-rich subfamily MELIPONINAE, the physiological basis of caste development was only reviewed extensively in the first two groups by de Wilde and Beetsma (1982). In the third group, the stingless bees, endocrine regulation of caste development has received far less attention, and only recently data on ecdysteroid titers and *corpora allata* (CA) activity in postembryonic development of queens and workers of one trigonine species have been published (Hartfelder 1986a, 1987).

A comparison of caste development in stingless bees with the regulatory features for caste differentiation in *Apis* and *Bombus* will therefore constitute the main topic of this review, especially since Strambi et al. (1984) recently presented a detailed analysis of ecdysteroid and juvenile hormone titers in the prepupal stage of two bumble bee species, and thus shed new light on the importance of the temporal sequence of the hormone peaks. Finally, I will attempt to integrate the trophogenic and endocrine aspects involved in caste differentiation in eusocial bees into a generalized picture of preimaginal development in holometabolous insects.

2 Caste Development in Stingless Bees

More conspiciously than in honey bees, colony homeostasis in stingless bees is closely associated with the oviposition process (Sakagami 1982). Shortly before the queen starts egg laying, newly constructed brood cells are mass-provisioned with larval food and closed immediately after oviposition has occurred. The larvae, thus, develop within a closed system with all the nutritional requirements necessary for larval, pupal and adult transformation present from the start.

Two different modes of caste development in stingless bees have been recognized for a long time (Ihering 1904; Schwarz 1948).

In the genus *Melipona*, queens emerge from cells which are indistinguishable - by size, shape or position within the brood comb - from those giving rise to workers or males. From the observation of an apparently fixed maximum ratio of queens to total female brood, Kerr (1946, 1950) proposed a genetic basis for caste development in this genus. After an extensive screening of environmental cues which could potentially modify such a mode of caste determination (Kerr et al. 1966), this hypothesis was later reformulated allowing for a trophogenic modulation of the genetic basis of caste development

(Kerr 1974). Yet, this "tropho-genetic" hypothesis has not been accepted unanimously - critical statements have been given by Velthuis (1976) and Darchen and Delage-Darchen (1977) - and the question still remains unsettled.

In the large trigonine group most species rear queens in cells which are distinctly larger than worker cells and contain significantly more larval food. These royal cells are located at the margin of the brood combs. The trophogenic basis for caste differentiation in the trigonines has been demonstrated experimentally for a considerable number of species (Darchen and Delage-Darchen 1971; Camargo 1972; Silva 1977).

3 Endocrine Control of Caste Formation in Stingless Bees

With respect to the endocrine control of caste differentiation, the two groups of stingless bees have revealed striking similarities, despite of the differences in the role played by the amount of larval food. Topical application of JH and JH-analogues on worker larvae from various species of the genus *Melipona*, and from different genera of the tribus TRIGONINI induced the formation of queen characters (Campos et al. 1975; Velthuis and Velthuis-Kluppell 1975; Campos 1977, 1978, 1979a, b; Campos et al. 1979, 1983; Bonetti 1982). In all these species the period of greatest sensitivity to exogenous JH was found to be within the spinning phase of the last larval instar. The conclusion drawn from these experiments was that an elevated JH titer in queen larvae during this critical period could be responsible for queen-specific gene activation.

The observation that in *Scaptotrigona postica depilis* the CA of queen larvae remain more active at the end of the feeding phase, and at the beginning of the spinning phase resume JH synthesis distinctly earlier than in worker larvae (Hartfelder 1987), contributed the endocrinological basis to this hypothesis. In Fig. 1 rates of JH synthesis of brain-*corpora cardiaca-corpora allata*-complexes and ecdysteroid titers are shown for the last two larval instars of queens and workers of *Scaptotrigona postica depilis*. In this graph, pupal ecdysis served as reference point for the time scale. Brain-CC-CA-complexes were incubated *in vitro* with ^{14}C-methyl-methionine as a JH precursor, and ecdysteroid titers were measured by a radioimmunoassay. Total ecdysteroids in haemolymph extracts were expressed as ecdysone equivalents. A caste-specific modulation of the ecdysteroid titer is apparent in the prepupal stages. Up till the end of the spinning phase, ecdysteroid titers in queens and workers were fluctuating in parallel fashion. CA activity in JH synthesis was measured by a radiochemical assay. This method has a double advantage. Firstly, it represents a direct approach for measuring

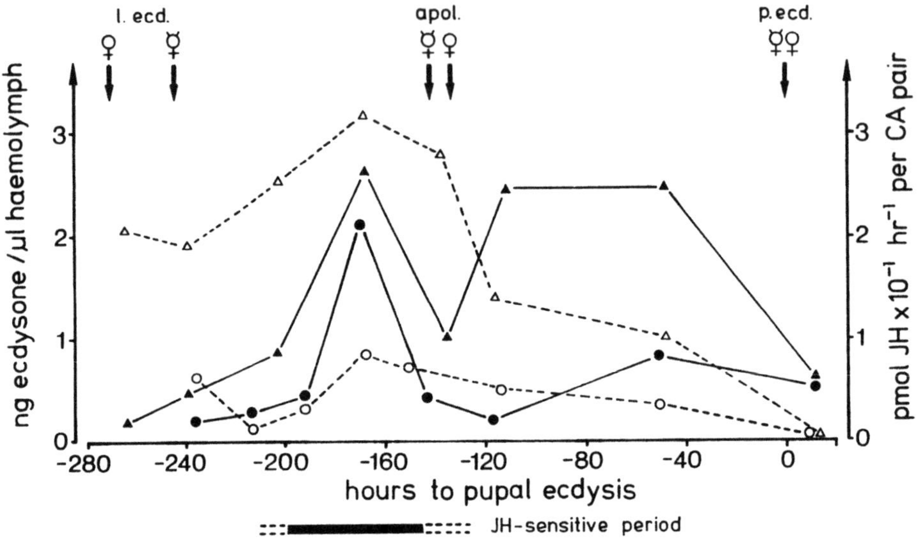

Fig. 1 - Ecdysteroid titers, expressed in ecdysone equivalents, in the haemolymph of queen (▲——▲) and worker larvae (●——●) of the stingless bee, *Scaptotrigona postica depilis*, during the last larval instar, and rates of JH biosynthesis for queen (△---△) and worker larvae (○---○). Arrows mark the events of larval ecdysis, apolysis, and pupal ecdysis for queens and workers. Below is indicated the JH-sensitive period for queen induction in the closely related species *Scaptotrigona xanthotricha* according to Campos et al. (1983).

rates of JH synthesis by *in vitro* incubation of the CA, and secondly the results are expressed in a quantitative manner. Earlier attempts to reach conclusions on CA activity, using measurements of CA volume and counts of cell number in the CA, have been carried out by Kerr et al. (1975) in *Melipona quadrifasciata* and by Bueno and Beig (1980) in *Scaptotrigona postica*. The activity curves, obtained by reliance on such parameters of indirect evidence for CA activity, are quite different from those based on direct radiochemical measurements. For the two castes of *Sc. postica* the volumes of the *corpora allata* showed a minimum in the prepupal phase. This is partly in accordance with my data since rates of JH synthesis gradually diminish towards pupal ecdysis. Shortly after pupal ecdysis, however, CA volumes of queen pupae increase rapidly while worker CA remain small. This caste-specific difference in CA volume is not reflected in the observed rates of JH synthesis by white-eyed pupae in *Scaptotrigona postica depilis*. A similar situation has been described for the changes in CA volume during the postembryonic development of honey bees. During pupal development the CA volumes in the two castes were distinctly different (Pflugfelder 1948; Lukoschus 1956) while JH titers - measured by physico-chemical methods (Rembold 1986) - remained at the detection limit in both castes until the late pharate adult stages.

In experiments which explicitly addressed the question of correlation between CA volume (or number of cells in the CA) and activity of JH synthesis, Tobe and Pratt (1975) showed that changes in spontaneous activity of JH synthesis in adult female *Schistocerca gregaria* were almost completely independent of changes in CA volume. Furthermore, Szibbo and Tobe (1981a, b) found that the cyclic changes in CA volume and cell number observed during the vitellogenic cycles in *Diploptera punctata* females could also not be correlated to JH synthesis by the CA. In preimaginal development CA volume and cell number are apparently related more directly to general growth than to rates of JH synthesis (Tobe and Stay 1985). This holds for hemimetabolous species like *Diploptera punctata* (Szibbo et al. 1982) as well as for holometabolous insects like *Manduca sexta* (Granger et al. 1979). Even from data on the ultrastructure of CA cells it is difficult to reach unequivocal conclusions on the state of CA activity (Wirtz 1973; Sedlak et al. 1983; Johnson et al. 1985).

4 Caste and Metamorphosis: the Role of JH in the Last Larval Instar

In the models proposed by de Wilde and Beetsma (1982) and Wheeler (1986), external stimuli relevant to caste development pass through a series of - still largely unknown - transformational steps and become manifest in a divergent activity of the endocrine system. At a later developmental stage the established differences in the endocrine system control the manifestation of alternative morphs. In a heuristically very stimulating paper Nijhout and Wheeler (1982) discussed the role of JH in this context. In their model which is mainly based on hormonal regulation of lepidopteran metamorphosis, the target organs have specific JH-sensitive periods for genes related to metamorphosis (i.e. larval/pupal and pupal/adult alternatives), and also for genes related to caste characters. During these JH-sensitive periods the presence or absence of a threshold JH titer determines which one of the respective alternatives will be expressed during the next moulting cycle.

The pattern of CA activity in penultimate and last larval instar *Scaptotrigona postica depilis* worker and queen larvae corresponds to the general basic scheme of CA activity in holometabolous insects. This scheme was largely elaborated from studies on regulation of metamorphosis in the tobacco hornworm (Granger et al. 1982). In *Manduca sexta* larvae JH titer (Fain and Riddiford 1975; Riddiford 1980a) and CA activity drop to a minimum in the middle of the last larval instar. The timing of this minimum coincides with decisive changes in behaviour. The larvae stop feeding and enter into a wandering stage in search for a suitable pupation site. During this wandering stage, JH synthesis recommences and reaches a maximum at the time of head capsule apolysis. Shortly afterwards, the activity of the CA stops again and the JH titer declines to "zero". This rapid disappearance of JH from the haemolymph is caused by the activation of JH-specific esterases (Sanburg et al. 1975; Vince and Gilbert 1977). The existence of two JH-sensitive periods in the last larval instar of *Manduca sexta* has been postulated by Williams (1976) and Riddiford (1980c). The first one is located at the end of the feeding phase before the onset of ecdysone secretion for the "commitment peak". During this period, a declining JH titer permits the formation of a pupal type cuticle after the following apolysis. During the second JH-sensitive period which starts about 48 hours after the "commitment peak", a newly increasing JH titer inhibits a precocious expression of adult characters (Riddiford 1980b).

In honey bees a comparatively large number of studies has attempted to elucidate the role of JH in caste differentiation processes by either mixing JH or JH-analogues with larval food or topically applying these compounds on 2.0 to 4.5 days old worker larvae (Wirtz 1973; Rembold et al. 1974; Asencot and Lensky 1976, 1984; Copijn et al. 1979; Dietz et al. 1979). This includes the larval instars from L3 up to the feeding phase of L5, when staged according to Rembold et al. (1980). Unfortunately, so far only one study addressed the problem of metamorphosis-regulating JH-sensitive periods (Zdárek and Haragsim 1974), and it would be stretching the interpretation of their data too far, if one were to try to deduce two separate metamorphosis-regulating JH-sensitive periods during the last larval instar of the honey bee. Nevertheless, it is possible to say that JH applied to feeding larvae had a stronger effect on the larval-pupal than on the pupal-adult transition. This latter transition was affected more seriously when JH had been applied to spinning larvae or prepupae. The JH-sensitive period for the queen-worker decision, thus, apparently may start before the metamorphosis-regulating periods. Such an interpretation would also explain the results of Wirtz (1973), Goewie and Beetsma (1976) and Copijn et al. (1979) who obtained the highest degrees of queen-likeness when they treated 3.0 - 3.5 days old worker larvae with JH at concentrations which minimally affected the regulation of metamorphosis, thus resulting in a minimal rate of mortality. When larvae were additionally fed with royal jelly, queen development could be induced by JH treatment well into the feeding phase of L5 (Zdárek and Haragsim 1974; Dietz et al. 1979), and therefore the sensitive period for caste development might at least partly overlap with the metamorphosis-regulating JH-sensitive period.

JH titers were distinctly higher in honey bee queen larvae than in worker larvae during the fourth and the early feeding phase of the fifth instar when measured either by the *Galleria* assay (Wirtz 1973), RIA (Lensky et al. 1978) or GC-MS (Rembold 1986). This caste-specific difference in JH titers coincides exactly with the above described JH-sensitive period for the induction of queen characters.

Different variations on the theme of endocrine regulation of caste development are played in bumble bees, where two species, *Bombus hypnorum* and *Bombus terrestris*, have been studied in detail by Röseler and collaborators. The pocket maker species *B. hypnorum* shows the more primitive traits in the control of polymorphism (Röseler 1975). Queens and workers are merely differentiated on the physiological level, i.e. they differ in their haemolymph protein spectra (Röseler and Röseler 1974). In this species queen development could be induced by topical application of JH to last instar larvae (Röseler 1976). The larvae were most sensitive in the early prepupal phase, when they responded to doses as low as 0.1 - 1.0 ug JH. Similar doses had also proven most effective for the induction of queen development in stingless bees when applied to cocoon spinning larvae (Bonetti 1982). The physiological basis underlying these queen induction phenomena by JH treatment of *B. hypnorum* prepupae was shown to lie in a caste-specific modulation of JH titers during this developmental phase (Röseler 1977). Using the *Galleria* bioassay for the quantification of JH in haemolymph, JH titers of queen and worker larvae were found to differ for the first time, approximately 2 days after the start of prepupal development, whereas before his stage they ran closely parallel. In a more detailed RIA analysis of JH titers in haemolypmph samples from individual bumble bee larvae during the course of prepupal development, Strambi et al. (1984) could show that only queen prepupae possessed a distinct JH peak about 55 hours after the beginning of prepupal development (i.e. approximately 15 hours before pupal ecdysis).

Contrary to the possibility of artificially inducing queen development in *Bombus hypnorum* by treatment with exogenous JH, identical experiments with the pollen storer species *Bombus terrestris* always produced negative results (Röseler and Röseler 1974). In this species, it is the egg-laying queen who controls the developmental pathways taken by her offspring, since during most of the year queens are only reared in her absence (Röseler 1970). Subsequent analyses showed that during most of prepupal development the JH titer in workers is even slightly higher than in queens (Röseler 1977). Unlike *B. hypnorum* workers, *B. terrestris* worker prepupae had a distinct JH peak approximately 30 hours before pupal ecdysis (Strambi et al. 1984). Queen prepupae had also a clear JH peak in their haemolymph titer, and as in *B. hypnorum* queens it occurred approximately 15 hours before pupal ecdysis. Thus, the JH peak in *B. terrestris* queens seems to be shifted more towards the end of prepupal development. As in stingless bees, nothing is known so far on the timing of metamorphosis regulating JH-sensitive periods in the last larval instar of bumble bees.

5 Caste-specific Modulation of Ecdysteroid Titers: Timing and Function

In contrast to the comparatively large number of studies dedicated to the elucidation of the role of JH in the regulation of caste polymorphism in social insects, it was not until recently that the second complex of morphogenetic hormones, the ecdysteroids, started to receive more attention. Already 30 years ago Lukoschus (1956) had pointed out drastic caste-specific differences in the volume of prothoracic gland nuclei of cocoon-spinning larvae and prepupae of the honey bee. He interpreted this phenomenon as an indication of a period of increased moulting hormone synthesis in queen larvae. His hypothesis was corroborated by RIA analysis of ecdysteroid contents in whole body extracts of honey bee larvae and pupae (Hagenguth and Rembold 1978). These authors found that in queen larvae ecdysteroids reached a higher level during the spinning and prepupal phase. In the prepupal phase of *Bombus hypnorum* and *Bombus terrestris* similar caste-specific differences in ecdysteroid titers were not observed (Strambi et al. 1984). However, in queens of both species the ecdysteroid peaks were shifted more towards pupal ecdysis when compared to the situation in worker larvae.

In stingless bees ecdysteroid titers were analysed for the trigonine species *Scaptotrigona postica depilis* (Fig. 1). Up till the end of the spinning phase in both castes the ecdysteroid titers fluctuate in a parallel fashion. With the beginning of the prepupal stage, the ecdysteroid titer in queens rapidly increases and remains at a much higher level as compared to workers. At pupal ecdysis the titers in both queens and workers drop to a minimum level. This caste-specific pattern of ecdysteroid titer modulation corresponds to the one observed in *Apis mellifera*, with the exception that in the analysis of haemolymph titers - as carried out in *Scaptotrigona postica depilis* (Hartfelder 1986a) - two peaks become clearly discernible in the last larval instar. This was not the case in the honey bee, where whole body extracts were used in a RIA analysis with the same antibody (Hagenguth and Rembold 1978).

A sequence of at least two ecdysteroid peaks in the last larval instar seems to constitute part of the general pattern of metamorphosis in holometabolous insects (Bollenbacher et al. 1975, 1978; Lafont et al. 1975; Calvez et al. 1976; Delachambre et al. 1980; Riddiford 1980a; Briers et al. 1983; Bainbridge and Bownes

1988). The first ecdysteroid peak in the last larval instar of *Manduca sexta* occurs for the first time in the absence of JH at this point of larval development. It commits the epidermal cells to the pupal program of cuticle synthesis (Riddiford 1978; 1980b, c). The second ecdysteroid peak then triggers the events related to ecdysis. At the moment of ecdysis itself, the edysteroid titer in the haemolymph of *M. sexta* has to drop to a minimum level before the secretion of eclosion hormone can start (Truman et al. 1981, 1983).

It is still too early to formulate a sound hypothesis on the meaning of a caste-specific modulation of ecdysteroid titers for the expression of caste characters. Since, however, this phenomenon was observed in all three groups of highly eusocial bees, studies on the effects of treatment with exogenous 20-hydroxy-ecdysone, ecdysone or makisterone A which recently has been identified as a major ecdysteroid compound in haemolymph of honey bee pupae (Feldlaufer et al. 1985) should be taken into consideration. It might also prove worthwhile to combine such experiments with topical applications of JH, because several studies on lepidopteran species carried out in the last few years demonstrated intercations between these two systems of morphogenetic hormones (Sehnal et al. 1986; Shaaya et al. 1986). The rising titer of JH at the beginning of prepupal development prevents the precocious expression of adult characters (Kiguchi and Riddiford 1978) and also stimulates ecdysone secretion by the prothoracic glands (Riddiford 1980a; Cymborowski and Zimowska 1984; Gruetzmacher et al. 1984; Whisenton et al. 1985). These results could prove of some importance for the interpretation of caste-specific modulation of hormone titers in eusocial bees, since they could explain the earlier rise and higher level of the ecdysteroid titer in the spinnning and prepupal phase of stingless bee queens as a consequence of the preceeding peak of differential CA activity. Such a model would also hold true for the situation in *Bombus terrestris* where the JH titer in prepupae increases in workers before queens, and so does the ecdysteroid titer (Strambi et al. 1984).

Such speculations on a synergistic or at least permissive role of ecdysteroids in the endocrine control of caste differentiation by JH, however, cannot be seen as a unifying hypothesis, and clearly have to be restricted to the eusocial bees. A quite different situation in the role of these two morphogenetic hormones was observed in the endocrine regulation of caste polymorphism in ants. In *Myrmica rubra* (Brian 1974) and in *Pheidole pallidula* (Passera 1982) topical applications of ecdysone on overwintered brood - which under normal conditions gives rise to a large number of queens - significantly hampered queen development. These results, indicating a suppressive effect of ecdysone on the expression of queen characters in ants, received support from analyses of ecdysteroid titers in egg-laying queens of *Pheidole pallidula* (Suzzoni et al. 1980) and in the final larval instar of *Plagiolepis pygmaea* (Suzzoni et al. 1983). In all ant species studied so far, ecdysone had an antagonistic effect to JH with respect to queen differentiation.

Caste formation in ants differs from bees in two important aspects. Firstly, overwintering plays a decisive role in all those temperate climate species in which queen-worker differentiation occurs during larval development. Wheeler (1986) has interpreted the necessity of overwintering for queen formation as a guarantee for the maintenance of bipotentiality, i.e. non-overwintering larvae lose their bipotential state and subsequently develop into workers. She thus clearly distinguished caste induction *per se* from maintenance of bipotentiality as a prerequisite of queen development. It surely should prove of interest to test whether this categorical difference has a physiological correlate, i.e. whether the effects of ecdysteroid treatment of last instar larvae in ants can be subsumed under aspects of the endocrine regulation of overwintering. It remains an open question to show that overwintering might be under the control of prothoracicotropic hormone, i.e. a situation similar to the endocrine regulation of larval diapause in many insect species (Bollenbacher and Granger 1985). Thus, for the elaboration of a unifying model of endocrine regulation of caste differentiation one will need to know more about temporal relationships of JH and ecdysteroid titers in overwintering, queen-biased larvae, and non-overwintering, worker-biased larvae.

The second major difference between bees and ants is the evolution of a polymorphic worker caste in the latter. According to Wheeler (1986), this was possible by shifting the queen-worker decision - and its corresponding endocrine implications - from the last larval instar to earlier stages. In *Pheidole pallidula* this decision occurs already during embryogenesis and is under the control of maternal hormones (Suzzoni et al. 1980). Thus, the evolutionary "old" JH-sensitive period for the queen-worker decision in the last larval instar of highly eusocial HYMENOPTERA could now acquire a new function in ants and serve as a developmental switch for the worker-soldier differentiation.

6 Synopsis of Trophogenic and Endocrine Control of Caste Differentiation

Hormone titers or activities of the respective endocrine glands in queen larvae and pupae of *Apis mellifera* (Wirtz 1973; Lensky et al. 1978; Hagenguth and Rembold 1978; Rembold 1986) and of *Scaptotrigona postica depilis* (Hartfelder 1986 a, 1987; Hartfelder and Engels 1987) generally are only obtainable by artificial queen rearing, i.e. by transfering young worker larvae to queen cells, and thus exposing them to a "queenly feeding program". But only very few studies have so far dealt with the question of how these caste-specific sequences in diet exert their effects on the endocrine system. All studies in this direction were carried out on *Apis mellifera* larvae, and generally remained in the realm of histological analysis (Lukoschus 1956; Wirtz 1973; Dogra et al. 1977; Goewie 1978; Ulrich and Rembold 1983). Model conceptions on the dynamics of interaction of the dietary sequence and the endocrine system in bees, thus, to a large extent have to fall back upon studies of lepidopteran species.

The consumption of larval food constitutes an integral part of the regulatory system of development in all groups of insects, and starvation experiments during certain larval periods of the tobacco hornworm, *Manduca sexta*, had significant effects on the activity of the *corpora allata* (Bhaskaran 1981; Cymborowski et al. 1982) and on the secretion of prothoracicotropic hormone (Gilbert et al. 1980). At this point where the problem of caste differentiation in eusocial bees is to be discussed under the heading of how the modulations of the two morphogenetic hormones, JH and ecdysone, might possibly be related to the nutritional condition of the larvae, two general hypotheses can be formulated:

a) The modulation occurs as a consequence of a specific haemolymph composition, which in itself reflects a caste-specific composition of larval food, or
b) the modulation results from the attainment of a critical larval size.

Hypothesis a) subsumes the classical "determinator" concept (Rhein 1933; Weaver 1955; Rembold and Hanser 1964; Rembold 1976) as well as the concept of a "balanced composition" of larval food (Haydak 1943; Weaver 1974; Rembold and Lackner 1981; Hartfelder 1986b; Wittmann and Engels 1987). Both these concepts are still at a problematic stage. The isolation of a determinator substance could only particularly be achieved (Rembold et al. 1974), and nowadays the queen inducing dietary compounds are considered to release an allatotropic effect of a more general nature (de Wilde 1976). The concept of "nutritional balance" on the other hand is of a very elusive logical structure and thus posseses a low heuristic value. One recent attempt to deal with this logical problem has resulted in a comparative chemical analysis of the larval food from several species of stingless bees (Hartfelder and Engels 1988), but also did not come to unambigous conclusions. Nevertheless, this train of thought should not be abandoned, since recent studies on several lepidopteran species postulated a surprising relationship between gamma-aminobutyric acid titers in the haemolymph and endocrine events related to metamorphosis (Jungreis and Omilianowski 1980; Jungreis et al. 1980; Jungreis 1982).

Significant amounts of this amino acid are also present in larval food of stingless bees (Hartfelder 1985).

Hypothesis b) would include the findings of Goewie (1978) on excitatory electrophysiological effects of sugars on chemoreceptors on the larval mouthparts of honey bees. Such a seemingly parsimonious and straightforward explanation of the first trophogenic steps in caste induction, however, is called into question in view of the analysis of the sugar composition of honey bee larval food (Brouwers 1984), showing exactly contrary results on the fructose/glucose ratio in royal jelly to the one which would have been predicted from Goewie's study. It is out of the question, however, that the sugar content in larval food constitutes an important phagostimulating factor in caste development (Asencot and Lensky 1976, 1984, 1988).

The concept of critical larval weight and its consequences for the control of metamorphosis events has received much more attention in the tobacco hornworm than in bees. In accurately timed starvation experiments Nijhout (1975) and Williams (1976) could show that for the start of metamorphosis a larva has to reach a critical weight at the end of the penultimate and in the last larval stage. Attainment of this critical size then modulates the activity of the CA. Yet, starvation does not affect the JH titer in a direct manner, but apparently does so via a distortion of body size parameters, which are associated with developmental control in the last larval instar and with the possibility to add a supernumerary larval instar (Safranek and Williams 1984).

A speculative translation of this concept to the phenomenon of caste polymorphism in eusocial bees, at least for stingless bees, could be formulated in the following manner. At the moulting event leading to the penultimate or to the last larval instar, the attainment of a critical size might determine whether in the last larval instar the activity of JH synthesis in the CA will show a queen-typical early increase, or will remain at the lower level of worker larvae. The *Scaptotrigona* situation would then differ from the *Manduca* model in that the critical weight in *Scaptotrigona* will not decide whether a supernumerary larval instar will be added, but whether queen or worker characters of the bipotent larvae will alternatively become expressed in the following steps of metamorphosis.

7 Nutritional Programs may affect the Temporal Sequence of Endocrine Events

In Fig. 2 event-o-grams for the honey bee and stingless bees are presented, depicting the sequences of trophogenic and endocrine events relevant to caste polymorphism. From these graphs it is

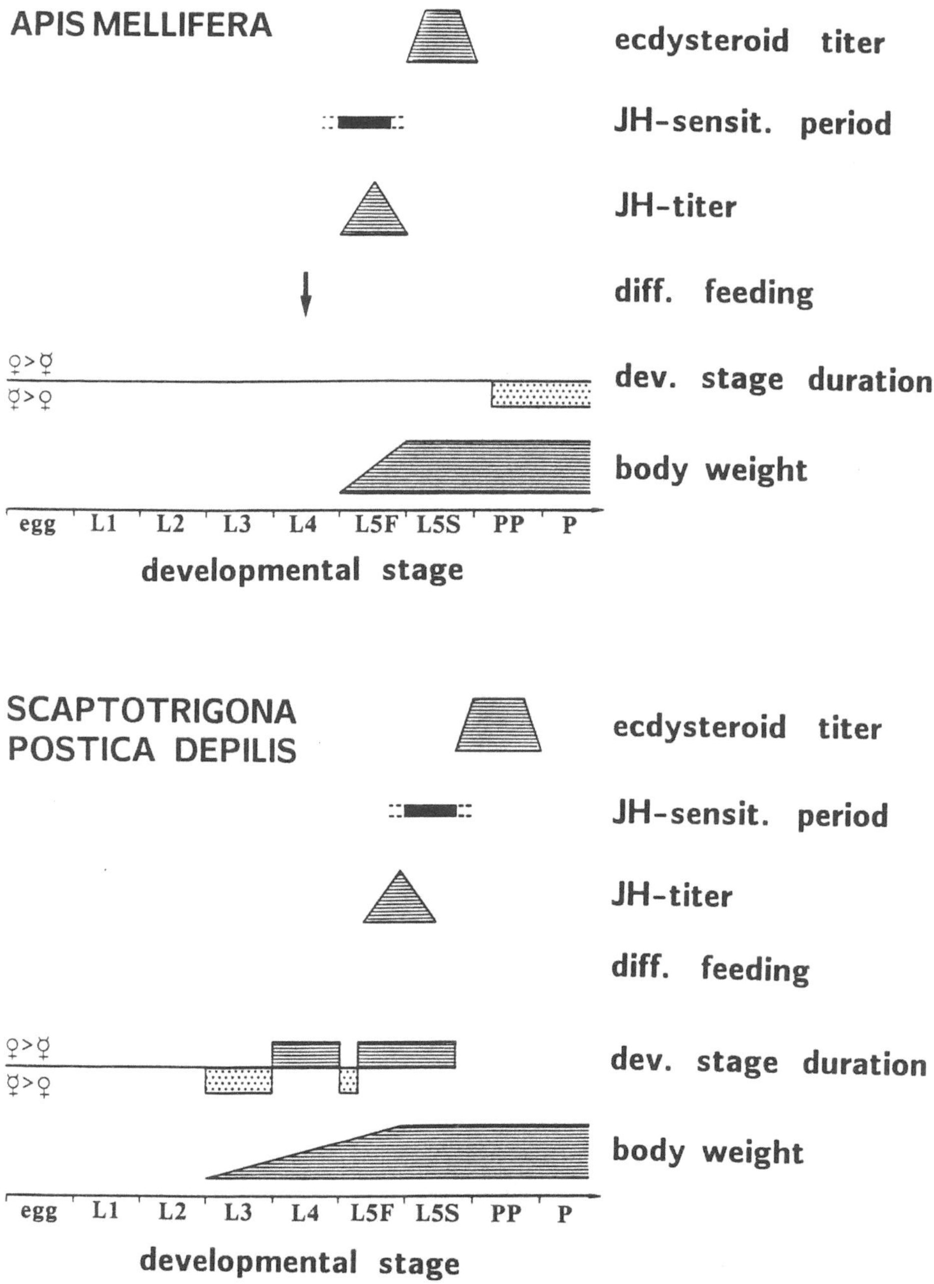

Fig. 2 - Temporal sequence of trophogenic and endocrine events, and their effects on growth rates (weight) and duration of developmental stages in caste development of the honey bee and a stingless bee. Striped bars indicate phases where values in queens are larger than in workers while stippled bars stand for the opposite situation. L_1-L_4 = first to fourth larval instar; the last larval instar is subdivided into a feeding phase = L_{5F}, a spinning phase = L_{5S}, and a prepupal phase = PP; P = pupal phase.

obvious that quite serious consequences for the control mechanisms of caste development may be entailed by the different modes of larval feeding, i.e. mass provisioning in stingless bees *versus* progressive and differential feeding in honey bees. Generally speaking, the mass provisioning in stingless bees results in a more diagonal event-o-gram wit an early divergence in body weight and early differences in the duration of the developmental phases, while the caste inducing endocrine events are located after the feeding period later in larval development (Hartfelder 1987). The event-o-gram for *Apis mellifera* on the other hand presents a more nested appearance (for a compilation of the data used in the construction of this event-o-gram see the reviews of Beetsma (1979) and de Wilde and Beetsma (1982)). A caste-specific peak of JH content at the beginning of the feeding period of L5 occurs soon after the nutritional switch in L4 (Rembold 1986). Interestingly, the honey bee is the only one example in eusocial bees where queen development is more rapid than worker development. This might have been achieved by moving the caste-relevant endocrine events to the beginning of the last larval instar, i.e. into a position of different temporal relation to the metamorphosis-regulating endocrine events, instead of maintaining them at the end of this instar.

The situation in bumble bees, especially in *Bombus hypnorum*, is of an intermediate type. Like in stingless bess, the duration of queen development is longer than that of workers, although only total durations for larval and pupal development are published, and a caste-specific fixation of the duration of postembryonic development was discernable only for *Bombus terrestris* (Röseler and Röseler 1974). In both species significantly higher frequencies of feeding acts of larvae by nurse bees were observed in the last days of the larval feeding period. The translation of this trophic signal into endocrine events, however, is visible later than in the honey bee and in the stingless bees, since Strambi et al. (1984) could show that in both bumble bee species hormone titers do not differ until the beginning of the prepupal phase. As already mentioned above, the two bumble bee species differ in an important aspect, since in *Bombus terrestris* a dominant egg-laying queen in a colony strictly controls the developmental pathway taken by her offspring, while this is not the case in *Bombus hypnorum* (Röseler 1970, 1976). The *B. terrestris* queen exerts her control very early in larval development, and as a result her interference exercises a sort of "epistatic" control over the following trophogenic and endocrine regulatory features of caste polymorphism.

8 References

Asencot M, Lensky Y (1976) The effect of sugars and juvenile hormone on the differentiation of the female honey bee larvae (*Apis mellifera* L.). Life Sciences 18:693-700

Asencot M, Lensky Y (1984) Juvenile hormone induction of "queenliness" on female honey bee (*Apis mellifera* L.) larvae reared on worker jelly and on stored royal jelly. Comp Biochem Physiol 78B:109-117

Asencot M, Lensky Y (1988) The effect of soluble sugars in stored royal jelly on the differentiation of female honey bee (*Apis mellifera* L.) larvae to queens. Insect Biochem 18:127-133

Bainbridge SP, Bownes M (1988) Ecdysteroid titers during *Drosophila* metamorphosis. Insect Biochem 18:185-197

Beetsma J (1979) The process of queen-worker differentiation in the honey bee. Bee World 60:24-39

Bhaskaran G (1981) Regulation of *corpus allatum* activity in last instar *Manduca sexta* larvae. In: Bhaskaran G, Friedman S, Rodriguez JG (eds) Topics in insect endocrinology and nutrition. Plenum Press, New York, pp 53-81

Bollenbacher WE, Granger NA (1985) Endocrinology of the prothoracicotropic hormone. In: Kerkut GA, Gilbert LI (eds) Comprehensive insect physiology, biochemistry and pharmacology. Pergamon Press, Oxford, vol 8, pp 109-151

Bollenbacher WE, Vedeckis VW, Gilbert LI, O'Connor JD (1975) Ecdysone titers and prothoracic gland activity during the larval-pupal development of *Manduca sexta*. Dev Biol 44:46-53

Bollenbacher WE, Zvenko H, Kumaran AK, Gilbert LI (1978) Changes in ecdysone content during postembryonic development of the wax moth, *Galleria mellonella*: The role of the ovary. Gen Comp Endocrinol 34:169-179

Bonetti AM (1982) Açao do hormônio juveníl sobre a expressao gênica em *Melipona* (Hymenoptera, Apidae, Meliponinae). Master thesis, Univ Sao Paulo, Ribeirao Preto

Brian MV (1974) Caste differentiation in *Myrmica rubra*: The role of hormones. J Insect Physiol 20:1351-1365

Briers T, van Beek E, De Loof A (1983) Ecdysteroid activity during metamorphosis and in male and female adults of four blowfly species. Comp Biochem Physiol 74A:521-524

Brooks RW, Roubik DW (1983) A halictine bee with distinct castes: *Halictus hesperus* (Hymenoptera: Halictidae) and its bionomics in Central Panamá. Sociobiol 7:263-282

Brouwers EVM (1984) The glucose-fructose ratio in the food of the honey bee larvae. J Apic Res 23:94-101

Bueno OC, Beig D (1980) Biometric study of the *corpora allata* in *Scaptotrigona postica* during postembryonic development. J Apicult Res 19:219-223

Calvez BM, Hirn M, De Reggi M (1976) Ecdysone changes in the haemolymph of two silkworms (*Bombyx mori* and *Phylosamia cynthia*) during larval and pupal development. FEBS letters 71:57-61

Camargo CA de (1972) Determinaçao de castas em *Scaptotrigona postica* Latreille. Rev Brasil Biol 32:133-138

Campos LA de O. (1977) O hormônio juveníl nas abelhas: Seu papel na diferenciaçao das castas e nos aspectos do contrôle social. Doctoral thesis, Univ Sao Paulo, Ribeirao Preto

Campos LA de O. (1978) Sex determination in bees. VI: Effect of a juvenile hormone analogue in males and females of *Melipona quadrifasciata* (Apidae). J Kansas Entomol Soc 51:228-234

Campos LA de O. (1979a) Determinaçao do sexo nas abelhas. XIV: Papel do hormônio juveníl na diferenciaçao das castas na subfamilia *Meliponinae* (Hymenoptera, Apidae). Rev Brasil Biol 39:965-971

Campos LA de O. (1979b) Determinaçao de sexo em abelhas. XIII: Determinaçao das castas em *Partamona cupira* (Hymenoptera, Apidae). Papel de hormônio juvenil. Ciência e Cultura 31:65-68

Campos LA de O., Velthuis HHW, Velthuis-Kluppell FM (1975) Juvenile hormone and caste determination in a stingless bee. Naturwiss. 62:98-99

Campos LA de O., Kerr WE, da Silva DLN (1979) Sex determination in bees. Relative action of genes *xa* and *xb* on sex determination in *Melipona* bees. Rev Brasil Genet 3:267-280

Campos LA de O., Drummond MS, Lacerda L de M. (1983) Sex determination in bees. 18: The role of juvenile hormones I, II and III in caste differentiation in *Scaptotrigona xanthotricha*. Ciência e Cultura 35:209-211

Copijn GM, Beetsma J, Wirtz P (1979) Queen differentiation and mortality after application of different juvenile hormone analogues to worker larvae of the honey bee (*Apis mellifera* L.). Proc Kon Ned Akad Wet 82C:29-42

Cymborowski B, Zimowska G (1984) Switchover in the sensitivity of the prothoracic glands to juvenile hormone in the cotton leafworm, *Spodoptera littoralis*. J Insect Physiol 30:911-918

Cymborowski B, Bogus M, Beckage NE, Williams CM, Riddiford LM (1982) Juvenile hormone titres and metabolism during starvation-induced supernumerary larval moulting of the tobacco hornworm, *Manduca sexta*. J Insect Physiol 28:129-135

Darchen R, Delage-Darchen B (1971) Le determinisme des castes chez les *Trigones*. Insect soc 18:121-134

Darchen R, Delage-Darchen B (1977) Sur le determinisme des castes chez les *Melipones*. Bull Biol 111:91-109

Delachambre J, Besson MT, Connat J-L, Delbecque JP (1980) Ecdysteroid titers and integumental events during the metamorphosis of *Tenebrio molitor*. In: Hoffmann JA (ed) Progress in ecdysone research. Elsevier, Amsterdam, pp 211-234

Dietz A, Hermann HR, Blum MS (1979) The role of exogenous JH I, JH III and anti-JH (Precocene II) on queen induction of 4.5-day-old worker honeybee larvae. J Insect Physiol 25:503-512

Dogra GS, Ulrich GM, Rembold H (1977) A comparative study of the endocrine system of the honey bee larvae under normal and experimental conditions. Z Naturforsch 32C:637-642

Fain MJ, Riddiford LM (1975) Juvenile hormone titers in the haemolymph during late larval development of the tobacco hornworm, *Manduca sexta* (L.). Biol Bull 149:506-521

Feldlaufer MF, Herbert EW, Svoboda JA, Thompson MJ, Luby WR (1985) Makisterone A. The major ecdysteroid from the pupa of the honey bee, *Apis mellifera*. Insect Biochem 15:597-600

Gilbert LI, Bollenbacher WE, Granger NA (1980) Insect endocrinology: Regulation of endocrine glands, hormone titer and hormone metabolism. Ann Rev Entomol 42:493-510

Goewie EA (1978) Regulation of caste differentiation in the honey bee (*Apis mellifera* L.). Med Landbouwhogeschool Wageningen 78-15:1-75

Goewie EA, Beetsma J (1976) Induction of caste differentiation in the honey bee (*Apis mellifera* L.) after topical application of JH-III. Proc Kon Ned Akad Wet 79C:466-469

Granger NA, Bollenbacher WE, Vince R, Gilbert LI, Baehr JC, Dray F (1979) *In vitro* biosynthesis of juvenile hormone by the larval *corpora allata* of *Manduca sexta*: Quantification by radioimmunoassay. Mol Cell Endocrinol 16:1-17

Granger NE, Niemiec SM, Gilbert LI, Bollenbacher WE (1982) Juvenile hormone synthesis in vitro by larval and pupal corpora allata of *Manduca sexta*. Mol Cell Endocrinol 28:587-604

Gruetzmacher MC, Gilbert LI, Granger NA, Goodman W, Bollenbacher WE (1984) The effect of juvenile hormone on prothoracic gland function during the larval-pupal development of *Manduca sexta*: An *in situ* and *in vitro* analysis. J Insect Physiol 30:331-340

Hagenguth H, Rembold H (1978) Kastenspezifische Modulation des Ecdysteroid-Titers bei der Honigbiene. Mitt dtsch Ges allg angew Entomol 1:296-298

Hartfelder K (1985) Composition of larval food in the stingless bee *Scaptotrigona postica*. Mitt dtsch Ges allg angew Ent 4:339-341

Hartfelder K (1986a) Caste differentiation in stingless bees: Activity of the *corpora allata* and ecdysteroid titers during the critical last larval instar. Adv Invertebr Reprod 4:518

Hartfelder K (1986b) Kastendifferenzierung bei Meliponen. Vergleichende Analysen von Larvenfutter unter dem Aspekt einer ausgewogenen Nahrungszusammensetzung. Apidol 17:361-364

Hartfelder K (1987) Rates of juvenile hormone synthesis control caste differentiation in the stingless bee *Scaptotrigona postica depilis*. Roux's Arch Dev Biol 166:522-526

Hartfelder K, Engels W (1987) Endokrine Regulation der Kastendifferenzierung bei der stachellosen Biene *Scaptotrigona depilis*. Verh Dtsch Zool Ges 80:166

Hartfelder K, Engels W (1988) The composition of larval food in stingless bees: Evaluating nutritional balance by chemosystematic methods. Insect soc 35: in press

Haydak M (1943) Larval food and development of castes in the honey bee. J econ Entomol 36:778-792

Ihering H von (1904) Biologie der Stachellosen Honigbienen Brasiliens. Zool Jb Abt Syst Geogr Biol 19:179-187

Johnson GD, Stay B, Rankin SM (1985) Ultrastructure of *corpora allata* of known activity during the vitellogenic cycle in the cockroach *Diploptera punctata*. Cell Tissue Res 239:317-327

Jungreis AM (1982) Evidence supporting a proposed association between hemolymph gamma-aminobutyric acid and neurosecretion in lepidopterous insects. Comp Biochem Physiol 71C:169-175

Jungreis AM, Omilianowski DR (1980) Gamma-aminobutyric acid and glutamic acid in *Manduca sexta*: Proposed roles in insect development. Comp Biochem Physiol 67C:173-185

Jungreis AM, McCaleb DC, Korman AK (1980) Hemolymph gamma-aminobutyric acid and its titers in the greater wax moth, *Galleria mellonella*: Changes during normal or supernumerary larval development and following injury. Comp Biochem Physiol 67C:167-172

Kerr WE (1946) Formaçao das castas no gênero *Melipona* (Illiger, 1806) - nota prévia. An Esc Sup Agric "Luiz de Queiroz" 3:299-312

Kerr WE (1950) Evolution of the mechanism of caste determination in the genus *Melipona*. Evolution 4:7-13

Kerr WE (1974) Sex determination in bees. III: Caste determination and genetic control in *Melipona*. Insect soc 21:357-368

Kerr WE, Stort AC, Montenegro MS (1966) Importância de alguns fatôres ambientais na determinaçao das castas do genero *Melipona*. An Acad Bras Ciências 38:149-168

Kerr WE, Akahira Y, de Camargo CA (1975) Sex determination in bees. IV: Genetic control of juvenile hormone production in *Melipona quadrifasciata* (Apidae). Genetics 81:749-756

Kiguchi K, Riddiford LM (1978) A role of juvenile hormone in the pupal development of the tobacco hornworm. J Insect Physiol 24:673-680

Knerer K, Atwood CE (1966) Polymorphism in some nearctic halictine bees. Science 152:1262-1263

Knerer G, Plateaux-Quénu C (1966) Sur le polymorphisme des femelles chez quelques *Halictinae*. Compt Rend Acad Sci (Paris) 263:2014-2017

Lafont R, Mauchamp B, Pennetier JL, Tarraux P, Dettys L, Delbecque JP (1975) alpha-ecdysone and beta-ecdysone levels in insect haemolymph: Correlation with developmental events. Experientia 31:1214-1242

Lensky Y, Baehr J-C, Pocheron P (1978) Dosages radio-immunologiques des ecdysones et des hormones juveniles au cours du développement post-embryonnaire chez les ouvrieres et les reines d'abeille (*Apis mellifica* L. var. *ligustica*). CR Acad Sci (Paris) 287D:821-824

Lukoschus F (1956) Untersuchungen zur Entwicklung der Kastenmerkmale bei der Honigbiene (*Apis mellifera*, L.). Z Morphol Ökol Tiere 45:157-197

Nijhout HF (1975) A threshold size for metamorphosis in the tobacco hornworm, *Manduca sexta* (L.). Biol Bull 149:214-225

Nijhout HF, Wheeler DE (1982) Juvenile hormone and the physiological basis of insect polymorphism. Quart Rev Biol 57:109-133

Passera L (1982) Endocrine regulation of caste determination in ants. In: Jaisson P (ed) Social insects in the tropics. Univ Paris-Nord, Paris, vol 1, pp 41-62

Pflugfelder O (1948) Volumetrische Untersuchungen an den *Corpora allata* der Honigbiene, *Apis mellifera*, L.. Biol Zentralbl 67:223-241

Rembold H (1976) The role of the determinator in caste formation in the honey bee. In: Lüscher M (ed) Phase and caste determination in insects. Pergamon Press, Oxford, pp 21-34

Rembold H (1986) Control of imaginal development by juvenile hormone. Adv Invertebr Reprod 4:59-68

Rembold H, Hanser G (1964) Über den Weiselzellenfuttersaft der Honigbiene. VIII: Nachweis des determinierenden Prinzips im Futtersaft der Königinnenlarve. Hoppe Seyler's Z Physiol Chem 316:206-212

Rembold H, Lackner B (1981) Rearing of honey bee larvae *in vitro*: Effect of yeast extract on queen differentiation. J Apic Res 20:165-171

Rembold H, Czoppelt C, Rao PJ (1974) Effect of juvenile hormone treatment on caste differentiation in the honey bee, *Apis mellifera*. J Insect Physiol 20:1193-1202

Rembold H, Lackner B, Geistbeck J (1974) The chemical basis of honey bee, *Apis mellifera*, caste formation. Partial purification of queen bee determinator from royal jelly. J Insect Physiol 20:307-314

Rembold H, Kremer J-P, Ulrich GM (1980) Characterization of postembryonic developmental stages of the female honey bee, *Apis mellifera*, L.. Apidol 11:29-38

Rhein W von (1933) Über die Entstehung des weiblichen Dimorphismus im Bienenstaate. W Roux Arch Entwicklungsmech Organ 129D:601-665

Riddiford LM (1978) Ecdysone-induced changes in cellular commitment of the epidermis of the tobacco hornworm, *Manduca sexta*, at the initiation of metamorphosis. Gen Comp Endocrinol 34:438-446

Riddiford LM (1980a) Interaction of ecdysteroids and juvenile hormone in the regulation of larval growth and metamorphosis of the tobacco hornworm. In: Hoffmann JA (ed) Progress in ecdysone research. Elsevier, Amsterdam, pp 409-433

Riddiford LM (1980b) The hormonal control of morphogenesis of a lepidopteran epidermal cell. In: Locke M, Smith DS (eds) Insect biology in the future. Academic Press, New York, pp 403-421

Riddiford LM (1980c) Insect endocrinology: Action of hormones at the cellular level. Ann Rev Physiol 42:511-528

Röseler P-F (1970) Unterschiede in der Kastendetermination zwischen den Hummelarten *Bombus hypnorum* und *Bombus terrestris*. Z Naturforsch 25B:543-548

Röseler P-F (1975) Die Kasten der sozialen Bienen. In: Lindauer M (ed) Informationsaufnahme und Informationsverarbeitung im lebenden Organismus. Akad Wiss Lit, Mainz, vol 3, pp 145-237

Röseler P-F (1976) Juvenile hormone and queen rearing in bumblebees. In: Lüscher M (ed) Phase and caste determination in insects. Pergamon Press, Oxford, pp 55-61

Röseler P-F (1977) Endocrine control of polymorphism in bumblebees. Proc VIII Congr IUSSI, Wageningen, pp 22-33

Röseler P-F, Röseler I (1974) Morphologische und physiologische Differenzierung der beiden Hummelarten *Bombus hypnorum* (L.) und *Bombus terrestris* (L.). Zool Jb Physiol 78:175-198

Safranek L, Williams CM (1984) Critical weights for metamorphosis in the tobacco hornworm, *Manduca sexta*. Biol Bull 167:555-567

Sakagami SF (1982) Stingless bees. In: Hermann HR (ed) Social insects, vol 3. Academic Press, New York, pp 361-423

Sakagami SF, Fukushima K (1961) Female dimorphism in a social halictine bee, *Halictus (Seladonia) aerius* (Smith). Japan J Ecol 11:118-124
Sanburg LL, Kramer KJ, Kezdy FJ, Law J, Oberlander H (1975) Role of juvenile hormone esterases and carrier proteins in insect development. Nature 253:266-267
Schwarz HF (1948) Stingless bees (*Meliponinae*) of the Western hemisphere. Bull Amer Mus Nat Hist 90:VI + 546 pp.
Sedlak BJ, Marchione L, Devorkin B, Davino R (1983) Correlations between endocrine gland ultrastructure and hormone titers in the fifth larval instar of *Manduca sexta*. Gen Comp Endocrinol 52:291-310
Sehnal F, Delbecque JP, Maróy P, Malá J (1986) Ecdysteroid titers during larval life and metamorphosis of *Galleria mellonella*. Insect Biochem 16:157-162
Shaaya E, Spindler-Barth M, Spindler K-D (1986) The effect of a juvenile hormone analogue on ecdysteroid titre during development and hnRNA formation in the moth, *Ephestia cautella*. Insect Biochem 15:181-185
Silva DLN da (1977) Estudos bionômicos em colonias mistas de *Meliponinae* (Hymenoptera, Apoidea). Bol Zool Univ Sao Paulo 2:7-106
Strambi A, Strambi C, Röseler P-F, Röseler I (1984) Simultaneous determination of juvenile hormone and ecdysteroid titers in the hemolymph of bumblebee prepupae (*Bombus hypnorum* and *Bombus terrestris*). Gen Comp Endocrinol 55: 83-88
Suzzoni J-P, Passera L, Strambi A (1980) Ecdysteroid titre and caste determination in the ant *Pheidole pallidula* (Nyl.) (Hymenoptera: Formicidae). Experientia 36:1228-1229
Suzzoni J-P, Passera L, Strambi A (1983) Ecdysteroid production during caste differentiation in larvae of the ant, *Plagiolepis pygmaea*. Physiol Entomol 8:93-96
Szibbo CM, Tobe SS (1981a) The mechanism of compensation in juvenile hormone synthesis following unilateral allatectomy in *Diploptera punctata*. J Insect Physiol 27:609-613
Szibbo CM, Tobe SS (1981b) Cellular and volumetric changes in relation to the activity cycle of the *corpora allata* of *Diploptera punctata*. J Insect Physiol 27:655-665
Szibbo CM, Rotin D, Feyereisen R, Tobe SS (1982) Synthesis and degradation of C_{16} juvenile hormone (JH III) during the final two stadia of the cockroach, *Diploptera punctata*. Gen Comp Endocrinol 48:25-32
Tobe SS, Pratt GE (1975) The synthetic activity and glandular volume of the *corpus allatum* during ovarian maturation in the desert locust, *Schistocerca gregaria*. Life Sciences 17:417-422
Tobe SS, Stay B (1985) Structure and regulation of the *corpus allatum*. Adv Insect Physiol 18:305-342
Truman JW, Taghert PH, Copenhaver PF, Tublitz NJ, Schwartz LM (1981) Eclosion hormone may control all ecdyses in insects. Nature 297:70-71
Truman JW, Rountree DB, Reiss SE, Schwartz LM (1983) Ecdysteroids regulate the release and action of eclosion hormone in the tobacco hornworm, *Manduca sexta*. J Insect Physiol 29:895-900
Ulrich GM, Rembold H (1983) Caste-specific maturation of the endocrine system in the female honey bee larva. Cell Tissue Res 203:49-55
Velthuis HHW (1976) Environmental, genetic and endocrine influences in stingless bee caste determination. In: M Lüscher (ed) Phase and caste determination in insects, Pergamon Press, Oxford, pp 63-70
Velthuis HHW, Velthuis-Kluppell FM (1975) Caste differences in a stingless bee, *Melipona quadrifasciata* Lep., influenced by juvenile hormone application. Proc Kon Ned Akad Wet 78(C):81-94
Vince RK, Gilbert LI (1977) Juvenile hormone esterase activity in precisely timed last instar larvae and pharate pupae of *Manduca sexta*. Insect Biochem 7: 115-120
Weaver N (1955) Rearing of honey bee larvae on royal jelly in the laboratory. Science 121:509-510
Weaver N (1974) Control of dimorphism in the female honey bee. 3. Balance of nutrients. J Apic Res 13:93-101
Wheeler DE (1986) Developmental and physiological determinants of caste in social Hymenoptera: Evolutionary implications. Amer Nat 128:13-34

Whisenton LR, Bowen MF, Granger NA, Gilbert LI, Bollenbacher WE (1985) Brain mediated 20-hydroxyecdysone regulation of juvenile hormone synthesis by the corpora allata of the tobacco hornworm, *Manduca sexta*. Gen Comp Endocrinol 58: 311-???

Wilde J de (1976) Juvenile hormone and caste differentiation in the honey bee (*Apis mellifera* L.). In: Lüscher M (ed) Phase and caste determination in insects. Pergamon Press, Oxford, pp 5-20

Wilde J de, Beetsma J (1982) The physiology of caste determination in insects. Adv Insect Physiol 16:167-246

Williams CM (1976) Juvenile hormone .... In retrospect and prospect. In: Gilbert LI (ed) The juvenile hormones. Plenum Press, New York, pp 1-14

Wirtz P (1973) Differentiation in the honey bee larvae. Med Landbouwhogeschool Wageningen, 73-5:1-155

Wittmann D, Engels W (1987) Welche Diät ergibt Arbeiterinnen bei *in vitro* Aufzucht von Honigbienen? Apidol 18:279-288

Zdárek J, Haragsim O (1974) Action of juvenoids on metamorphosis of the honey-bee *Apis mellifera*. J Insect Physiol 20:209-221

LIST OF CONTRIBUTORS

Prof. Dr. Alfred **Buschinger**
Zoologisches Institut der Technischen Hochschule Darmstadt,
Schnittspahnstrasse 3, D-6100 Darmstadt, West Germany

Prof. Dr. Robin M. **Crewe**
Department of Zoology, University of Witwatersrand,
1 Jan Smuts Avenue, Johannesburg, South Africa

Prof. Dr. Wolf **Engels**
Zoologisches Institut der Universität Tübingen,
Auf der Morgenstelle 28, D-7400 Tübingen, West Germany

Dr. Klaus **Hartfelder**
Zoologisches Institut der Universität Tübingen,
Auf der Morgenstelle 28, D-7400 Tübingen, West Germany

Dr. Cor G.J. **van Honk**
Laboratorium voor Vergelijkende Fysiologie, Rijksuniversiteit
Utrecht, Jan van Galenstraat 40, NL-3572 LA Utrecht, The Netherlands

Prof. Dr. Vera L. **Imperatriz-Fonseca**
Departamento de Écologia Geral, Instituto de Biociências,
Universidade de Sao Paulo, Caixa Postal 11.461, 05508 Sao Paulo, Brazil

Prof. Dr. Charles D. **Michener**
Departments of Entomology and of Systematics & Ecology,
and Snow Entomological Museum, University of Kansas,
Lawrence, Kansas 66045-2106, U.S.A.

Prof. Dr. Charles **Noirot**
Laboratoire de Zoologie, Université de Dijon,
6, boulevard Gabriel, F-21.000 Dijon, France

Prof. Dr. Bernhard **Rensch**
Zoologisches Institut der Westfälischen Wilhelms-Universität,
Badestrasse 9, D-4400 Münster, West Germany

Prof. Dr. Peter-Frank **Röseler**
Zoologisches Institut II der Universität Würzburg,
Röntgenring 10, D-8700 Würzburg, West Germany

Prof. Dr. Friedrich **Ruttner**
Institut für Bienenkunde der Universität Frankfurt,
Karl-von-Frisch-Weg 2, D- 6370 Oberursel, West Germany

Prof. Dr. Alain **Strambi**
CNRS - Laboratoire de Neurobiologie, Neuroendocrinologie des Insectes,
31, chemin Joseph Aiguier, F-13.402 Marseille Cédex 9, France

Prof. Dr. Hayo H.W. **Velthuis**
Vakgroep Vergelijkende Fysiologie, Ethologie & Socio-ecologie,
Rijksuniversiteit Utrecht, Postbus 80.086, NL-3508 TB Utrecht, The Netherlands

MIX
Papier aus verantwortungsvollen Quellen
Paper from responsible sources
FSC® C105338

If you have any concerns about our products,
you can contact us on
ProductSafety@springernature.com

In case Publisher is established outside the EU,
the EU authorized representative is:
Springer Nature Customer Service Center GmbH
Europaplatz 3, 69115 Heidelberg, Germany

Printed by Libri Plureos GmbH
in Hamburg, Germany